INTERNATIONAL SERIES IN
NATURAL PHILOSOPHY
GENERAL EDITOR: D. TER HAAR

VOLUME 110

RELATIVISTIC ASTROPHYSICS

Other Pergamon Titles of Interest

Books
CLARK & STEPHENSON:
The Historical Supernovae

GINZBURG
Theoretical Physics and Astrophysics

GURZADYAN:
Flare Stars

KONDRATYEV & HUNT:
Weather and Climate on Planets

REDDISH:
Stellar Formation

SAHADE & WOOD:
Interacting Binary Stars

SOLOMON & EDMUNDS:
Giant Molecular Clouds in the Galaxy

*Journals**
Chinese Astronomy and Astrophysics
Planetary and Space Science
Vistas in Astronomy
**Free specimen copy available on request.*

A full list of titles in the International Series in Natural Philosophy follows the index.

NOTICE TO READERS

Dear Reader

If your library is not already a standing order customer or subscriber to this series, may we recommend that you place a standing or subscription order to receive immediately upon publication all new volumes published in this valuable series. Should you find that these volumes no longer serve your needs your order can be cancelled at any time without notice.
The Editors and the Publisher will be glad to receive suggestions or outlines of suitable titles, reviews or symposia for consideration for rapid publication in this series.

ROBERT MAXWELL
Publisher at Pergamon Press

RELATIVISTIC ASTROPHYSICS

By

MAREK DEMIAŃSKI

University of Warsaw, Poland

Translated by

ANTONI POL

PERGAMON PRESS
OXFORD · NEW YORK · TORONTO
SYDNEY · PARIS · FRANKFURT

PWN — POLISH SCIENTIFIC PUBLISHERS
WARSZAWA

U.K.	Pergamon Press Ltd., Headington Hill Hall, Oxford OX3 0BW, England
U.S.A.	Pergamon Press Inc., Maxwell House, Fairview Park, Elmsford, New York 10523, U.S.A.
CANADA	Pergamon Press Canada Ltd., Suite 104, 150 Consumers Rd., Willowdale, Ontario M2J 1P9, Canada
AUSTRALIA	Pergamon Press (Aust.) Pty. Ltd., P.O. Box 544, Potts Point, N.S.W. 2011, Australia
FRANCE	Pergamon Press SARL, 24 rue des Ecoles, 75240 Paris, Cedex 05, France
FEDERAL REPUBLIC OF GERMANY	Perg mon Press GmbH, 6242 Kronberg-Taunus, Hammerweg 6, Federal Republic of Germany

First English edition 1985
Translation from the Polish orginal
Astrofizyka relatywistyczna
published in 1978 by Państwowe Wydawnictwo Naukowe

Cover design by Zygmunt Ziemka

British Library Cataloguing in Publication Data
Demiański, Marek
Relativistic astrophysics. – (International series in natural philosophy; v. 110)
1. Astrophysics
I. Title II. Series
523'. Cl QB460
ISBN 0-08-025042-4

Printed in Poland

Preface

The last fifteen years have brought a number of important astronomical discoveries. During this period quasars, relic radiation, pulsars, X-ray sources and what is probably a black hole were all observed for the first time. Theoretical research also made a great leap forward. John Wheeler and a group of young collaborators became interested in the last stages of the evolution of stars. Their studies gave rise to the black hole concept and initiated the intense development of relativistic astrophysics. Even before the microwave background was discovered, a group led by Jakov Zeldovich began to advocate the hot model of the universe and created a complete cosmological theory explaining the successive stages of the development of the universe from a singular state to the formation of galaxies and stars. In this way, by combining interest in the evolution of the universe and in systems dominated by gravitational effects, relativistic astrophysics emerged as a separate field of research.

The short history of the development of relativistic astrophysics abounds in magnificent achievements and frustrating difficulties. Subtle mathematical considerations have permitted predicting the properties of black holes. On the other hand, we still do not know what physical processes are responsible for the observed radiation from quasars and pulsars.

The range of astrophysical observation has increased considerably over the last decade. It now covers almost the whole spectrum of electromagnetic radiation. Studies of cosmic radiation also supply interesting information, and recently new possibilities have arisen for an improvement in neutrino detection methods and a substantial increase in the sensitivity of gravitational wave detectors. Success in these last two undertakings will bring a new stream of data on processes occurring in stellar interiors and in regions of space where the gravitational field is very strong. It will then be possible to investigate the processes involving black holes more closely and to learn what happens in the interiors of stars and in the central regions of large stellar clusters.

Theoretical studies have also led to some surprising results. Until recently, thermonuclear reactions, which release about 1% of the rest-mass energy, were considered the most effective source of energy. Accretion of matter onto a rotating black hole turns out to be a much more efficient process, in which as much as 50% of the rest-mass energy can be released. Stephen Hawking has recently proved that black holes emit particles. This is a very important result since it concerns the quantum processes in strong gravitational fields and may lead to the long-sought-for links between quantum mechanics and general relativity theory.

The literature of relativistic astrophysics already includes a few monographs. Those which merit particular distinction are **Relativistic Astrophysics** (vol. 1: **Stars and Relativity**, vol. 2: **Structure and Evolution of the Universe**) by Zeldovich and Novikov, and **Gravitation** by Misner, Thorne and Wheeler. My intention is to present, in a more concise form, those achievements of relativistic astrophysics which are already well established by observation and have stood the test of time. Of necessity, many special results are omitted; for example no mention is made of variational principles for rotating stars and only a fragmentary treatment is given of homogeneous anisotropic cosmological models. The book is devoted to relativistic astrophysics and is not an exposition of general relativity. It is assumed that the reader knows the elements of the General Theory of Relativity up to the level of Landau and Lifshitz's classic book, **The Classical Theory of Fields.**

In writing this book, I used original papers and review articles. Basic references are given at the end of each chapter and are often restricted to review articles.

To all who helped me in the writing of the book with their advice, criticism and encouragement, and particularly to Dr. Marek Abramowicz for preparing a draft of the chapter on rotating stars, and to the Physics Editors of the Polish Scientific Publishers, for their patience and understanding, I extend heart-felt thanks.

MAREK DEMIAŃSKI

Contents

CHAPTER 1

GRAVITATIONAL FIELD

1.1. Newton's Theory of Gravitation

Classical celestial mechanics and models of stars are based on the assumption that bodies attract each other according to Newton's theory of gravitation. The fundamental quantity describing the gravitational field in this theory is the gravitational potential ϕ. The distribution of mass is represented by the density of matter ρ as a function of position; the gravitational potential ϕ is then determined from the Poisson equation

$$\Delta\phi = 4\pi G\rho, \tag{1.1}$$

where $G = 6.67 \cdot 10^{-8}\ \mathrm{cm^3\ g^{-1} s^{-2}}$ is the gravitational constant. For the potential ϕ to be determined uniquely, boundary conditions must be given. For a bounded distribution of mass it is usually assumed that far from the sources, at infinity, the gravitational potential vanishes as $1/r$. The Poisson equation can then be solved uniquely and ϕ can be written in the form

$$\phi(\boldsymbol{r}, t) = -G\int \frac{\rho(\boldsymbol{r}', t)}{|\boldsymbol{r}-\boldsymbol{r}'|}\mathrm{d}^3\boldsymbol{r}'. \tag{1.2}$$

Far from the sources, for $|\boldsymbol{r}| = r \gg R$, where R characterizes the size of the region filled with matter, $\dfrac{1}{|\boldsymbol{r}-\boldsymbol{r}'|}$ can be represented by the series

$$\frac{1}{|\boldsymbol{r}-\boldsymbol{r}'|} = \frac{1}{r} - x'^a\left(\frac{1}{r}\right)_{,a} + \frac{1}{2}x'^a x'^b\left(\frac{1}{r}\right)_{,ab} + \ldots \tag{1.3}$$

The small Latin indices $a, b, c, \ldots$ run over the values $1, 2, 3$, and repeated indices denote summation, e.g. $x^a p_a = x^1 p_1 + x^2 p_2 + x^3 p_3$. A comma denotes partial differentiation, e.g. $r_{,a} = \dfrac{\partial r}{\partial x^a}$. Substituting this expansion into (1.2) gives

$$\phi(\boldsymbol{r}, t) = -\frac{GM}{r} + G\left(\frac{1}{r}\right)_{,a}\int x'_a \rho(\boldsymbol{x}', t)\,\mathrm{d}^3\boldsymbol{x}' - \frac{1}{2}G\left(\frac{1}{r}\right)_{,ab}\int x'_a x'_b \rho(\boldsymbol{x}', t)\,\mathrm{d}^3\boldsymbol{x}' + \ldots, \tag{1.4}$$

where $M = \int \rho\,\mathrm{d}^3\boldsymbol{x}$. In the centre-of-mass system this reads

$$\phi(\boldsymbol{r}, t) = -\frac{GM}{r} - \frac{1}{6}GQ^{ab}\left(\frac{1}{r}\right)_{,ab}, \tag{1.5}$$

where

$$Q^{ab} = \int \rho(x', t)(3x'^a x'^b - x'^c x'_c \delta^{ab}) d^3x' \tag{1.6}$$

denotes the quadrupole moment of the mass distribution.

The equations of motion for a test particle can be found from Newton's second law, which takes the form

$$m\ddot{x}_a = -m\phi_{,a}. \tag{1.7}$$

Owing to the equality of gravitational and inertial mass, an experimentally confirmed fact, gravitational forces are locally indistinguishable from inertial forces. A gravitational field is generally an inhomogeneous field, and by passing to a noninertial reference frame one can eliminate it only along an arbitrary curve.

Now consider two test particles moving in a gravitational field described by a potential ϕ. If their positions are x^a and $x^a + \delta x^a$, respectively, the relative acceleration of the particles is given by

$$m\delta\ddot{x}^a = -mK^a{}_b\,\delta x^b, \tag{1.8}$$

where $K^a{}_b = \phi^{,a}{}_{,b}$. In this case the acting force is called the *tidal force*; as can be seen from the form of $K^a{}_b$, it is of quadrupole character. It is this force produced by the gravitational field of the Moon that gives rise to tides.

1.2. Gravitational Field in the General Theory of Relativity

Newton's theory of gravitation requires the concepts of absolute time and absolute space. In this theory gravity propagates at an infinite velocity.

The Special Theory of Relativity introduced the concept of space-time. In this four-dimensional space both time and space lose their absolute character, and become dependent on the choice of observer. In addition, the speed of light is the maximum possible velocity at which information can be transmitted.

Having formulated the principles of the Special Theory of Relativity in 1905, Albert Einstein began to work on a relativistic gravitational theory. A general outline of such a theory was set out by Einstein in 1916. The theory is known as the General Theory of Relativity (G.T.R.). Created by pure deduction, the General Theory of Relativity is one of the most beautiful theories in physics. The two cornerstones on which G.T.R. rests are the Principle of Equivalence and the Principle of General Covariance. The Principle of Equivalence states that locally, in small regions of space-time, the influence of a gravitational field can be eliminated by passing to a noninertial reference frame. Gravitational fields produced by bounded distributions of matter differ from inertial forces in their global properties. Such fields vanish at infinity, while the apparent inertial forces acting in noninertial reference frames grow boundlessly at infinity or, at best, tend to some constant value.

The Principle of Equivalence is a consequence of the observed equality of gravitational and inertial mass. Galileo's observations implied proportionality of the two masses. Their equality was later confirmed by Eötvos and, with greater precision, by Dicke. The most precise measurements, carried out by Braginsky, established this equality to the accuracy of 10^{-12}.

The Principle of General Covariance says that the laws of physics should be formulated in such a way that they may be invariant under arbitrary coordinate transformations in space-time.

We do not propose to expound general relativity, but we will recall some general facts, to be used further on.

The distance between two neighbouring points in space-time is defined as

$$ds^2 = g_{\alpha\beta}\, dx^\alpha dx^\beta, \tag{1.9}$$

where $g_{\alpha\beta}$ is a symmetric tensor † with signature $(+, -, -, -)$, called the metric tensor, and small Greek indices run from 0 to 3. A test particle moves in space-time along a timelike geodesic, i.e. a curve $x = x(s)$ such that

$$\frac{d^2x^\alpha}{ds^2} + \Gamma^\alpha_{\beta\rho} \frac{dx^\beta}{ds} \frac{dx^\rho}{ds} = 0, \qquad \frac{dx^\alpha}{ds} \cdot \frac{dx_\alpha}{ds} = 1, \tag{1.10}$$

where

$$\Gamma^\alpha_{\beta\rho} = \tfrac{1}{2} g^{\alpha\nu}(g_{\beta\nu,\rho} + g_{\rho\nu,\beta} - g_{\beta\rho,\nu}) \tag{1.11}$$

is a Christoffel symbol. Equation (1.10) of the geodesic has a simple geometrical meaning. It implies that the tangent vector $u^\alpha = \dfrac{dx^\alpha}{ds}$ satisfying the equation

$$u^\beta \nabla_\beta u^\alpha = u^\alpha{}_{;\beta} u^\beta = \dot{u}^\alpha = (u^\alpha{}_{,\beta} + \Gamma^\alpha_{\beta\rho} u^\rho) u^\beta = 0 \tag{1.12}$$

is parallel-displaced along the geodesic. Geometrically, the parameter s is interpreted as the arc length. From the physical point of view, it is the proper time.

Photons and other particles moving with the speed of light travel in space-time along null geodesics. These are geodesics along which $ds = 0$. The arc length is no longer a good parameter here. To describe null geodesics we introduce an affine parameter λ. Let $x^\alpha = x^\alpha(\lambda)$; the parameter λ is called an affine parameter if the tangent vector $k^\alpha = \dfrac{dx^\alpha}{d\lambda}$, which is a null vector ($k^\alpha k^\beta g_{\alpha\beta} = k^\alpha k_\alpha = 0$), satisfies the equation $k^\alpha{}_{;\beta} k^\beta = 0$. It can be seen that the affine parameter is not determined uniquely and transformations $\lambda \to a\lambda + b$, where a and b are constants, are allowed.

The metric tensor $g_{\alpha\beta}$ is related to the energy-momentum tensor $T_{\alpha\beta}$ of the matter distribution by Einstein's field equations

$$G_{\alpha\beta} = R_{\alpha\beta} - \frac{1}{2} g_{\alpha\beta} R = \frac{8\pi G}{c^4} T_{\alpha\beta}, \tag{1.13}$$

† More precisely, $g_{\alpha\beta}$ is a symmetric covariant tensor of rank two, i.e. an object which under a change f the coordinate system $x'^\mu = x'^\mu(x^\sigma)$ is transformed according to the formula $g_{\alpha\beta} = \dfrac{\partial x'^\mu}{\partial x^\alpha} \dfrac{\partial x'^\nu}{\partial x^\beta} g_{\mu\nu}$. At every fixed point of space-time, $g_{\alpha\beta}$ is a symmetric tensor whose diagonal form has one positive and three negative elem nts.

where $R=R^{\alpha}_{\alpha}$ is the curvature scalar, $R_{\alpha\beta}=R^{\sigma}{}_{\alpha\sigma\beta}$ the Ricci tensor and $R^{\alpha}{}_{\beta\gamma\delta}$ the curvature tensor. $G_{\alpha\beta}$ is called the Einstein tensor. Specific forms of the energy-momentum tensor for a perfect fluid and an electromagnetic field will be given in Chapter 2.

Just as the Poisson equation in Newton's theory of gravitation does not determine the gravitational potential uniquely, the Einstein equations in the General Theory of Relativity do not determine the metric uniquely. They have to be supplemented with boundary conditions. When we consider a spatially bounded distribution of matter, we assume that, asymptotically, the metric tensor $g_{\alpha\beta}$ approaches the Minkowski metric. Such a gravitational field, or rather such a geometry, is said to be asymptotically flat.

1.3. Local Inertial Frames

While observing the motion of the planets or stars, we refer the results of observation to an inertial frame. In the first case we can take as an inertial frame a frame with the origin at the centre of the Sun, in the second case a frame with its origin at the centre of the Galaxy. According to the Principle of Equivalence, the General Theory of Relativity only allows local inertial frames.

Consider an arbitrary observer O moving in space-time and suppose that his world-line, i.e. the trajectory of his motion, is given by $x^{\alpha}=x^{\alpha}(s)$. As one of the basis vectors of a local reference frame we can choose the observer's velocity four-vector $u^{\alpha}=\frac{dx^{\alpha}}{ds}$. We remember that as s is proper time, the velocity four-vector is a unit timelike vector because $u^{\alpha}u_{\alpha}=1$. The fact that this vector is chosen as the timelike vector of our basis is symbolically denoted by $\boldsymbol{e}_{\hat{0}}=\boldsymbol{u}$.† In the plane perpendicular to the velocity four-vector $\boldsymbol{u}$ we choose three spacelike unit vectors $\boldsymbol{e}_{\hat{a}}$ such that $\boldsymbol{e}_{\hat{a}}\cdot\boldsymbol{e}_{\hat{b}}=-\delta_{\hat{a}\hat{b}}$, where $\boldsymbol{e}_{\hat{a}}\cdot\boldsymbol{e}_{\hat{b}}$ denotes the scalar product with respect to the metric in space-time. Such a system of vectors $\boldsymbol{e}_{\hat{0}}, \boldsymbol{e}_{\hat{a}}$ is called a local reference frame of the observer O.

If, for example, a photon with wave four-vector k^{α} is received by our observer, the observer will assign to it a wave vector $k_{\hat{a}}=\boldsymbol{k}\cdot\boldsymbol{e}_{\hat{a}}$ and a wavelength λ given by $\boldsymbol{k}\cdot\boldsymbol{e}_{\hat{0}}=\frac{2\pi}{\lambda}$.

We now need a prescription defining the way in which such a basis, chosen at one point, will be propagated along the observer's world-line. What matters here is that the resulting local reference frame should approximate to an inertial frame as close as possible. When the observer moves with a non-zero acceleration, we are not able to eliminate all inertial forces; however, we can choose the propagation mode so that the spatial axes of the frame do not rotate. The simplest way would be to move the basis vectors parallelly along the observer's world-line. In general, however, the world-line is not a geodesic and therefore the transport of the tangent vector along the world-line is not parallel, and in order to pass to a system in which $\boldsymbol{e}_{\hat{0}}$ will coincide with the velocity four-vector it is necessary to per-

† The basis vectors are denoted by $\boldsymbol{e}_{\hat{0}}, \boldsymbol{e}_{\hat{1}}, \boldsymbol{e}_{\hat{2}}, \boldsymbol{e}_{\hat{3}}$; they satisfy the orthogonality relation $\boldsymbol{e}_{\hat{\alpha}}\cdot\boldsymbol{e}_{\hat{\beta}} = e^{\alpha}{}_{\hat{\alpha}}e^{\beta}{}_{\hat{\beta}}g_{\alpha\beta}=\eta_{\hat{\alpha}\hat{\beta}}$, where $\eta_{\hat{\alpha}\hat{\beta}}=\text{Diag}\,||1,-1,-1,-1||$.

form a Lorentz transformation. Suppose we have

$$\frac{D}{ds} e_{\hat{\alpha}} = -e_{\hat{0}}(\dot{u} \cdot e_{\hat{\alpha}}) + \dot{u}(e_{\hat{0}} \cdot e_{\hat{\alpha}}), \tag{1.14}$$

where $\dot{u}$ denotes the observer's four-acceleration $(u^\beta \nabla_\beta)u$, and $\frac{D}{ds} e_{\hat{\alpha}} = (u^\beta \nabla_\beta) e_{\hat{\alpha}}$. The index $\hat{\alpha}$ numbers the basis vectors. Then it is readily seen that when the observer moves along a geodesic and $\dot{u}=0$, the basis vectors are parallelly transported. A basis obeying formula (1.14) is said to be Fermi–Walker transported. Note that by setting $\hat{\alpha}=0$ in (1.14), we simply get the identity $\dot{u} = \frac{D}{ds} e_{\hat{0}}$, while for the spatial vectors of the basis we have

$$\frac{D}{ds} e_{\hat{a}} = -e_{\hat{0}}(\dot{u} \cdot e_{\hat{a}}). \tag{1.15}$$

It remains to check that Fermi–Walker transport preserves the orthogonality relations between the basis vectors,

$$e_{\hat{\alpha}} \cdot e_{\hat{\beta}} = \eta_{\hat{\alpha}\hat{\beta}}, \tag{1.16}$$

where $\eta_{\hat{\alpha}\hat{\beta}}$ is the Minkowski metric expressed in Cartesian coordinates. Using (1.14), one can easily find that

$$\frac{D}{ds}(e_{\hat{\alpha}} \cdot e_{\hat{\beta}}) = \left(\frac{D}{ds} e_{\hat{\alpha}}\right) \cdot e_{\hat{\beta}} + e_{\hat{\alpha}} \cdot \left(\frac{D}{ds} e_{\hat{\beta}}\right) = 0, \tag{1.17}$$

which implies that the angles between the basis vectors are preserved. More generally, Fermi–Walker transport has the property of preserving scalar products.

Physically, a local inertial reference frame can be realized by means of a system of three gyroscopes suspended at the centre of mass and oriented along three perpendicular directions. In order to measure vectorial or tensorial quantities, we usually project them onto the basis vectors $e_{\hat{\alpha}}$, obtaining "physical components". As an example, let us take the electromagnetic field tensor $F_{\mu\nu}$. In a local inertial system we assign to it the components $F_{\hat{\alpha}\hat{\beta}} = e^\mu{}_{\hat{\alpha}} F_{\mu\nu} e^\nu{}_{\hat{\beta}}$, and use them to determine the local properties of the field. Specifically, the electric field is given by $E^{\hat{\alpha}} = F^{0\hat{\alpha}}$ and the magnetic field by $B^{\hat{\alpha}} = \epsilon^{\hat{0}\hat{\alpha}\hat{\beta}\hat{\gamma}} F_{\hat{\beta}\hat{\gamma}}$.† The complete system of Maxwell's equations can be written in the form

$$F_{[\hat{\alpha}\hat{\beta};\hat{\gamma}]} = 0, \quad F^{\hat{\alpha}\hat{\beta}}{}_{;\hat{\beta}} + F^{\gamma\hat{\beta}}(e^{\hat{\alpha}}{}_{\alpha} e^{\alpha}{}_{\hat{\gamma};\hat{\beta}} + \delta^{\hat{\alpha}}{}_{\gamma} e^{\beta}{}_{\hat{\beta};\beta}) = \frac{4\pi}{c} j^{\hat{\alpha}}. \tag{1.18}$$

Similarly, one can write all known laws of physics in a locally inertial reference frame.

1.4. Spherically Symmetric Gravitational Fields

The spherically symmetric gravitational field plays a fundamental role in astrophysical considerations. Gravitational fields of stars and clusters of stars are very well approximated by spherically symmetric fields.

† $\epsilon^{\hat{\alpha}\hat{\beta}\hat{\gamma}\hat{\delta}}$ is a completely antisymmetric Levi-Civita symbol such that $\epsilon^{\hat{0}\hat{1}\hat{2}\hat{3}} = 1$.

Intuitively, the concept of spherical symmetry is clear. If a field is spherically symmetric, there should exist a coordinate system in which the field strength depends only on the distance from a certain distinguished point – the centre of symmetry. To formulate the concept of a spherically symmetric gravitational field mathematically, let us first recall the definition of the group of rotations in a three-dimensional Euclidean space R^3. The group of rotations in R^3 is the three-parameter group of transformations under which a given point, for convenience chosen to be the origin of the coordinate system, is invariant. In Cartesian coordinates the generators of this group are

$$X_1=z\frac{\partial}{\partial y}-y\frac{\partial}{\partial z},\quad X_2=x\frac{\partial}{\partial z}-z\frac{\partial}{\partial x},\quad X_3=y\frac{\partial}{\partial x}-x\frac{\partial}{\partial y}. \tag{1.19}$$

In the spherical coordinates,

$$\begin{aligned} x&=r\sin\theta\cos\varphi,\\ y&=r\sin\theta\sin\varphi,\\ z&=r\cos\theta, \end{aligned} \tag{1.20}$$

the generators become

$$\begin{aligned} X_1&=\sin\varphi\frac{\partial}{\partial\theta}+\cot\theta\cos\varphi\frac{\partial}{\partial\varphi},\\ X_2&=-\cos\varphi\frac{\partial}{\partial\theta}+\cot\theta\sin\varphi\frac{\partial}{\partial\varphi},\\ X_3&=-\frac{\partial}{\partial\varphi}. \end{aligned} \tag{1.21}$$

They satisfy the well-known commutation rules

$$[X_a, X_b]=\epsilon_{abc}X_c, \tag{1.22}$$

where ϵ_{abc} is a completely antisymmetric tensor such that $\epsilon_{123}=1$.

Now consider a space-time with a metric tensor $g_{\alpha\beta}(x^\mu)$. If we perform an infinitesimal coordinate transformation $x^\alpha\to x^\alpha+\xi^\alpha(x^\mu)$, the question arises under what conditions the form of the metric will remain unchanged. Supposing the transformed metric to be the same function as the original metric, from the transformation law $g_{\alpha\beta}(x'^\mu)=g_{\rho\sigma}(x^\mu+\xi^\mu)\dfrac{\partial x'^\rho}{\partial x^\alpha}\dfrac{\partial x'^\sigma}{\partial x^\beta}$ one can derive the following condition for ξ^μ:

$$\xi_{\alpha;\beta}+\xi_{\beta;\alpha}=2\xi_{(\alpha;\beta)}=0, \tag{1.23}$$

where the parantheses denote symmetrization over the indices they enclose. The above condition is known as Killing's equation and any vector ξ_α that satisfies it is said to be a Killing vector. The Killing vectors generate a group of isometries, i.e. coordinate transformations under which the metric tensor is form-invariant.

A spherically symmetric gravitational field is a space-time in which the three-parameter group of rotations acts as a group of isometries. In other words, there exist three linearly independent Killing vectors $\xi_{\hat{a}}$† such that

$$[\xi_{\hat{a}}, \xi_{\hat{b}}] = \epsilon_{\hat{a}\hat{b}\hat{c}}\xi_{\hat{c}}. \tag{1.24}$$

The condition of spherical symmetry permits a considerable simplification in the form of the line element. Space-time can be parameterized by means of arbitrary coordinate systems; in the case under consideration the choice of time t and spherical coordinates suggests itself. Assuming that the Killing vectors are given by (1.21) and solving equations (1.23), we eventually obtain the general form of the metric as

$$ds^2 = A\,dt^2 + 2D\,dt\,dr - B\,dr^2 - C(d\theta^2 + \sin^2\theta\,d\varphi^2), \tag{1.25}$$

where A, B, C and D are arbitrary functions of t and r. The most general coordinate transformation that leaves the form of the metric unchanged is given by $t = g(t', r')$, $r = f(t', r')$; its Jacobian should of course be non-zero. The functions f and g can be chosen so as to eliminate the mixed term $dt\,dr$ from the line element (1.25) and to make the function C equal to r^2. Then, we have

$$ds^2 = e^{\nu}c^2dt^2 - e^{\lambda}dr^2 - r^2(d\theta^2 + \sin^2\theta\,d\varphi^2), \tag{1.26}$$

where ν and λ are arbitrary functions, dependent only on t and r.

There is still room for arbitrary changes of the time coordinate

$$t = f(t). \tag{1.27}$$

Einstein's field equations for metric (1.26) are reduced to the following system:

$$\frac{8\pi G}{c^4}T_0^0 = -e^{-\lambda}\left(\frac{1}{r^2} - \frac{\lambda'}{r}\right) + \frac{1}{r^2}, \tag{1.28}$$

$$\frac{8\pi G}{c^4}T_1^1 = -e^{-\lambda}\left(\frac{\nu'}{r} + \frac{1}{r^2}\right) + \frac{1}{r^2}, \tag{1.29}$$

$$\frac{8\pi G}{c^4}T_2^2 = \frac{8\pi G}{c^4}T_3^3 = -\frac{1}{2}e^{-\lambda}\left(\nu'' + \frac{\nu'^2}{2} + \frac{\nu' - \lambda'}{r} - \frac{\nu'\lambda'}{2}\right) + \frac{1}{2}e^{-\nu}\left(\ddot{\lambda} + \frac{\dot{\lambda}^2}{2} - \frac{\dot{\lambda}\dot{\nu}}{2}\right), \tag{1.30}$$

$$\frac{8\pi G}{c^4}T_0^1 = -e^{-\lambda}\frac{\dot{\lambda}}{r}; \tag{1.31}$$

a dot denotes differentiation with respect to $x^0 = ct$, and a prime denotes differentiation with respect to r.

The system becomes particularly simple in empty space. If we set the energy-momentum tensor to zero, the result will be

$$e^{-\lambda}\left(\frac{\nu'}{r} + \frac{1}{r^2}\right) - \frac{1}{r^2} = 0, \tag{1.32}$$

† More precisely, we should speak of Killing vector fields satisfying Killing's equation (1.23).

$$e^{-\lambda}\left(\frac{\lambda'}{r}-\frac{1}{r^2}\right)+\frac{1}{r^2}=0, \tag{1.33}$$

$$\dot{\lambda}=0; \tag{1.34}$$

equation (1.30) with the right-hand side equal to zero is automatically satisfied. Adding equations (1.32) and (1.33), we find that

$$\lambda'+\nu'=0, \tag{1.35}$$

and therefore

$$\lambda+\nu=f(t). \tag{1.36}$$

The freedom of choice of the time coordinate, which allows us to add an arbitrary function of time to ν, can now be used to eliminate $f(t)$. Without any loss of generality, we can assume that

$$\lambda+\nu=0. \tag{1.37}$$

Thus in the case of a spherically symmetric gravitational field in empty space it is always possible to choose a coordinate system in which the metric components are time-independent. To put it another way, in an empty space-time, the existence of the three Killing vectors of the group of rotations implies the existence of a fourth independent Killing vector, connected with time-translations. This fourth Killing vector is a timelike vector. For a spherically symmetric gravitational field in empty space, the time coordinate can be chosen so as to make this vector equal to $\frac{\partial}{c\partial t}$. There exists, therefore, a family of spacelike hypersurfaces to which the timelike Killing vector field is orthogonal.† A space-time having this property is said to be static. If a space-time admits a timelike Killing vector ξ_α which is not orthogonal to any family of spacelike hypersurfaces, i.e. $\xi_{[\alpha;\beta}\xi_{\gamma]}\neq 0$, it is called a stationary space-time. From what we have said it follows that a spherically symmetric gravitational field in empty space is static. This statement is known as Birkhoff's theorem. Equation (1.33) can now be integrated easily, and we get

$$e^{-\lambda}=e^{\nu}=1-\frac{r_g}{r}. \tag{1.38}$$

where r_g is a constant.

The boundary conditions are satisfied automatically because for large r, $e^{-\lambda}=e^{\nu}\to 1$ and the metric is asymptotically flat. Far from the mass distribution, where the gravitational field is weak, predictions of general relativity should coincide with those of Newton's theory. As we shall see later, this is so when $r_g=2GM/c^2$; r_g is called the gravitational or the Schwarzschild radius of the system. Here M denotes the mass equivalent to the total energy,

† A vector field ξ determines a family of hypersurfaces if $\xi_{[\alpha;\beta}\xi_{\gamma]}=0$. The Killing vector $\xi=\frac{\partial}{c\partial t}$ satisfies this condition.

hence to the sum of the rest-mass energy, internal energy and (negative) gravitational potential energy.

The line element finally takes the form

$$ds^2 = \left(1 - \frac{r_g}{r}\right)c^2 dt^2 - \frac{dr^2}{1 - r_g/r} - r^2(d\theta^2 + \sin^2\theta\, d\varphi^2). \tag{1.39}$$

This solution of Einstein's equations was for the first time obtained by Schwarzschild in 1916. It describes a gravitational field in empty space, produced by an arbitrary spherically symmetric distribution of matter. It also describes a gravitational field outside moving matter if the motion is spherically symmetric.

The Schwarzschild metric written in the coordinates (t, r, θ, φ) is singular on the surface $r = r_g$. We shall look at the properties of this surface a little later. Now let us only note that the length of the Killing vector $\xi_{\hat{t}} = \frac{\partial}{c\partial t}$ is $\xi_{\hat{t}} \cdot \xi_{\hat{t}} = 1 - \frac{r_g}{r}$. In the region $r > r_g$ it is a timelike vector and therefore the space-time is static there. When $r = r_g$, $\xi_{\hat{t}}$ is a null vector, and when $r < r_g$, it is spacelike and the space-time is neither static nor stationary there. Metric (1.39) has a real singularity at $r = 0$.

1.5. Motion of Test Particles in a Spherically Symmetric Gravitational Field

We can gain a deeper insight into the differences between Newton's theory and the General Theory of Relativity by examining motions of test particles in a spherically symmetric gravitational field. Test particles move along timelike geodesics. If u^α is the velocity four-vector of a test particle, the momentum is given by $p^\alpha = mcu^\alpha$ (m is the particle mass) and therefore

$$p^\alpha p_\alpha = m^2 c^2. \tag{1.40}$$

The motion of the particle can be found by using conservation laws. To this end, we first show that every Killing vector is associated with a constant of the motion, often known as a first integral. Let ξ_α be a Killing vector and p^α the momentum four-vector of the test particle, so that $\frac{D}{ds} p^\alpha = 0$. Differentiating $\xi_\alpha p^\alpha$ we find that

$$\frac{D}{ds}(\xi_\alpha p^\alpha) = \xi_\alpha \frac{D}{ds} p^\alpha + p^\alpha \xi_{\alpha;\beta} u^\beta = 0. \tag{1.41}$$

The right-hand side vanishes because of Killing's equation and the equations of motion of the test particle (the second term can be rewritten as $p^\alpha u^\beta \xi_{\alpha;\beta} = mcu^\alpha u^\beta \xi_{\alpha;\beta} = mcu^\alpha u^\beta \xi_{(\alpha;\beta)} = 0$). Thus $\xi_\alpha p^\alpha = \text{const}$.

Since a spherically symmetric gravitational field admits four Killing vectors, there will exist four first integrals. In Newton's theory, angular momentum is conserved in a spherically symmetric gravitational field. This is so also in general relativity. The motion of a test particle is therefore a plane motion, and, with no loss in generality, we can assume that it takes place in the plane $\theta = \pi/2$. Then there only remain the two independent first integrals $E/c=\xi_{\hat{t}}\cdot p$ and $J=-\xi_{\hat{\varphi}}\cdot p$. The normalization condition (1.40) takes the form

$$\frac{(E/c)^2}{1-r_g/r}-\left(1-\frac{r_g}{r}\right)p_r^2-\frac{J^2}{r^2}=m^2c^2\,; \tag{1.42}$$

remembering that

$$p_r=g_{rr}p^r=-\frac{mc}{1-\dfrac{r_g}{r}}\frac{dr}{ds}=-\frac{E}{c\left(1-\dfrac{r_g}{r}\right)^2}\frac{dr}{cdt}, \tag{1.43}$$

we finally get

$$\frac{1}{1-\dfrac{r_g}{r}}\frac{dr}{cdt}=\frac{mc^2}{E}\sqrt{\left(\frac{E}{mc^2}\right)^2-1+\frac{r_g}{r}-\frac{J^2}{m^2c^2r^2}+\frac{J^2r_g}{m^2c^2r^3}}\,. \tag{1.44}$$

A number of important conclusions can be drawn from this equation (Mielnik and Plebański, 1962). Denote the function occurring under the root by $F(r)$. The test particle will move on a circular orbit $r=r_0=\text{const}$ if $F(r_0)=0$ and $\frac{\partial F}{\partial r}(r_0)=0$. The condition $\frac{\partial F}{\partial r}(r_0)=0$ implies that the particle can move on a circular orbit if its angular momentum J with respect to the centre exceeds $\sqrt{3}mcr_g$. The radius of the orbit of a particle moving with the critical angular momentum $J=\sqrt{3}mcr_g$ is $r=3r_g$, and the energy of the particle is equal to $E=\sqrt{\tfrac{8}{9}}mc^2$. When the angular momentum is greater than $\sqrt{3}mcr_g$, two circular orbits

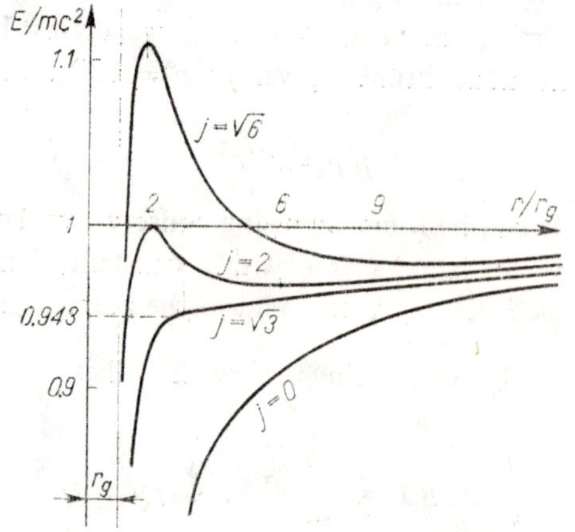

FIG. 1.1. The effective potential of the motion of a test particle in a Schwarzschild space-time for different values of angular momentum j measured in mcr_g units (adapted from Zeldovich and Novikov, 1971).

are possible. The orbit corresponding to the minimum of the function $F(r)$ is stable; the minimum radius of such an orbit is $3r_g$. If J/mcr_g is denoted by j, the radii of the circular orbits are given by

$$r=r_gj(j\mp\sqrt{j^2-3}), \tag{1.45}$$

and the energy of particles moving on these orbits can be found from

$$\left(\frac{E}{mc^2}\right)^2 = 1 - \frac{j \pm \sqrt{j^2-3}}{27j}\left[6 - j(j \pm \sqrt{j^2-3})\right]. \tag{1.46}$$

The upper signs, both in (1.45) and in (1.46), correspond to the unstable orbit.

As $j \to \infty$, the radius of the unstable orbit approaches $\frac{3}{2}r_g$ and the particle energy grows to infinity. Thus, in a spherically symmetric Schwarzschild field there are no circular orbits

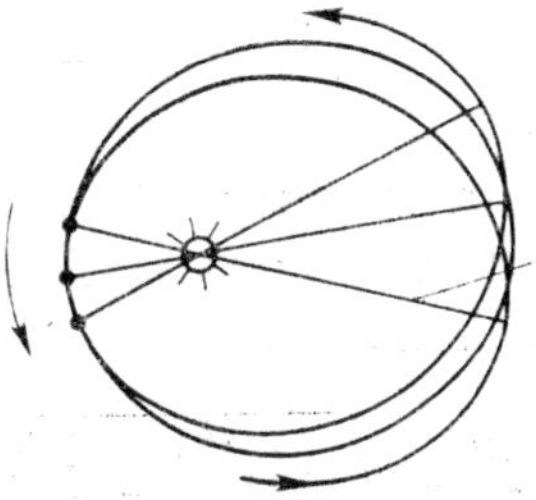

FIG. 1.2. Perihelion motion of Mercury in the gravitational field of the Sun.

with radii less than $\frac{3}{2}r_g$. Let us recall that the Newtonian theory imposes no restrictions on the radii of circular orbits.

The path of the particle can be found, just as in Newton's theory, from the law of conservation of angular momentum. We have

$$J = -\xi_{\hat{\varphi}} \cdot p = -p_\varphi = r^2 \frac{d\varphi}{dr} \frac{E}{c\left(1 - \frac{r_g}{r}\right)} \frac{dr}{cdt}, \tag{1.47}$$

which gives

$$\varphi = \int \frac{J dr}{r^2 mc \sqrt{\left(\frac{E}{mc^2}\right)^2 - 1 + \frac{r_g}{r} - \frac{J^2}{m^2c^2r^2} + \frac{J^2 r_g}{m^2c^2r^3}}}. \tag{1.48}$$

Depending on the values of the total energy E and the angular momentum J, the equation $F(r)=0$ may have two roots $r_{\min}$ and $r_{\max}$ and the particle path is bound and elliptical, or only one or no roots and then the path is unbound. When a particle moving on a bound orbit completes one full cycle, first approaching the centre and then returning to the most distant point, its radius vector will be displaced through the angle

$$\delta\varphi = -2mc \frac{\partial}{\partial J} \int_{r_{\min}}^{r_{\max}} \sqrt{\left(\frac{E}{mc^2}\right)^2 + 1 - \frac{r_g}{r} - \frac{J^2}{m^2c^2r^2} + \frac{J^2 r_g}{m^2c^2r^3}}\, dr - 2\pi. \tag{1.49}$$

Regarding the last term under the root as a small correction, we obtain

$$\delta\varphi = \frac{3}{2} \frac{\pi r_g^2 m^2 c^2}{J^2} = \frac{6\pi G^2 M^2 m^2}{c^2 J^2}, \tag{1.50}$$

where we have used the fact that $r_g = 2GM/c^2$. This slow rotation of the orbit is known as the perihelion motion.

Now examine the simplest kind of motion, namely radial motion where $\varphi = \text{const}$ and $J=0$. Applying the law of conservation of energy to a particle which at an initial moment rested at $r=r_0$, we find $E = mc^2\sqrt{1-r_g/r_0}$, and, using (1.44), we get

$$\frac{dr}{dt} = c\left(1-\frac{r_g}{r}\right)\sqrt{1-\frac{1-\dfrac{r_g}{r}}{1-\dfrac{r_g}{r_0}}}. \tag{1.51}$$

The result is surprising. From the point of view of a distant observer, the radial velocity of a particle falling freely towards the centre will decrease to zero as $r \to r_g$. Note that for r_0 and r much greater than r_g we obtain the well-known Newtonian formula

$$\frac{dr}{dt} = \sqrt{2GM\left(\frac{1}{r}-\frac{1}{r_0}\right)}. \tag{1.52}$$

Suppose a set of observers is placed at rest along the particle path. Each observer measures the local velocity of the particle at the instant it passes him by. In their local inertial reference frames $ds^2 = c^2 d\tau^2 - dR^2$, where $c\,d\tau = \sqrt{1-\frac{r_g}{r}}\,c\,dt$, and $dR = \frac{dr}{\sqrt{1-r_g/r}}$,† so the velocity they will assign to the particle will be

$$v_r = \frac{dR}{d\tau} = \frac{1}{1-\dfrac{r_g}{r}}\frac{dr}{dt} = c\sqrt{1-\frac{1-\dfrac{r_g}{r}}{1-\dfrac{r_g}{r_0}}}. \tag{1.53}$$

From the point of view of local resting observers, the particle moves faster and faster, and as it approaches $r=r_g$ its velocity increases to the speed of light. That a far-away observer sees the velocity of the particle fall to zero is due to time dilation in the gravitational field; to the time interval $\Delta\tau$ measured by a local observer he will assign the time interval $\Delta t = \Delta\tau/\sqrt{1-\frac{r_g}{r}}$.

We may now ask how long it takes for the particle to fall from $r=r_0$ to r_g as measured by remote clock. For this purpose we calculate

$$\Delta t = \int_{r_0}^{r} dt = \int_{r_0}^{r} \frac{1}{c\left(1-\dfrac{r_g}{r}\right)}\left(1-\frac{1-\dfrac{r_g}{r}}{1-\dfrac{r_g}{r_0}}\right)^{-\frac{1}{2}} dr. \tag{1.54}$$

† We are considering radial motions, therefore $\theta = \text{const}$ and $\varphi = \text{const}$.

The integral is divergent as $r \to r_g$, so *from the point of view of a remote observer the particle approaches the surface $r = r_g$ asymptotically and will only reach it after an infinite time.* The proper time measured by an observer moving with the particle is given by

$$\Delta\tau = \frac{1}{c}\int_{r_0}^{r} ds = \frac{1}{c}\int_{r_0}^{r} \frac{dr}{\sqrt{\frac{r_g}{r} - \frac{r_g}{r_0}}} = \frac{1}{c}\sqrt{\frac{r_0}{r_g}}\left(r_0 \arctan\sqrt{\frac{r_0}{r} - 1} - r\sqrt{\frac{r_0}{r} - 1}\right) \tag{1.55}$$

and when $r \to r_g$, $\Delta\tau$ is *finite.*

The equations of motion for photons can be obtained by taking the limit $m \to 0$ and putting $Jc/E = l$ in (1.44) and in the law of conservation of angular momentum; l is interpreted as an impact parameter. In this way we get the equations

$$\frac{dr}{dt} = c\left(1 - \frac{r_g}{r}\right)\sqrt{1 - \frac{l^2}{r^2}\left(1 - \frac{r_g}{r}\right)}, \tag{1.56}$$

$$\frac{d\varphi}{dt} = \frac{lc}{r^2}\left(1 - \frac{r_g}{r}\right), \tag{1.57}$$

from which we can easily find the trajectory

$$\frac{dr}{d\varphi} = \frac{r^2}{l}\sqrt{1 - \frac{l^2}{r^2}\left(1 - \frac{r_g}{r}\right)}. \tag{1.58}$$

If we neglect the relativistic term containing r_g, we obtain an equation of a straight line, so we should expect since in a flat Minkowski space-time photons propagate along straight lines. The trajectory of a light ray passing near a mass, where general relativity cannot be neglected, is bent. Carrying out calculations similar to those for perihelion motion, we find that the asymptotes of the trajectory of a light ray sent from infinity with impact parameter l form the angle

$$\delta\varphi = \frac{2r_g}{l}. \tag{1.59}$$

A light ray or radio wave passing near the sun's limb should be deflected by $\delta\varphi = 1.75$ arc sec. This prediction is presently confirmed by observations with the accuracy of 1 percent.

The fact that light rays are deflected in gravitational fields leads to the following interesting possibility. If a galaxy or a quasar G lies close enough to the line of sight of another more distant galaxy or a quasar S then the gravitational field of the galaxy G acts as a gravitational lense and at least two images of S appear (Refsdal ,1964). The number of images and the luminosity of images depend on the geometry and on the mass distribution in G. The luminosity of images can be enhanced manyfold and several images could appear (Bourassa and Kantowski, 1975).

Recently two examples of a double (Walsh *et al.*, 1979) and a triple quasars (Weymann *et al.*, 1980) were discovered. There is enough observational evidence to believe that they are manifestation of the gravitional lense effect.

There is one more important conclusion concerning the behaviour of photons in a gravitational field to be drawn from the energy conservation law. Remembering that the momentum of a photon is $h\nu/c$, we can write this law as

$$\frac{\nu}{\nu_0}=\sqrt{\frac{1-\dfrac{r_g}{r_0}}{1-\dfrac{r_g}{r}}} \tag{1.60}$$

or, using the wavelength, in the equivalent from

$$\frac{\lambda_0}{\lambda}=\sqrt{\frac{1-\dfrac{r_g}{r_0}}{1-\dfrac{r_g}{r}}}, \tag{1.61}$$

where λ_0 and λ are the wavelengths assigned to the photon by observers at the distances of r_0 and r from the centre, respectively. If $r>r_0$, then $\lambda_0<\lambda$, which means that the wave-

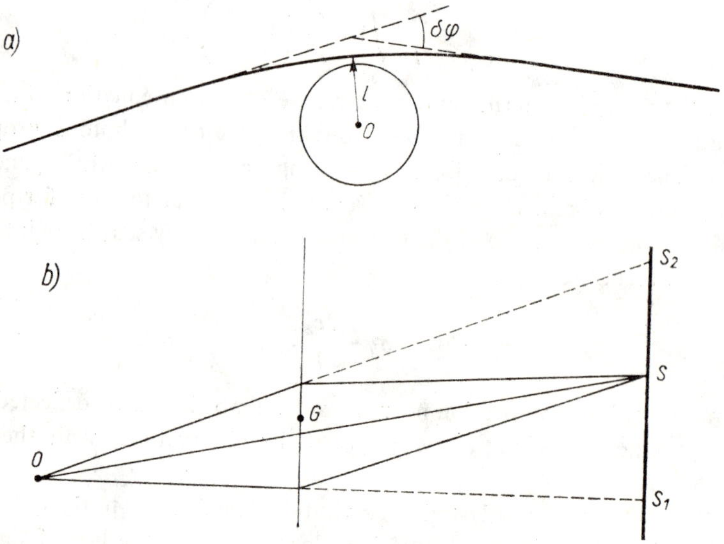

FIG. 1.3. a) Deflection of light in the gravitational field of the Sun. b) Gravitational dense effect.

length of the photon is shifted towards the red part of the spectrum. In the first approximation, the change in the wavelength is proportional to the difference between the gravitational potentials at the emission and the observation points. Indeed, if r_0 and r r_g, then

$$\frac{\lambda_0}{\lambda}=1-\frac{r_g}{2r_0}+\frac{r_g}{2r}\,. \tag{1.62}$$

1.6. Extension of the Schwarzschild Metric

When we examined the motion of test particles in a spherically symmetric Schwarzschild field we noticed that the surface $r=r_g$ is, in a sense, a boundary. From the point of view of an observer at infinity, even particles moving radially towards the centre reach this surface only asymptotically after an infinite time. On the other hand, the proper time measured along the trajectory of such a particle is finite. The natural question arises whether

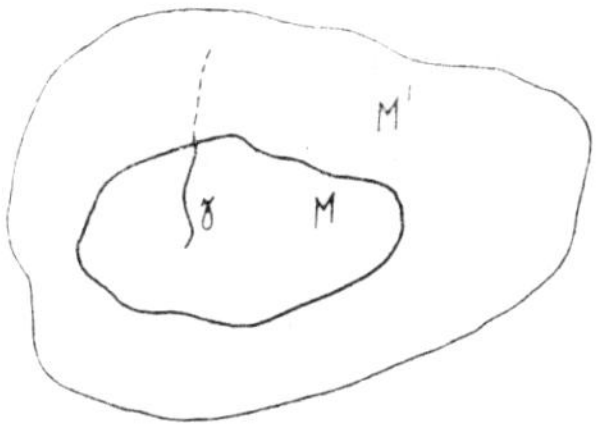

FIG. 1.4. Extension of a manifold M. The curve γ can be extended to the manifold M'.

the surface $r=r_g$ is a real boundary of the space-time or a restriction imposed by the choice of coordinates. The behaviour of the world-lines of test particles and that of scalar polynomials of the curvature tensor suggest that $r=r_g$ is not a real boundary and that the space-time can be extended.

First of all we need to state exactly what is meant by extension. Given two space-times M and M', we shall call M' an extension of M if there exists a metric-preserving embedding (a one-to-one mapping) of M into M' which maps M onto a proper submanifold of M'. The general extension problem is very difficult and it is not known whether it admits a unique solution. However, if the space-time to be extended has some additional properties, for example if it has a group of symmetries, then finding its extension may be much simpler and, indeed, is possible in the cases of interest.

If the coordinates used as a parametrization of M cover only some part of it, say a region U, so there exist incomplete geodesics determining the regular boundary of U, then, leaving aside mathematical subtleties, the extension problem is reduced to finding a new coordinate system in which the geodesics can be extended beyond the boundary of the region U. This coordinate system transformation will in general be singular on the boundary.

An example of a partial extension of the Schwarzschild metric is provided by the following procedure. Instead of the coordinates (t, r, θ, φ), we use, after Eddington and Finkelstein, the coordinates (u, r, θ, φ) or (v, r, θ, φ), where

$$u=ct-r-r_g\ln\left(\frac{r}{r_g}-1\right)=ct-r^*, \tag{1.63}$$

and

$$v=ct+r+r_s\ln\left(\frac{r}{r_g}-1\right)=ct+r^*, \tag{1.64}$$

are retarded and advanced null coordinates respectively. In the coordinates (u, r, θ, φ)

the line element takes the form

$$ds^2=\left(1-\frac{r_g}{r}\right)du^2+2dudr-r^2d\Omega^2, \tag{1.65}$$

and expressed in the coordinates (v, r, θ, φ) it becomes

$$ds^2=\left(1-\frac{r_g}{r}\right)dv^2-2dvdr-r^2d\Omega^2, \tag{1.66}$$

where $d\Omega^2=d\theta^2+\sin^2\theta d\varphi^2$ is the line element on the unit sphere. Both forms of the metric are regular at $r=r_g$.

The light cones for the two metrics are shown in Fig. 1.5 and Fig. 1.6. A test particle moving on a timelike geodesic in the space-time described by the line element (1.66) can reach region T from region R. Regions R and T are fundamentally different. In region R the curves r=const are timelike and a test particle may remain at a constant distance from the centre. This is not possible in region T; moreover, all null geodesics, even those running radially from the centre, are bent and reach the singularity $r=0$ after a finite proper time. It follows that once a test particle enters region T, it must hit the centre $r=0$ after a finite proper time. In region T, not only test particles but also light signals will be caught

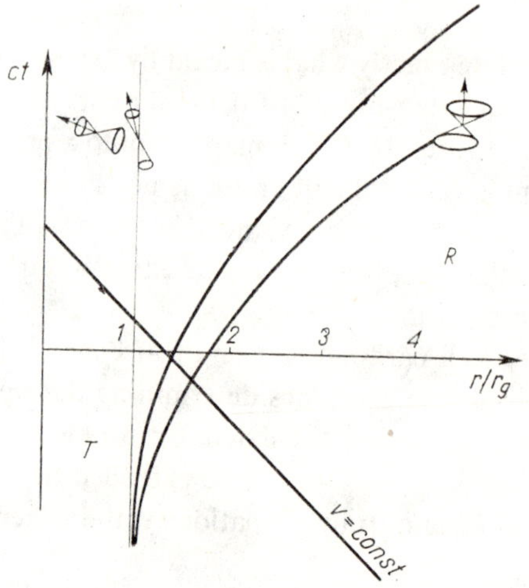

FIG. 1.5. Schwarzschild space-time in advanced coordinates. The light cones and the future direction are indicated. No signal can pass from region T to region R.

by gravitation. No information, therefore, can penetrate from region T to region R. It is not without justification, therefore, that the surface $r=r_g$ is called a *horizon.*

In the coordinate system (u, r, θ, φ) the situation can be thought of as time-reversed. Region T' is also nonstationary; this time, however, test particles and light signals are repelled from the centre and leave region T' after a finite proper time. A particle or a light signal that finds itself outside region T' can never return to it. The surface $r=r_g$ plays the role of a horizon again since no particle or light can penetrate from region R' to region T'.

To gain a better understanding of the properties of Schwarzschild space-time, let us consider its 2-dimensional subspaces. An interesting example is the subspace $t=\text{const}$, $\theta=\pi/2$, whose line element is

$$-ds^2=\frac{dr^2}{1-\frac{r_g}{r}}+r^2d\varphi^2. \tag{1.67}$$

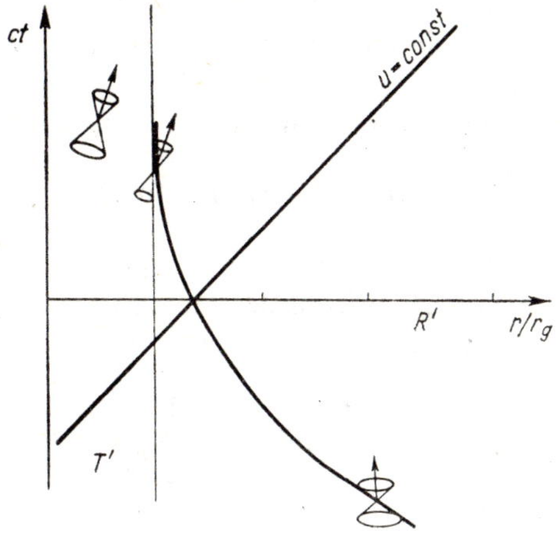

FIG. 1.6. Schwarzschild space-time in retarded coordinates. The light cones and the future directions are indicated. No signal can pass from region R' to region T'.

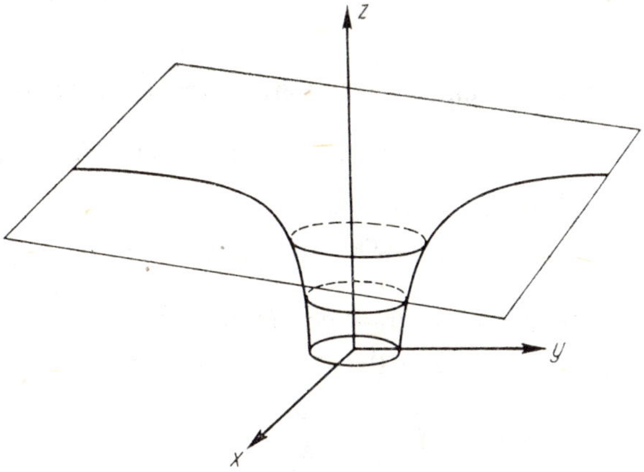

FIG. 1.7. Embedding of the subspace $t=\text{const}$, $\theta=\pi/2$ of Schwarzschild space-time into Euclidean space.

For large r this subspace is locally similar to a 2-dimensional Euclidean plane, and when $r\to r_g$, it bends. Let us embed this surface into a 3-dimensional Euclidean space, in which we introduce for convenience the cylindrical coordinates $\rho^2=x^2+y^2$, $\tilde{\varphi}=\arctan y/x, z$.

The line element of the Euclidean space written in the cylindrical coordinates reads

$$dl^2 = d\rho^2 + \rho^2 d\tilde{\varphi}^2 + dz^2. \tag{1.68}$$

It is easy to check that

$$\rho = r, \quad \varphi = \tilde{\varphi}, \quad z = 2\sqrt{r_g(r - r_g)}, \tag{1.69}$$

is the desired embedding of the surface t=const, $\theta = \pi/2$ for $r \geqslant r_g$. It is shown in Fig. 1.7. In order to extend the metric we would like to extend this image to the region $r < r_g$. The surface t=const, $\theta = \pi/2$, is spacelike for $r > r_g$ and timelike for $r < r_g$. The extension thus requires a new coordinate system.

After Kruskal we introduce new coordinates U, V, θ, φ, to make the line element read

$$ds^2 = -f(dU^2 - dV^2) - r^2 d\Omega^2, \tag{1.70}$$

where $f > 0$ is a function dependent only on $U^2 - V^2 = g(r)$ and regular for $r \to r_g$. In view of spherical symmetry, it is enough to consider the surface t, r, whose line element in Schwarzschild coordinates is given by

$$ds^2 = \left(1 - \frac{r_g}{r}\right) c^2 dt^2 - \frac{dr^2}{1 - \frac{r_g}{r}}. \tag{1.71}$$

Comparing (1.70) and (1.71), we find that

$$U \pm V = h_{\pm}(ct \pm r^*), \tag{1.72}$$

$$-fh'_- h'_+ = 1 - \frac{r_g}{r}, \tag{1.73}$$

where $r^* = r + r_g \ln\left(\frac{r}{r_g} - 1\right)$, h_+ and h_- are arbitrary functions of their arguments and a prime denotes differentiation with respect to those arguments. The functions h_+ and h_- will satisfy the condition $U^2 - V^2 = g(r)$ if

$$h_+(w) = Ae^{Cw}, \quad h_-(w) = Be^{-Cw}, \tag{1.74}$$

where A, B ($A \cdot B > 0$) and C are arbitrary constants.

From equation (1.73) we can find the function f:

$$f = \frac{r - r_g}{rABC^2 e^{2Cr^*}}. \tag{1.75}$$

It will be regular for $r \to r_g$ if $C = \frac{1}{2r_g}$. Then

$$f = \frac{4r_g^3}{ABre^{r/r_g}}. \tag{1.76}$$

By putting the constants A and B equal to unity we set a scale on the coordinate axes U

and V. Finally, we get

$$U-V=\left(\frac{r}{r_g}-1\right)^{\frac{1}{2}} e^{-\frac{ct-r}{2r_g}},$$

$$U+V=\left(\frac{r}{r_g}-1\right)^{\frac{1}{2}} e^{\frac{ct+r}{2r_g}} \qquad (1.77)$$

$$f=4(r_g^3/r)e^{-r/r_g}.$$

Solving this system of equations for U and V in the region I, $r>r_g$, we obtain

$$U=(r/r_g-1)^{\frac{1}{2}}e^{r/2r_g}\cosh\frac{ct}{2r_g},$$

$$V=(r/r_g-1)^{\frac{1}{2}}e^{r/2r_g}\sinh\frac{ct}{2r_g}, \qquad (1.78)$$

$$\frac{U}{V}=\coth\frac{ct}{2r_g},$$

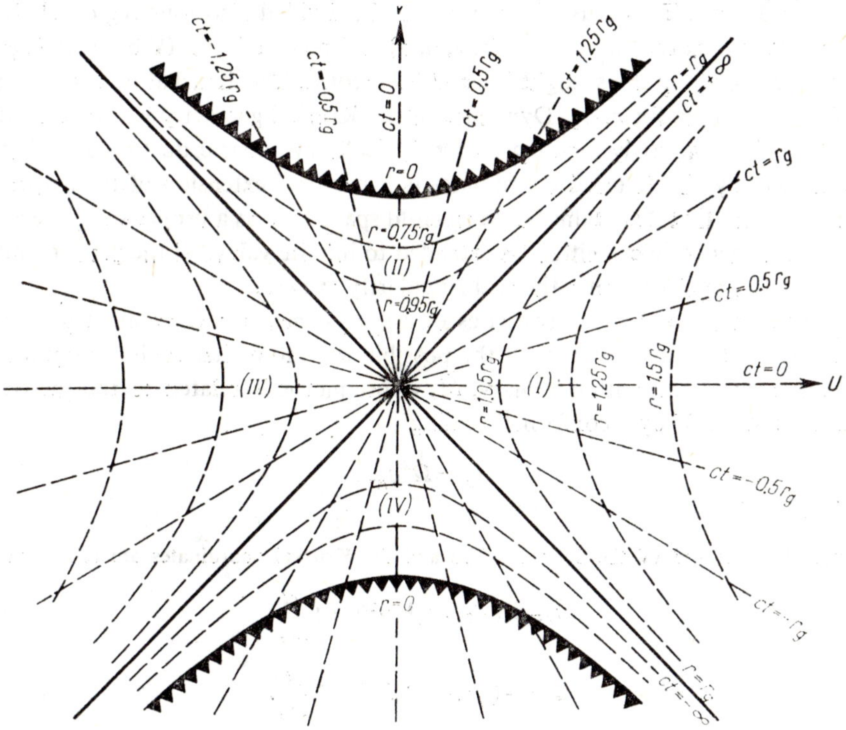

FIG. 1.8. Maximal extension of the Schwarzschild space-time in Kruskal coordinates (Kruskal diagram). The Schwarzschild coordinates cover either region I or region III only. Using the Finkelstein coordinates, one can cover either regions I and II or regions I and IV. Each point of the diagram represents a two-sphere (adapted from Misner, Thorne and Wheeler, 1973).

whereas in the region II $r>r_g$

$$U=(1-r/r_g)^{\frac{1}{2}}e^{r/2r_g}\sinh\frac{ct}{2r_g},$$
$$V=(1-r/r_g)^{\frac{1}{2}}e^{r/2r_g}\cosh\frac{ct}{2r_g}, \tag{1.79}$$
$$\frac{U}{V}=\tanh\frac{ct}{2r_g}.$$

An interpretation of Schwarzschild coordinates is presented in Fig. 1.8. In the plane U, V, in the so-called Kruskal diagram, the lines $t=\text{const}$ are straight lines passing through the origin, and the curves $r=\text{const}$ are represented by rectangular hyperbolas when $r>r_g$ and by conjugate hyperbolas when $r<r_g$. Schwarzschild coordinates cover only part of the Kruskal diagram (region I or III).

Using the Kruskal coordinates, we can find the extension of the surface $t=\text{const}$, $\theta=\pi/2$. Embedding this surface into a three-dimensional Euclidean space gives the surface shown in Fig. 1.9. It is invariant under the transformation $z\to -z$. The Schwarzschild coordinates cover only one half of it, the upper or the lower one.

Finally, some remarks on the interpretation of the Schwarzschild space-time in the Kruskal coordinates. The Kruskal diagram can be divided into four regions.† Regions I and II are covered by the coordinates (v, r, θ, φ) and regions I and IV by (u, r, θ, φ). Light signals propagate along the straight lines $U\pm V=\text{const}$. The Kruskal metric is not static. The field is static only for $r>r_g$. Dynamics of the Kruskal geometry will be considered in more detail when we analyse collapse of spherically symmetric dust clouds. The space-time described by the Kruskal coordinates is a maximal extension of the Schwarzschild metric in the sense that it contains Schwarzschild space-time as a proper subspace and every timelike or null geodesic can either be extended to infinite values of the proper time or the affine parameter, or else reaches the real singularity at $r=0$.

In order to describe global properties of a space-time, it is convenient to compactify it by adding points at infinity. One of the possible ways to realize such a programme is to pass to a new space-time with boundary, $\tilde{M}$, whose metric is related to the metric of the physical space-time M by a conformal scaling

$$\tilde{g}_{\mu\nu}=\Omega^2 g_{\mu\nu}, \tag{1.80}$$

† In regions III and IV the Schwarzschild and the Kruskal coordinates are related as follows:

$$U=-(r/r_g-1)^{\frac{1}{2}}e^{r/2r_g}\cosh\frac{ct}{2r_g},\qquad \text{III}$$
$$V=-(r/r_g-1)^{\frac{1}{2}}e^{r/2r_g}\sinh\frac{ct}{2r_g},$$

$$U=-(1-r/r_g)^{\frac{1}{2}}e^{r/2r_g}\sinh\frac{ct}{2r_g},\qquad \text{IV}$$
$$V=-(1-r/r_g)^{\frac{1}{2}}e^{r/2r_g}\cosh\frac{ct}{2r_g}.$$

where $\Omega^2>0$ is a conformal factor. If Ω is a regular function on $\tilde{M}$ and $\Omega=0$ on the boundary $\mathscr{I}$ of the space-time $\tilde{M}$, with $\nabla_\mu\Omega\neq0$ on $\mathscr{I}$, we say that $\tilde{M}$ is a compactified image of the space-time M.

In the case of a Schwarzschild space-time, if we assume $\Omega=1/r$, we obtain a manifold which in Kruskal coordinates is shown in Fig. 1.10 (a Penrose diagram). Here I^0 represents

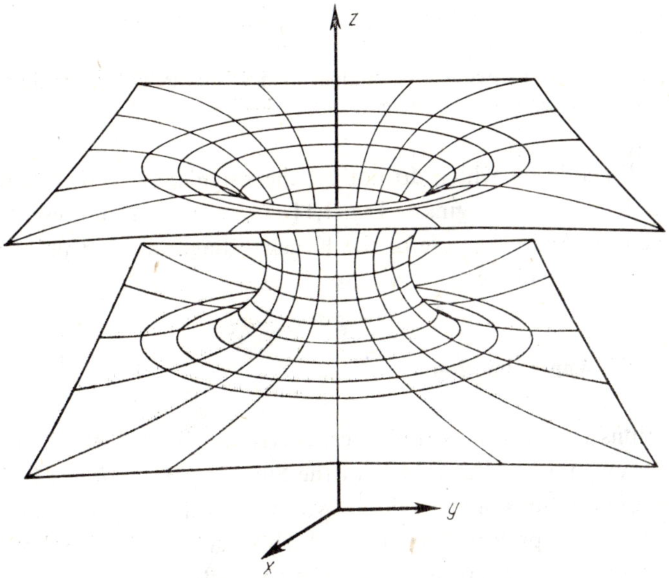

FIG. 1.9. The subspace $t=$const, $\theta=\pi/2$ of the extended Schwarzschild space-time, embedded in the Euclidean space.

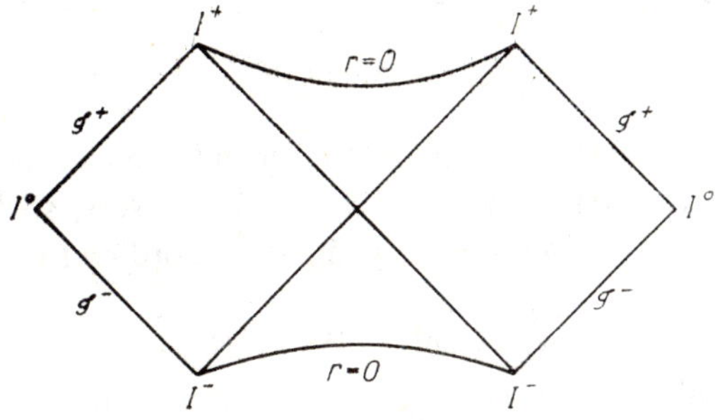

FIG. 1.10. Compactified maximal extension of the Schwarzschild space-time.

spacelike infinity, I^+ and I^- represent timelike infinities in the future and in the past, respectively, and $\mathscr{I}^+$ and $\mathscr{I}^-$ are null infinities in the future and in the past. I^+ is the point of asymptotic convergence of those future-directed timelike geodesics which do not reach the singularity. I^- is the origin, in the infinite past, of timelike geodesics. I^0 is both the starting and terminal point of spacelike geodesics. $\mathscr{I}^+$ is the set of the limit points of those

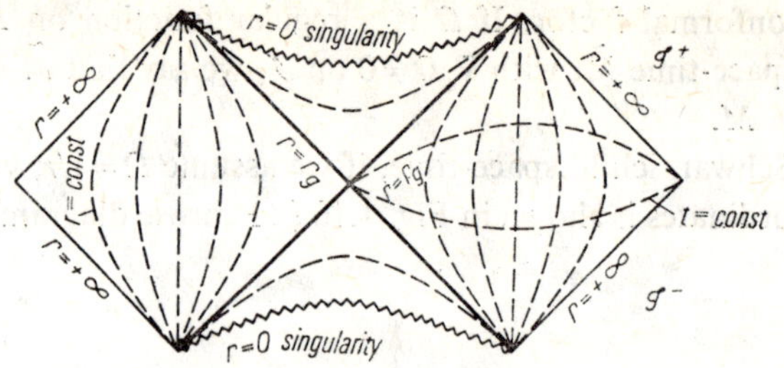

FIG. 1.11. Surfaces t=const and r=const in a compactified Schwarzschild space-time. Every point of the diagram represents a two-sphere.

future-directed null geodesics which do not hit the singularity, and $\mathscr{I}^-$ the set of points from which the null geodesics originate. Fig. 1.11 shows the lines representing the surfaces t=const and r=const. As in the case of the Kruskal diagram, each point on the Penrose diagram represents a two-sphere.

1.7. The Weak-Field Approximation

In most situations we meet in astrophysics, corrections introduced by general relativity are very small, for example on the surface of the Sun the ratio $r_g/R_\odot$ is $4.24 \cdot 10^{-6}$. Instead of applying complex equations of general relativity we can then content ourselves with small corrections. The zeroth approximation from which to start is flat Minkowski space-time.

We choose a coordinate system so as to make the metric read

$$g_{\mu\nu} = \eta_{\mu\nu} + h_{\mu\nu}, \tag{1.81}$$

where $\eta_{\mu\nu}$ is the Minkowski metric tensor in Cartesian coordinates and $h_{\mu\nu}$ are small corrections. The coordinate system is not defined uniquely by those conditions: we can perform transformations

$$x^\mu \to \Lambda^\mu_\nu x^\nu + \xi^\mu(x^\nu). \tag{1.82}$$

The first term is a Lorentz transformation and the second a small correction. Under Lorentz transformations the metric corrections $h_{\mu\nu}$ behave like tensors, so $h'_{\mu\nu} = \Lambda^\alpha_\mu \Lambda^\beta_\nu h_{\alpha\beta}$, whereas under the transformation $x^\mu \to x^\mu + \xi^\mu(x^\nu)$ they change according to

$$h'_{\mu\nu} = h_{\mu\nu} - \xi_{\mu;\nu} - \xi_{\nu;\mu}. \tag{1.83}$$

Note that in the vectorial and tensorial quantities of zero or first order in h, we raise and lower indices using the Minkowski metric, for example $h^\mu_\nu = \eta^{\mu\sigma} h_{\sigma\nu}$.

We now insert the metric tensor (1.81) in the Einstein equations, remembering that the flat space-time has been furnished with Cartesian coordinates, and ignoring non-linear terms in $h_{\mu\nu}$. We obtain

$$h_{\mu\alpha,\nu}{}^{,\alpha} + h_{\nu\alpha,\mu}{}^{,\alpha} - h_{\mu\nu,\alpha}{}^{,\alpha} - h_{,\mu\nu} - \eta_{\mu\nu}(h_{\alpha\beta,}{}^{\alpha,\beta} - h_{,\beta}{}^{,\beta}) = \frac{16\pi G}{c^4} T_{\mu\nu}, \tag{1.84}$$

where $T_{\mu\nu}$ is the energy-momentum tensor and $h = h^\alpha_\alpha$. The field equations simplify if we introduce quantities $\psi_{\mu\nu}$ defined as

$$\psi_{\mu\nu}=h_{\mu\nu}-\tfrac{1}{2}\eta_{\mu\nu}h\,; \tag{1.85}$$

then

$$-\psi_{\mu\nu,\alpha}{}^{,\alpha}-\eta_{\mu\nu}\psi_{\alpha\beta,}{}^{\alpha,\beta}+\psi_{\mu\alpha,}{}^{\alpha}{}_{,\nu}+\psi_{\nu\alpha,}{}^{\alpha}{}_{,\mu}=\frac{16\pi G}{c^4}\,T_{\mu\nu}\,. \tag{1.86}$$

In the linear approximation the curvature tensor has the form

$$R_{\alpha\mu\beta\nu}=\tfrac{1}{2}(h_{\alpha\nu,\mu\beta}+h_{\mu\beta,\nu\alpha}-h_{\mu\nu,\alpha\beta}-h_{\alpha\beta,\mu\nu}) \tag{1.87}$$

which, as can be verified easily, is invariant under transformations (1.83), so that the field equations are also invariant. By a suitable choice of the functions ξ^μ we can eliminate four components of $h_{\mu\nu}$ or, alternatively, we can impose on $h_{\mu\nu}$ four independent differential conditions analogous to the gauge conditions used in electrodynamics. Here we shall assume that

$$\psi^{\nu}_{\mu,\nu}=0, \tag{1.88}$$

simplifying the field equations to

$$\psi_{\mu\nu,\alpha}{}^{\alpha}=-\frac{16\pi G}{c^4}\,T_{\mu\nu}\,. \tag{1.89}$$

The imposed gauge conditions still leave some freedom of choice of the coordinate system: they are satisfied when ξ_μ is an arbitrary solution of the equation $\xi_{\mu,\alpha}{}^{,\alpha}=0$.

We now move on to find the linear approximation of the metric for a spherically symmetric gravitational field. There are two ways: either to solve equation (1.89) or to *linearize the Schwarzschild metric.* We choose the second way. Expanding the Schwarzschild line element with respect to r_g/r, to within linear terms we get

$$ds^2=(1-r_g/r)c^2dt^2-(1+r_g/r)dr^2-r^2d\Omega^2; \tag{1.90}$$

if we change the coordinate r to $r\sqrt{1+\frac{r_g}{r}}$ and restrict ourselves to terms linear in r_g/r again, then

$$ds^2=(1-r_g/r)c^2dt^2-(1+r_g/r)[dr^2+r^2(d\theta^2+\sin^2\theta\, d\varphi^2)], \tag{1.91}$$

and on passing to Cartesian coordinates we finally obtain

$$ds^2=(1+2\phi/c^2)c^2dt^2-(1-2\phi/c^2)(dx^2+dy^2+dz^2), \tag{1.92}$$

whence

$$h_{00}=\frac{2\phi}{c^2}, \quad h_{ij}=\frac{2\phi}{c^2}\,\delta_{ij}, \tag{1.93}$$

where ϕ denotes the Newtonian potential $\phi=-GM/r$. It remains to check whether the gauge condition is satisfied. By simple calculation we find $\psi_{00}=4\phi/c^2$ and $\psi_{ij}=0$, and therefore $\psi^{\nu}_{\mu,\nu}=0$, since ϕ is independent of time. Thus the line element (1.92) is indeed the sought-for linear approximation of the metric for a spherically symmetric gravitational field.

If, instead, we solve equation (1.89) with the energy-momentum tensor of a stationary rotating sphere of mass M and total angular momentum J, we obtain the metric†

$$ds^2=\left(1+\frac{2\phi}{c^2}\right)c^2dt^2-\left(1-\frac{2\phi}{c^2}\right)(dx^2+dy^2+dz^2)+\frac{4GJ}{c^3r^3}\,c\,dt(x\,dy-y\,dx). \quad (1.94)$$

In the weak-field approximation, the gravitational field generated by a rotating spherical body depends only on the total mass M and the total angular momentum J of the body.

1.8. Stationary Gravitational Fields. The Kerr Metric

In astrophysics we often deal with gravitational fields in which by a suitable choice of coordinates one can make the metric tensor time-independent. Such gravitational fields are called stationary fields. Stationary gravitational fields are produced by masses at rest or masses moving relative to an observer resting at infinity in such a way that the density of matter and the velocity of particles in any small volume of space are time-independent. The line element describing a stationary gravitational field can be written in the form

$$ds^2=g_{00}c^2dt^2+2g_{0i}c\,dt\,dx^i+g_{ik}dx^i dx^k, \quad (1.95)$$

where g_{00}, g_{0i} and g_{ik} are functions of the spatial coordinates x^k only. Metric (1.95) is invariant under the coordinate transformation

$$\begin{aligned} t &\to t+T(x^i), \\ x^i &\to X^i(x^k). \end{aligned} \quad (1.96)$$

Stationary gravitational fields can easily be distinguished in an invariant manner. Every such field admits a timelike Killing vector field ξ^α. One can choose a coordinate system so that $\xi=\frac{\partial}{c\,\partial t}$. In such a system, Killing's equations are reduced to $\frac{\partial}{\partial t}g_{\alpha\beta}=0$. If the timelike Killing vector field is proportional to a gradient, i.e. if

$$\xi_{[\alpha;\beta}\xi_{\gamma]}=0, \quad (1.97)$$

the space-time is called static. In static gravitational fields the Killing vector field is orthogonal to spacelike hypersurfaces. They can be chosen as constant-time hypersurfaces.

Knowing the Killing vector field, we can construct the vector ω^α defined as

$$\omega^\alpha=\epsilon^{\alpha\beta\gamma\delta}\xi_\beta\,\xi_{\gamma;\delta}; \quad (1.98)$$

the vanishing of the vector ω^α is a necessary condition for the vector field ξ^α to be proportional to a gradient. If ω^α is different from zero, the gravitational field is stationary. Later we shall see that the vector ω^α is associated with rotation.

Particularly interesting stationary gravitational fields are those which are axisymmetric. Gravitational fields generated by stationary rotating stars are examples of such fields.

† With the coordinate system chosen so that $\mathbf{J}=J\mathbf{e}_z$.

Carter has shown that the line element of the gravitational field generated by a stationary axisymmetric star with purely rotational fluid motions† can be put in the form

$$ds^2 = e^{2\nu}c^2\,dt^2 - e^{2\psi}(d\varphi - \tilde{\omega}\,dt)^2 - e^{2\mu}(d\rho^2 + dz^2), \tag{1.99}$$

where ρ, z and φ are cylindrical coordinates and ψ, ν, $\tilde{\omega}$ and μ are functions of ρ and z.

By using Einstein's equations for empty space it can be shown that with suitable coordinates $e^{\psi} = \rho e^{-\nu}$. However, if our aim is to match up the external and the internal solutions, we have to assume that $e^{\psi} = \rho B e^{-\nu}$, $B = \text{const}$.

The vacuum field equations can be written more simply if the line element is represented as

$$ds^2 = f(c\,dt - \omega\,d\varphi)^2 - \frac{1}{f}\left[e^{2\gamma}(d\rho^2 + dz^2) + \rho^2\,d\varphi^2\right], \tag{1.100}$$

where the functions, f, ω and γ depend only on ρ and z. They can be expressed in terms of the functions ν, μ and $\tilde{\omega}$:

$$\begin{aligned} e^{2\gamma} &= e^{2\mu}(e^{2\nu} - \tilde{\omega}^2\rho^2 e^{-2\nu}/c^2), \\ f &= e^{2\nu} - \tilde{\omega}^2\rho^2 e^{-2\nu}/c^2, \\ \omega &= \frac{\tilde{\omega}\rho^2/c}{e^{4\nu} - \tilde{\omega}^2\rho^2/c^2}. \end{aligned} \tag{1.101}$$

The field equations can now be written in the simple form‡

$$f\nabla^2 f = \nabla f \cdot \nabla f - \rho^{-2} f^4 \nabla\omega \cdot \nabla\omega, \tag{1.102}$$

$$\nabla \cdot (\rho^{-2} f^2 \nabla\omega) = 0, \tag{1.103}$$

and γ is determined by the equations

$$\begin{aligned} 4\gamma_{,\rho} &= \rho\left((\ln f)_{,\rho}^2 - (\ln f)_{,z}^2\right), \\ 2\gamma_{,z} &= \rho(\ln f)_{,\rho}(\ln f)_{,z}. \end{aligned} \tag{1.104}$$

Equation (1.103) will be satisfied identically if we assume that

$$\phi_{,\rho} = \rho^{-1} f^2 \omega_{,z}, \qquad \phi_{,z} = -\rho^{-1} f^2 \omega_{,\rho}. \tag{1.105}$$

The condition for this system to be integrable is

$$(\rho f^{-2}\phi_{,\rho})_{,\rho} + (\rho f^{-2}\phi_{,z})_{,z} = 0. \tag{1.106}$$

† Let $T_{\mu\nu}$ be the energy-momentum tensor and $\xi_\alpha = \frac{\partial}{c\partial t}$, $\eta_\alpha = \frac{\partial}{\partial\varphi}$ the Killing vectors. Carter's assumptions will be satisfied if $\xi_\alpha T^{\alpha[\beta}\xi^\lambda\eta^{\rho]} =$ ` and $\eta_\alpha T^{\alpha[\beta}\xi^\lambda\eta^{\rho]} = 0$.

‡ Here the differential operators and scalar products refer to 3-dimensional Euclidean space parametrized by cylindrical coordinates, for example

$$\nabla \cdot \mathbf{u} = \frac{1}{\rho}(\rho u_\rho)_{,\rho} + u_{z,z} + \frac{1}{\rho} u_{\varphi,\varphi}.$$

In terms of the functions f and ϕ, the field equations read

$$f\nabla^2 f=\nabla f\cdot\nabla f-\nabla\phi\cdot\nabla\phi, \tag{1.107}$$

$$f\nabla^2\phi=2\nabla f\cdot\nabla\phi. \tag{1.108}$$

Introducing a new complex function ξ, the Ernst potential, given by

$$f+i\phi=\frac{\xi-1}{\xi+1}, \tag{1.109}$$

we can write the field equations as one complex equation

$$(\xi\xi^*-1)\nabla^2\xi=2\xi^*\nabla\xi\cdot\nabla\xi. \tag{1.110}$$

The problem of finding a stationary, axisymmetric solution of Einstein's equations in empty space has thus been reduced to solving the Ernst equation (1.110). The equation has to be supplemented with boundary conditions. If we are interested in a gravitational field of a bounded distribution of matter, then, asymptotically, as

$$(\rho^2+z^2)^{\frac{1}{2}}\to\infty,\quad f+i\phi\to 1-\frac{2GM}{c^2(\rho^2+z^2)^{\frac{1}{2}}}-\frac{2iJzG}{c^3(\rho^2+z^2)^{\frac{3}{2}}}, \tag{1.111}$$

where M and J are the total mass and the total angular momentum of the system, respectively.

Among stationary, axisymmetric, vacuum solutions of Einstein's equations, the solution given by Kerr in 1963 is of particular interest. If instead of the coordinates ρ and z we introduce the spheroidal coordinates

$$\begin{aligned} z&=\kappa x\cdot y,\\ \rho&=\kappa(x^2-1)^{\frac{1}{2}}(1-y^2)^{\frac{1}{2}}, \end{aligned} \tag{1.112}$$

where x and y are dimensionless with $1\leqslant x<\infty$, $|y|\leqslant 1$, and $\kappa=\text{const}$, it can readily be verified that

$$\xi=px+iqy \tag{1.113}$$

is a solution of equation (1.108), provided that $p^2+q^2=1$. The constants p and q can be chosen so that $p=\left(1-\dfrac{a^2c^4}{M^2G^2}\right)^{\frac{1}{2}}$, $q=\dfrac{ac^2}{MG}$, where $a=J/Mc$. Passing to coordinates r and θ such that

$$\begin{aligned} x&=\left(\frac{G^2M^2}{c^4}-a^2\right)^{-\frac{1}{2}}\left(r-\frac{GM}{c^2}\right),\\ y&=\cos\theta, \end{aligned} \tag{1.114}$$

we can write the Kerr metric in the form

$$\begin{aligned} ds^2=c^2dt^2&-\frac{2GMr}{c^2(r^2+a^2\cos^2\theta)}(c\,dt-a\sin^2\theta\,d\varphi)^2-(r^2+a^2)\sin^2\theta\,d\varphi^2-\\ &-(r^2+a^2\cos^2\theta)\left(d\theta^2+\frac{dr^2}{r^2-\dfrac{2GMr}{c^2}+a^2}\right). \end{aligned} \tag{1.115}$$

Note that when $a=0$, this metric reduces to the Schwarzschild metric. The physical interpretation of the parameters M and a can be given by analysing the form of the metric far from the sources. When $r\to\infty$, the line element (1.115) can be written as

$$ds^2=\left(1-\frac{2GM}{c^2r}\right)c^2\,dt^2-\left(1+\frac{2GM}{c^2r}\right)(dx^2+dy^2+dz^2)+\frac{4GMa}{c^2r^3}\,c\,dt(x\,dy-y\,dx)\,. \quad (1.116)$$

Comparing this expression with the form of the metric describing a gravitational field of a bounded distribution of matter far from the sources, we come to the conclusion that M is the total mass of the system and Mac is the total angular momentum.

Using the weak-field approximation, Hernandez has shown that a very simple relationship exists between the multipole moments in the Kerr space-time, namely all odd multipole moments vanish, whereas the even ones are

$$Q_{2l}=(-1)^l Ma^{2l}. \quad (1.117)$$

In view of these relationships, only a very special distribution of matter may be the source of the Kerr metric.

A space-time described by the Kerr metric is stationary in the region where $g_{00}>0$. The equation $g_{00}=0$ has two roots

$$r_{E\pm}=\frac{GM}{c^2}\pm\sqrt{\frac{G^2M^2}{c^4}-a^2\cos^2\theta}\,. \quad (1.118)$$

We call the surface $r=r_{E+}$ the stationary limit surface. It is not a horizon and particles may cross to either side of it. We shall examine this more closely when considering the motion of test particles in Kerr space-time.

In a stationary, axisymmetric space-time, the horizon may only be a null, regular, stationary surface of revolution, whose generators coincide with the orbits of the Killing vector. Suppose $f(r,\theta)=\text{const}$ is such a surface. Then

$$\left(r^2-\frac{2GMr}{c^2}+a^2\right)\left(\frac{\partial f}{\partial r}\right)^2+\left(\frac{\partial f}{\partial\theta}\right)^2=0\,. \quad (1.119)$$

It follows that the horizon must be the surface

$$r_\pm=\frac{GM}{c^2}\pm\sqrt{\left(\frac{GM}{c^2}\right)^2-a^2}. \quad (1.120)$$

It is necessary to check whether the vector tangent to this surface can be normalized to coincide with the Killing vector. It is easy to verify that a linear combination of Killing vectors is also a Killing vector. Therefore, the vector tangent to the hypersurface $r=r_+$ should be proportional to $\boldsymbol{l}=\frac{\partial}{c\,\partial t}+\Omega_H\frac{\partial}{c\,\partial\varphi}$, where Ω_H is a constant. It follows from eqn. (1.119) that the vector normal to the horizon, $\boldsymbol{k}$, is isotropic and orthogonal to the Killing vector $\boldsymbol{l}$. If it is possible to choose Ω_H so as to make $\boldsymbol{l}$ isotropic on the horizon, then since $\boldsymbol{l}$ is orthogonal to $\boldsymbol{k}$, the two vectors will have to be proportional. From the isotropy condition for the vector $\boldsymbol{l}$ on the surface $r=r_+$ we find $\Omega_H = ac/(r_+^2+a^2)$. The surface $r=r_+$ is therefore the horizon. From (1.120), the horizon exists only when $\left(\frac{GM}{c^2}\right)^2 - a^2 \geqslant 0$.

To gain further insight into the properties of space-time outside the horizon, notice that the Killing vector $\frac{\partial}{c\,\partial t}$ which is timelike for $r>r_{E+}$ becomes null at the stationary limit surface and is spacelike for $r_+ \leqslant r < r_{E+}$. Therefore in the region $r<r_{E+}$ no particle can be stationary with respect to an observer at infinity. In the region $r_+<r<r_{E+}$ at each point off the rotation axis there is a linear combination of the Killing vectors $\frac{\partial}{c\,\partial t}$ and $\frac{\partial}{\partial \varphi}$ which is timelike, but no such combination exists for $r_{E-}<r<r_+$.

The Kerr metric is singular in the region where $r^2+a^2\cos^2\theta=0$. In a coordinate system which becomes Cartesian when $M=a=0$, this region is represented by a ring of radius a in the equatorial plane.

Just as in the case of the Schwarzschild metric, the space-time described by the Kerr metric is not complete. Completion of the Kerr metric is a complicated problem and will not be dealt with here. We shall restrict ourselves to mentioning the main results. Fig. 1.12

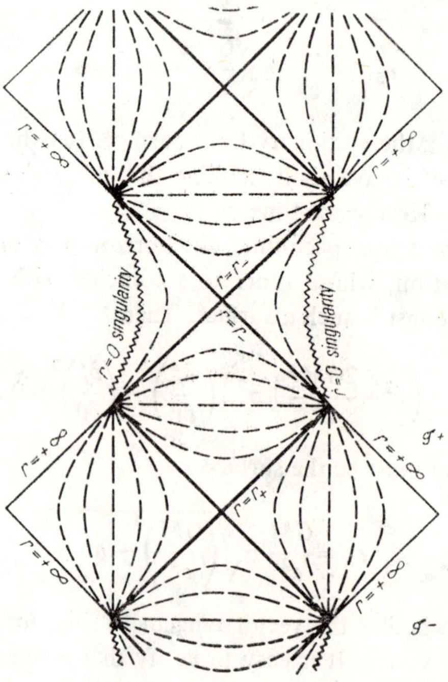

Fig. 1.12. Compactified maximal Kerr space-time. The surfaces r=const are indicated (adapted from Carter in *Black Holes*, 1973).

shows a conformal diagram of the maximal extension of the Kerr metric. It can be seen that the diagram is composed of elements of three types. The first type represents the region between infinity and the horizon, the second represents the region between r_+ and r_-, and the third the region between r_- and singularity. The global structure of the Kerr metric is much richer than that of the Schwarzschild metric. An ingoing observer who crossed the surface of the horizon may again find himself in the region between the horizon and infinity. This, however, will be a different asymptotically flat region of space-time, and the observer will no longer have any causal influence on the phenomena that occur in the region from which he started.

The Kerr metric plays a very important part in astrophysics, being the metric which describes gravitational field outside rotating black hole.

Among other stationary axisymmetric solutions of Einstein's equations, the class of solutions recently given by Tomimatsu and Sato deserves special attention. They have found a general algorithm which permits construction of solutions of equation (1.110) in cylindrical coordinates when the function ξ is a quotient of two polynomials:

$$\xi_n = \frac{\text{polynomial of degree } n^2}{\text{polynomial of degree } n^2-1}. \tag{1.121}$$

The corresponding solution is characterized by constants κ, p and q, with

$$\kappa = \frac{GMp}{c^2 n}, \quad p^2+q^2=1. \tag{1.122}$$

The space-time described by such a metric is asymptotically flat. As in the case of the Kerr metric, there are simple algebraic relations between the higher multipole moments, for example the quadrupole moment is expressed by the formula

$$Q = Ma^2 + \frac{n^2-1}{3n^2} M \left[\left(\frac{GM}{c^2} \right)^2 - a^2 \right]. \tag{1.123}$$

The physical significance of these metrics is obscure.

To end with, let us mention that in recent years a number of authors, using different methods, have obtained a general solution of the Ernst equation, which may represent the most general vacuum stationary axisymmetric and asymptotically flat space-time (Geroch, 1972; Belinsky and Zakharov, 1978; Hoenselaers, Kinnersley and Xanthopoulos, 1979; and others).

1.9. Motion of Test Particles and Light Signals in Kerr Space-Time

The problem of finding the null and timelike geodesics along which photons and test particles move in Kerr space-time may appear hopelessly difficult. As we know, Kerr space-time admits only two Killing vectors, which relate to the laws of conservation of energy and the axial component of angular momentum.

Investigating the Hamilton–Jacobi equation in Kerr space-time, Carter noticed that it was separable and so the problem of geodesics was reduced to quadratures.

As in special relativity, we define the momentum of a test particle as

$$p^\alpha = mc \frac{\mathrm{d}x^\alpha}{\mathrm{d}s}, \tag{1.124}$$

where m is the particle mass and s the arc length. The particle momentum is a timelike vector, its length being constant and equal to

$$p_\alpha p_\beta g^{\alpha\beta} = m^2 c^2. \tag{1.125}$$

Timelike geodesics can be found by putting

$$p_\alpha = -S_{,\alpha} \tag{1.126}$$

and solving the Hamilton–Jacobi equation

$$S_{,\alpha} S_{,\beta} g^{\alpha\beta} = m^2 c^2. \tag{1.127}$$

In Kerr space-time this equation takes the form

$$\left(1+\frac{2Mr(r^2+a^2)}{\Sigma\Delta}\right)\left(\frac{\partial S}{\partial t}\right)^2 + 4\frac{Mra}{\Sigma\Delta}\frac{\partial S}{\partial t}\frac{\partial S}{\partial \varphi} - \frac{\Delta}{\Sigma}\left(\frac{\partial S}{\partial r}\right)^2 - \frac{1}{\Sigma}\left(\frac{\partial S}{\partial \theta}\right)^2 - \frac{\Sigma-2Mr}{\Sigma\Delta\sin^2\theta}\left(\frac{\partial S}{\partial \varphi}\right)^2 = m^2, \tag{1.128}$$

where $\Sigma = r^2 + a^2\cos^2\theta$, $\Delta = r^2 - 2Mr + a^2$ and the system of units is chosen so that $G = c = 1$.

Assuming that the generating function S is of the form

$$S = -Et + \Phi\cdot\varphi + R(r) + \Theta(\theta). \tag{1.129}$$

we can write equation (1.128) as a system of two equations

$$\left(\frac{d\Theta}{d\vartheta}\right)^2 = Q + a^2(E^2 - m^2)\cos^2\theta - \Phi^2\cot^2\theta \tag{1.130}$$

$$\Delta^2\left(\frac{dR}{dr}\right)^2 = [r^2\Delta + 2Mr(r^2+a^2)]E^2 - 4MraE\Phi - (r^2-2Mr)\Phi^2 - \Delta(m^2r^2+Q). \tag{1.131}$$

By equating the derivatives of the function S with respect to the parameters E, Φ and Q

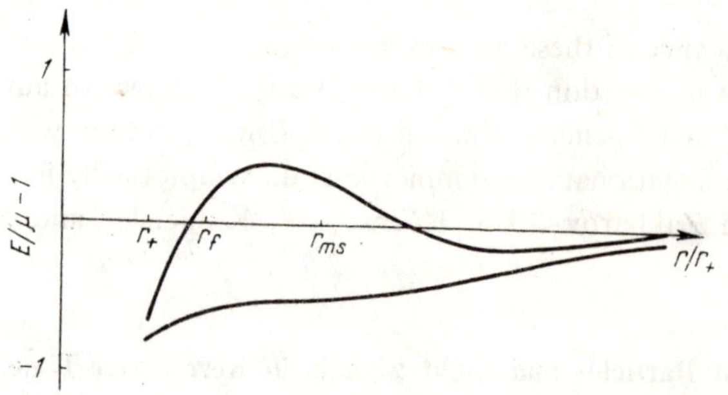

FIG. 1.13. Effective potential of the motion of a test particle in the Kerr space-time for two different values of the angular momentum of the particle (adapted from Bardeen in *Black Holes*, 1973).

to constants, we eventually find the equation of the geodesics:

$$t = \int^r \frac{[r^2\Delta + 2Mr(r^2+a^2)]E - 2ra\Phi}{\Delta(B(r))^{1/2}}\,dr + \int^\theta \frac{a^2E\cos^2\theta}{(A(\theta))^{1/2}}\,d\theta \tag{1.132}$$

$$\varphi = \int^r \frac{(r^2-2Mr)\Phi + 2MraE}{\Delta(B(r))^{1/2}}\,dr + \int^\theta \frac{\Phi\cot^2\theta}{(A(\theta))}\,d\theta \tag{1.133}$$

$$\int^{r}\frac{dr}{(B(r))^{\frac{1}{2}}}=\int^{\theta}\frac{d\theta}{(A(\theta))^{\frac{1}{2}}} \tag{1.134}$$

$A(\theta)$ and $B(r)$ denote the right-hand sides of eqns. (1.130) and (1.131).

In the general case, analytical determination of the timelike geodesics is not possible since the integrals occurring in formulae (1.132)–(1.134) reduce to elliptic integrals (Carter, 1968; de Felice, 1968; Bardeen, Press and Teukolsky, 1972). However, analysis of equations (1.130) and (1.131) provides a great deal of information about the motion of test particles. Equation (1.131) implies that, given constant E and Q, motion is possible only for the values of r for which $B(r)$ is non-negative. Setting $B(r)$ equal to zero, we can determine the relationship $E/m=V(r, \Phi, Q)$. The function $V(r, \Phi, Q)$ describes the effective radial potential. By simple algebraic calculations we find the two roots

$$V_{\pm}(r, \Phi, Q)=\frac{2\hat{r}\hat{a}\hat{\Phi}\pm\sqrt{D}}{\hat{r}^4+\hat{a}^2(\hat{r}^2+2\hat{r})}, \tag{1.135}$$

where

$$D=\hat{r}\hat{\Delta}\{\hat{r}^3\hat{\Phi}^2+(\hat{r}^2+\hat{Q})[\hat{r}^3+\hat{a}^2(\hat{r}+2)]\}. \tag{1.136}$$

To simplify the discussion, we have introduced dimensionless quantities $\hat{r}, \hat{a}, \hat{E}, \hat{\Phi}$ and $\hat{Q}$, defined by $r=\hat{r}M$, $a=\hat{a}M$, $E=\hat{E}m$, $\Phi=\hat{\Phi}Mm$, $Q=\hat{Q}M^2m^2$.

From formula (1.136), D is positive in the region between infinity and the horizon. Therefore, motion is possible when

$$\hat{E}>V_{+}(\hat{r}, \hat{\Phi}, \hat{Q}), \tag{1.137}$$

or when

$$\hat{E}<V_{-}(\hat{r}, \hat{\Phi}, \hat{Q}). \tag{1.138}$$

It follows from the form of equations (1.130) and (1.131) that motions corresponding to the parameters E, Φ, Q and $-E$, $-\Phi$, Q differ only in the sign of the velocity four-vector.

As in the case of the Schwarzschild metric, it can be shown (Wilkins, 1972) that if $\hat{E}^2\geqslant 1$ and $0\leqslant\hat{a}\leqslant 1$, the motion is unbound.

For general motion, we need to consider equation (1.130) again. Introduce the potential $W(\theta, \hat{\Phi}, Q)$ given by

$$W^2=1+\hat{a}^{-2}(\hat{\Phi}^2\sin^{-2}\theta-\hat{Q}\cos^{-2}\theta). \tag{1.139}$$

Equation (1.130) can now be written in the form

$$\left(\frac{d\hat{\Theta}}{d\theta}\right)^2=\hat{a}^2\cos^2\theta(\hat{E}^2-W^2), \tag{1.140}$$

from which one can conclude that the motion is possible only when

$$\hat{E}^2\geqslant W^2. \tag{1.141}$$

It follows from (1.139) that if $Q<0$ then $W^2(\theta)>1$, which implies that the motion is unbound in this case. For bound motion, $Q\geqslant 0$. Equations (1.139) and (1.131) imply that when $Q=0$, the motion takes place in the equatorial plane. A typical shape of the potential $W^2(\theta)$ with $Q>0$ is presented in Fig. 1.14. One can see from it that either the motion is confined to the equatorial plane and then $Q=0$, or, if $Q>0$, the test particle trajectory intersects the equatorial plane.

In the following, we shall restrict ourselves to considering circular orbits lying in the equatorial plane (Bardeen, 1973). We shall assume, therefore, that $Q=0$. Motion in a circular orbit $r=r_0$ will be possible if

$$B(r_0)=0 \quad \text{and} \quad \frac{\partial B}{\partial r}(r_0)=0. \tag{1.142}$$

A circular orbit will be stable when $\frac{\partial^2 B}{\partial r^2}(r_0)<0$ and unstable when $\frac{\partial^2 B}{\partial r^2}(r_0)>0$.

Solving the equations (1.142) for $\hat{E}$ and $\hat{\Phi}$, we obtain

$$\hat{E}=\frac{\hat{r}^{\frac{3}{2}}-2\hat{r}^{\frac{1}{2}}\pm\hat{a}}{\hat{r}^{\frac{3}{4}}(\hat{r}^{\frac{3}{2}}-3\hat{r}^{\frac{1}{2}}\pm2\hat{a})^{\frac{1}{2}}}, \tag{1.143}$$

$$\hat{\Phi}=\pm\frac{\hat{r}^{2}\pm2\hat{a}\hat{r}^{\frac{1}{2}}+\hat{a}^{2}}{\hat{r}^{\frac{3}{4}}(\hat{r}^{\frac{3}{2}}-3\hat{r}^{\frac{1}{2}}\pm2\hat{a})^{\frac{1}{2}}}. \tag{1.144}$$

For large values of $\hat{r}$ ($\hat{r}\gg1$), a particle moving on a circular orbit is bound and $\hat{E}\to1$ as $\hat{r}\to\infty$. Owing to the coupling between the angular momentum of the particle and that of the central body, the binding energy is different from the value it would take in a Schwarzschild field. More precisely, if the particle moves around the centre on a direct orbit, the binding energy is greater than in the spherically symmetric case, while for retrograde orbits

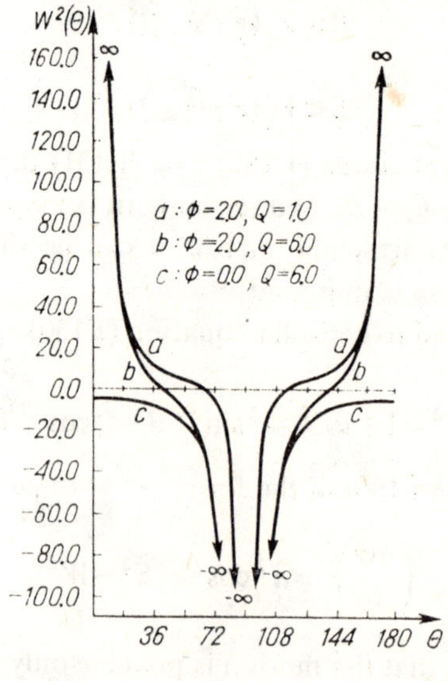

FIG. 1.14. Effective angular potential of the motion of a test particle in Kerr space-time for different values of the angular momentum of the particle (adapted from Wilkins 1972).

this energy is smaller. Note that when $a\neq0$, $V_+(\hat{r}, \hat{\Phi}, \hat{Q})$ becomes negative for retrograde orbits ($a\Phi<0$) sufficiently close to the horizon, regardless of whether or not these orbits lie in the equatorial plane. In this region particles with positive energy in the local observer's frame can have negative energy with respect to infinity. Such orbits may only exist in the region where $g_{00}<0$.

The following circular orbits are of particular interest: the circular photon orbit r_{ph}, the marginally bound orbit r_{mb} and the marginally stable orbit r_{ms}. The radius of the photon orbit can be obtained from the limit condition

$$\hat{r}^{\frac{3}{2}}-3\hat{r}^{\frac{1}{2}}\pm 2\hat{a}=0, \tag{1.145}$$

because in this case $\hat{E}\rightarrow\infty$ and $\hat{\Phi}\rightarrow\infty$. When $\hat{a}=0$, $\hat{r}_{ph}=3$, which corresponds to $r_{ph}=\frac{3}{2}r_g$. When $\hat{a}=1$ and the angular momentum of the central body reaches the maximum value

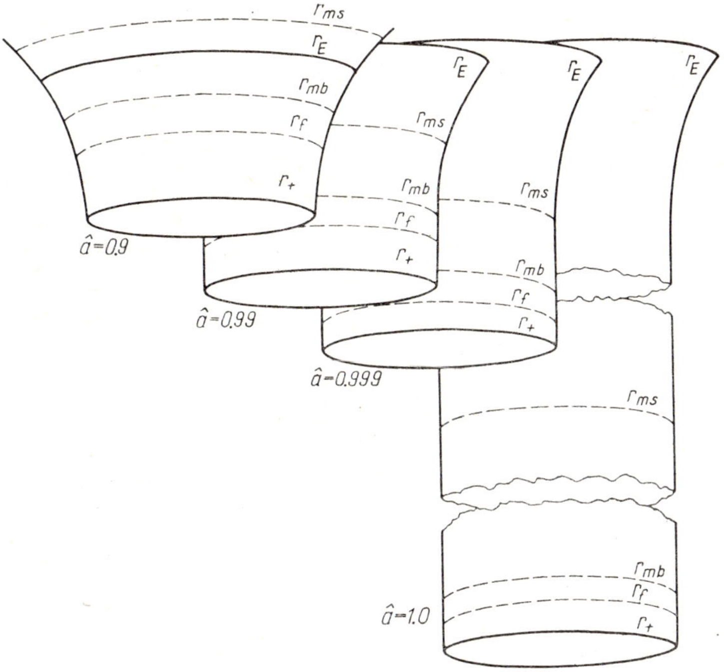

FIG. 1.15. Subspaces $t=$const, $\theta=\pi/2$ of Kerr space-time embedded in Euclidean space. As the parameter $|a|$ increases, the proper distance between the marginally stable circular orbit and the marginally circular photon orbits also increases. When $a^2\rightarrow(GM/c^2)^2$, this distance grows to infinity (adapted from Bardeen, Press and Teukolsky, 1972).

for which the horizon still exists, $\hat{r}_{ph-}=4$ and $\hat{r}_{ph+}=1$. The minimum radius at which the particle is still gravitationally bound can be determined from the condition $\hat{E}=1$; hence we get

$$\hat{r}_{mb}=2\mp\hat{a}+2\sqrt{1\mp\hat{a}}\,. \tag{1.146}$$

The radius of the marginally stable orbit can be found from the conditions

$$B(r_{ms})=0,\quad \frac{\partial B}{\partial r}(r_{ms})=0 \quad\text{and}\quad \frac{\partial^2 B}{\partial r^2}(r_{ms})=0. \tag{1.147}$$

After considerable algebraic manipulations, one gets

$$\hat{r}_{ms}=3+(3\hat{a}^2+Z^2)^{\frac{1}{2}}\mp\{(3-Z)[3+Z+2(3\hat{a}^2+Z^2)^{\frac{1}{2}}]\}^{\frac{1}{2}}, \tag{1.148}$$

where

$$Z=1+(1-\hat{a}^2)^{\frac{1}{3}}[(1+\hat{a})^{\frac{1}{3}}+(1-\hat{a})^{\frac{1}{3}}]. \tag{1.149}$$

When $\hat{a}=0$, $r_{ms}=6$, which gives $r_{ms}=3r_g$, in agreement with the result obtained in section 5, where we discussed motions in the Schwarzschild field.

Note that, just as in the case of the Schwarzschild metric, both for direct and retrograde orbits $\hat{r}_{ms}>\hat{r}_{mb}>\hat{r}_{ph}$. Dependence of the radii of these orbits on the parameter $\hat{a}$ is illustrated in Fig. 1.15.

Literature

BARDEEN, J. M., PRESS, W. H. and TEUKOLSKY, S. (1972) *Astrophys. J.* **178**, 347.

BARDEEN, J. M. (1973) *Timelike and Null Geodesics in the Kerr Metric*, in *Black Holes*, eds.: DeWitt, C., DeWitt, B., Gordon and Breach, N. Y.

BELINSKY, V. A. and ZAKHAROV, V. E. (1978) *Sov. Phys. JETP*, **48**, 985.

BOURASSA, R. R. and KANTOWSKI, R. (1975) *Astrophys. J.* **195**, 13.

CARTER, B. (1968) *Phys. Rev.* **174**, 1559.

CARTER, B. (1973) *Black Hole Equilibrium States*, in *Black Holes*, eds.: DeWitt, C., DeWitt, B., Gordon and Breach, N. Y.

ERNST, F. J. (1968) *Phys. Rev.*, **167**, 1175.

ERNST, F. J. (1968) *Phys. Rev.* **168**, 1415.

DE FELICE, F. (1968) *Nuovo Cimento* **57B**, 351.

GEROCH, R. (1972) *J. Math. Phys.* **13**, 394.

HAWKING, S. W. and ELLIS, G. F. R. (1973) *The Large Scale Structure of Space-Time*, Cambridge University Press, Cambridge.

HOENSELAERS, C., KINNERSLEY, W. and XANTHOPOULOS, B. C. (1979) *Phys. Rev. Lett.* **42**, 481.

KERR, R. P. (1963) *Phys. Rev. Lett.* **11**, 237.

KRUSKAL, M. (1960) *Phys. Rev.* **119**, 1743.

LANDAU, L. D. and LIFSHITZ, E. M. (1962) *The Classical Theory of Fields*, Addison–Wesley, Reading, Mass.

MIELNIK, B. and PLEBAŃSKI, J. (1962) *Acta Phys. Polon.* **21**, 239.

MISNER, C. W., THORNE, K. S., and WHEELER, J. A. (1973) *Gravitation*, Freeman and Comp., San Francisco.

PENROSE, R. (1964) *Conformal Treatment of Infinity*, in *Relativity, Gravitation and Topology*, eds.: DeWitt, B., DeWitt, C., Gordon and Breach, N. Y.

REFSDAL, S. (1964) *Mon. Not. Roy. Astron. Soc.* **128**, 307.

TOMIMATSU, A. and SATO, H. (1972) *Phys. Rev. Lett.*, **29**, 1344.

WALSH, D., CARSWELL, R. F. and WEYMANN, R. J. (1979) *Nature* **279**, 381.

WEYMANN, R. J., LATHAM, D., ANGEL, J. R. P., GREEN, R. F., LIEBERT, L. W., TURNSHEK, D. A., TURNSHEK, D. E. and TYSON, J. A. (1980) *Nature* **285**, 641.

WILKINS, D. C. (1972) *Phys. Rev.* **D5**, 814.

ZELDOVICH, Ya. B. and NOVIKOV I. D. (1971) *Relativistic Astrophysics, Vol. 1: Stars and Relativity*, The University of Chicago Press, Chicago.

CHAPTER 2

ELEMENTS OF RELATIVISTIC THERMODYNAMICS AND HYDRODYNAMICS

2.1. Relativistic Transformation Properties of Thermodynamic Quantities

A local observer comoving with matter can describe its thermodynamic state using such quantities as pressure p, temperature T, specific entropy per particle s, internal energy density ε, etc. In the proper frame comoving with matter, the thermodynamic state is governed by the laws of thermodynamics in their usual form. The standard form of the first law of thermodynamics is

$$đQ = \mathrm{d}\frac{\varepsilon}{n} - đL, \tag{2.1}$$

where $1/n$ is the molecular volume of the substance, $đL$ the work done on the system, and $đQ$ the amount of heat transferred to the system.

The second law of thermodynamics for quasistatic processes has the form

$$\mathrm{d}\frac{s}{n} = \frac{đQ}{T}, \quad \frac{đQ}{T} \geqslant 0. \tag{2.2}$$

It is not difficult to develop phenomenological thermodynamics and statistical mechanics in the proper frame. Difficulties arise when we attempt to give an invariant formulation of thermodynamics, i.e. such that its laws may hold in arbitrary local inertial frames. This formulation is usually based on a discussion of special examples which are used as a heuristic justification of a particular choice of certain fundamental quantities as invariants (Pauli, 1958; Ott, 1963; Møller, 1967). A consistent solution to the problem can be given on the basis of the generalized statistical mechanics proposed by Bergmann.

In this approach, the fundamental quantity is the density matrix ρ. We define the entropy S of the system as

$$S = -k\,\mathrm{Tr}(\rho \ln \rho), \tag{2.3}$$

where k is Boltzmann's constant and ρ is a positive-definite operator such that $\mathrm{Tr}(\rho) = 1$. A system with the given mean values $\langle A^j \rangle$ of the observables A^j is said to be in the state of thermodynamic equilibrium if its entropy is maximum for a given density matrix ρ satisfying

$$\mathrm{Tr}(\rho A^j) = \langle A^j \rangle. \tag{2.4}$$

In other words, in this state

$$\delta S = 0, \quad \delta^2 S < 0, \tag{2.5}$$

for all variations $\delta\rho$ satisfying the conditions

$$\begin{aligned}\delta\,\mathrm{Tr}(\rho A^j)&=0,\\ \delta\,\mathrm{Tr}(\rho)&=0.\end{aligned} \tag{2.6}$$

We shall use these conditions to determine the form of the operator ρ. In a representation of the density matrix, conditions (2.5) and (2.6) take the form

$$\begin{aligned}\delta S&=-k\sum_a \delta\rho_a(\ln\rho_a+1)=0,\\ &\sum_a A^j_a\,\delta\rho_a=0,\\ &\sum_a \delta\rho_a=0.\end{aligned} \tag{2.7}$$

Applying the method of Lagrange multipliers, we find that S is maximum when ρ_a is given by

$$\ln\rho_a+\sum_j\beta_j A^j_a+\lambda+1=0, \tag{2.8}$$

where β_j and λ are the Lagrange multipliers. Thus, we have

$$\rho_a=Z^{-1}\,e^{-\sum_j\beta_j A^j_a}. \tag{2.9}$$

The distribution function Z is defined by

$$Z(\beta_j,\gamma)=\mathrm{Tr}\big(e^{-\sum_j\beta_j A^j}\big), \tag{2.10}$$

where γ represents the set of all external parameters involved in the definition of A^j. The density matrix given by (2.9) corresponds to the canonical distribution in classical statistical mechanics.

For condition (2.4) to be satisfied, the values of the parameters β_j must be determined from

$$\langle A^j\rangle=-\frac{\partial\ln Z}{\partial\beta_j}. \tag{2.11}$$

If the system is in thermodynamic equilibrium, its entropy takes the value

$$S=k\Big(\sum_j\beta_j\langle A^j\rangle+\ln Z\Big). \tag{2.12}$$

In order that the density matrix ρ may be invariant under a linear transformation of A^j, β_j must transform by the transformation inverse to $A^j\to A^{j\prime}$.

The work performed on the system is connected with the change of the mean values $\langle A^j\rangle$ due to variation of the parameter γ. The adiabatic change of $\langle A^j\rangle$

$$\delta_{\mathrm{ad}}\langle A^j\rangle=\Big\langle\frac{\partial A^j}{\partial\gamma}\Big\rangle\delta\gamma \tag{2.13}$$

is called the generalized work done on the system. The generalized heat flux is the difference between the total variation of the mean value $\langle A^j \rangle$ and the generalized work:

$$\delta_Q \langle A^j \rangle = \delta \langle A^j \rangle - \delta_{\mathrm{ad}} \langle A^j \rangle . \tag{2.14}$$

Combining (2.12) and (2.11) gives

$$\delta S = k \sum_j \beta_j \delta_Q \langle A^j \rangle . \tag{2.15}$$

In thermodynamics, a change of the observable value $\langle A^j \rangle$ is usually denoted by $\mathrm{d}U^j$, where U^j is a thermodynamic function or a state variable; $đL^j$ denotes the work performed on the system and $đQ^j$ the generalized heat flux.

Processes which preserve the thermodynamic equilibrium obey the first law of thermodynamics

$$\mathrm{d}U^j = đQ^j + đL^j \tag{2.16}$$

and the second law of thermodynamics

$$\mathrm{d}S = k \sum_j \beta_j đQ^j . \tag{2.17}$$

The general formulae of relativistic thermodynamics will be obtained if the symmetry group of the theory is assumed to be the Lorentz group.

Let us apply this general formalism (in special relativity) to a system whose centre of mass moves along a world-line having unit timelike tangent vector u^α, and whose momentum four-vector is p^α. In the proper frame moving with the centre of mass, both the velocity and the momentum four-vectors have only the time component different from zero. Assume that $A^i \to p^\alpha$. In the proper frame, $p^\alpha = (\mathscr{E}/c, 0, 0, 0)$, where $\mathscr{E}$ denotes the total proper energy of the system. The first law of thermodynamics takes in this frame the well-known form

$$\mathrm{d}\mathscr{E} = đQ + đL . \tag{2.18}$$

The heat $đQ$ transforms therefore as the time component of a four-vector:

$$đQ' = \frac{đQ}{\sqrt{1 - \dfrac{v^2}{c^2}}} . \tag{2.19}$$

In order that the second law of thermodynamics may be invariant under Lorentz transformations, the quantities β_μ must transform as components of a four-vector; assume that

$$\beta_\mu = \beta u_\mu . \tag{2.20}$$

In the proper frame we then obtain

$$\mathrm{d}S = k \beta đQ \tag{2.21}$$

and comparing this expression with (2.2), we can see that $\beta = 1/kT$. Under a change of the inertial frame, the temperature T transforms according to

$$T = T' \sqrt{1 - \frac{v^2}{c^2}} . \tag{2.22}$$

For an arbitrary space-time, in the proper frame, we shall have

$$dS = k\beta\sqrt{g_{00}}\, đQ, \tag{2.23}$$

so that

$$\beta = \frac{1}{kT\sqrt{g_{00}}}.$$

Thus in the general case, the condition of thermal equilibrium reduces to

$$T\sqrt{g_{00}} = \text{const}. \tag{2.24}$$

For a weak field $g_{00} \approx 1 + 2\phi/c^2$, so approximately

$$T = \text{const}\left(1 - \frac{\phi}{c^2}\right). \tag{2.25}$$

The Newtonian gravitational potential ϕ is negative, and therefore the equilibrium state temperature will be higher in regions where $|\phi|$ is greater. If we pass to the limit as $c \to \infty$, the thermal equilibrium condition becomes

$$T = \text{const}. \tag{2.26}$$

In a similar way, one can show that the thermal equilibrium condition for a system of variable number of particles also reduces to (2.24) and that

$$\mu\sqrt{g_{00}} = \text{const}, \tag{2.27}$$

where μ is the chemical potential defined by

$$\mu = \left(\frac{\partial \varepsilon}{\partial n}\right)_{s,v}. \tag{2.28}$$

2.2. The Energy-Momentum Tensor

In order to find a complete solution of Einstein's equations, we need to determine the form of the energy-momentum tensor $T_{\alpha\beta}$. We shall assume that in the cases we are interested in, the matter can to a good approximation be considered to be a continuous medium.

The momentum flux through an element of the body surface is simply the force acting on this element. If we denote the stress tensor by S_{ab}, then $S_{ab}d\Sigma^b$ is the force acting on the surface element $d\Sigma^b$. We assume that Pascal's law holds in the frame in which the given element of the body is at rest, i.e. the pressure acting on this element is identical in all directions. Sometimes, especially in cosmological considerations, we shall assume that pressure is anisotropic and varies depending on direction. Choosing a reference frame which moves with matter and, at the same time, is the proper frame of the stress tensor, we get $S_{ab}d\Sigma^b = p_a d\Sigma_a$ (no summation over a). Assuming Pascal's law, we have $S_{ab}d\Sigma^b = pd\Sigma_a$.

In the proper frame, the components T^{0a} representing the energy density flux are equal to zero. The component T^{00} is the energy density of the system ε, i.e. the total energy referred to a unit of proper volume, or volume calculated in the frame in which the given element is at rest. In this way we find the form of the energy-momentum tensor in the proper frame to be

$$T^{\alpha\beta}=\begin{bmatrix}\varepsilon & 0 & 0 & 0\\ 0 & p & 0 & 0\\ 0 & 0 & p & 0\\ 0 & 0 & 0 & p\end{bmatrix}. \tag{2.29}$$

The general expression for the energy-momentum tensor can now easily be found by requiring that it be of form (2.29) in the frame in which the velocity four-vector u^α is $(1,0,0,0)$. It is easy to check that in an arbitrary coordinate system the energy-momentum tensor is given by

$$T^{\alpha\beta}=(\varepsilon+p)\,u^\alpha u^\beta-pg^{\alpha\beta}. \tag{2.30}$$

For the energy-momentum tensor $T^{\alpha\beta}$ to describe the behaviour of matter which is the source of a gravitational field, it is necessary that its divergence vanish,

$$T^{\alpha\beta}{}_{;\beta}=0, \tag{2.31}$$

because this is the integrability condition for Einstein's equations. Relations (2.31) are very important: it will be seen later that they contain the equations of motion.

The energy-momentum tensor of form (2.30) can describe the behaviour of real matter only when $\varepsilon\geqslant 0$. This restriction follows from the physical interpretation of ε as the density of energy.

Another example is the energy-momentum tensor of an electromagnetic field $F_{\alpha\beta}$:

$$T^{\alpha\beta}=\frac{1}{4\pi}\left(F^{\alpha\rho}F_\rho{}^\beta+\tfrac{1}{4}g^{\alpha\beta}F_{\rho\kappa}F^{\rho\kappa}\right). \tag{2.32}$$

One of its important algebraic properties is the vanishing of its trace: $T^\alpha{}_\alpha=0$.

2.3. Kinematic Quantities Characterizing the Motion of a Medium

Each particle of a medium is described by a world-line given in a local coordinate system as

$$x^\alpha=x^\alpha(y^a,s), \tag{2.33}$$

where y^a is the position of the particle at an initial moment and s is the proper time calculated along the world-line. Then

$$u^\alpha=\frac{\partial x^\alpha}{\partial s}=\dot{x}^\alpha,\qquad u^\alpha u_\alpha=1 \tag{2.34}$$

is the unit vector tangent to the curve. An infinitesimal displacement of the particle (infinitesimal variation of the world-line) is described by

$$\delta x^\alpha=\frac{\partial x^\alpha}{\partial y^a}\delta y^a, \tag{2.35}$$

$$(\delta x^\alpha)^{\cdot}=\frac{\partial}{\partial s}\delta x^\alpha=\frac{\partial}{\partial s}\left(\frac{\partial x^\alpha}{\partial y^a}\right)\cdot\delta y^a=\frac{\partial u^\alpha}{\partial y^a}\delta y^a=u^\alpha{}_{;\beta}\,\delta x^\beta. \tag{2.36}$$

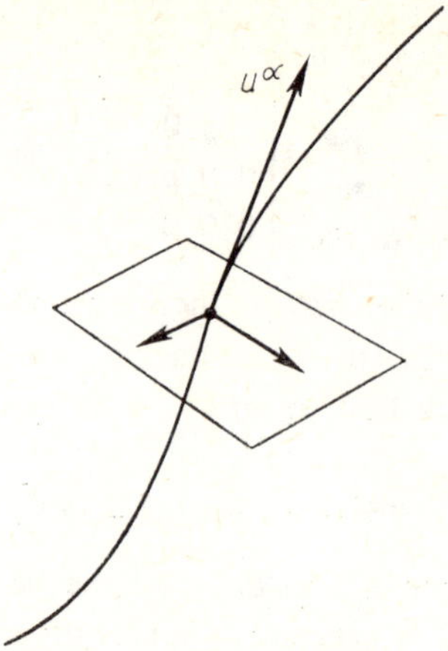

FIG. 2.1. A local spacelike subspace orthogonal to the observer's velocity four-vector.

The tensor

$$p^\alpha{}_\beta = \delta^\alpha{}_\beta - u^\alpha u_\beta \tag{2.37}$$

is the projection operator onto the local plane perpendicular to u^α:

$$p^\alpha{}_\beta u^\beta = 0, \quad p^\alpha{}_\beta p^\beta{}_\gamma = p^\alpha{}_\gamma. \tag{2.38}$$

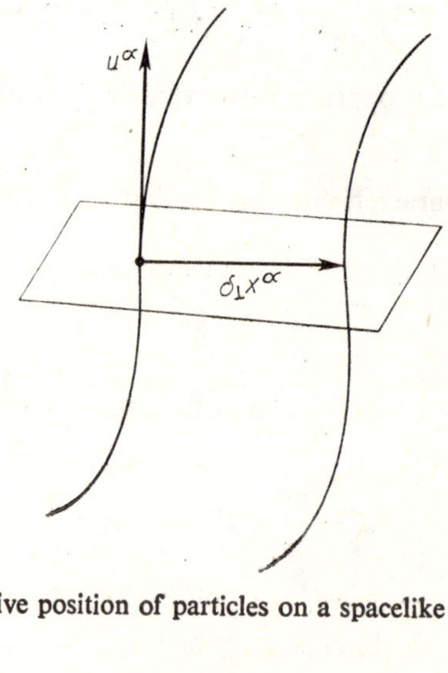

FIG. 2.2. Projection of the relative position of particles on a spacelike plane orthogonal to the particle velocity four-vector.

The orthogonal projection of the infinitesimal displacement vector δx^β,

$$\delta_\perp x^\alpha = \bar{\bar{p}}^\alpha{}_\beta \, \delta x^\beta \tag{2.39}$$

is the position vector of the particle $y^a + \delta y^a$ relative to the particle y^a, and its Fermi derivative

$$v^{\alpha}=p^{\alpha}{}_{\beta}(\delta_{\perp}x^{\beta})^{\cdot} \tag{2.40}$$

is the *relative velocity* of these particles.

The invariant acceleration four-vector of a particle is given by the spacelike vector

$$\dot{u}^{\alpha}=u^{\alpha}{}_{;\beta}u^{\beta}, \quad \dot{u}^{\alpha}u_{\alpha}=0. \tag{2.41}$$

It can be readily verified that

$$v^{\alpha}=u^{\alpha}{}_{;\beta}\delta_{\perp}x^{\beta}. \tag{2.42}$$

As in classical hydrodynamics, we shall write, after Ehlers, the displacement tensor $u^{\alpha}{}_{;\beta}p^{\beta}{}_{\gamma}$

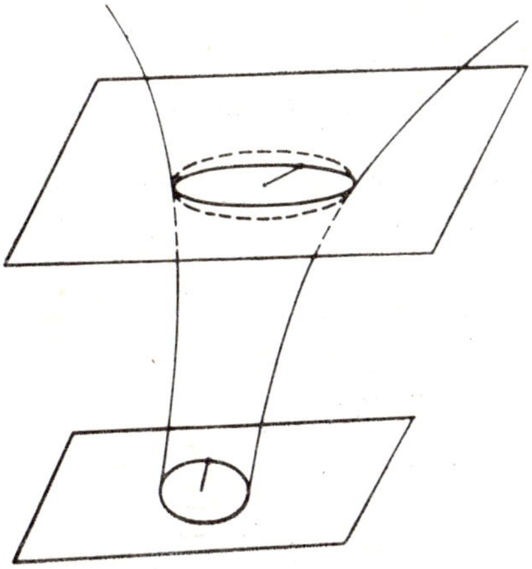

FIG. 2.3. An infinitesimal circle in a plane orthogonal to a congruence propagates along this congruence in such a wayt hat it is deformed to an ellipse in proportion to the shear σ and rotated in proportion to the rotation parameter ω; its area increases or decreases depending on the value of the expansion θ.

as a sum of irreducible components

$$u_{\alpha;\rho}p^{\rho}{}_{\beta}=\omega_{\alpha\beta}+\sigma_{\alpha\beta}+\tfrac{1}{3}\theta p_{\alpha\beta}, \tag{2.43}$$

where

$$\omega_{(\alpha\beta)}=\sigma_{[\alpha\beta]}=0, \quad \sigma^{\alpha}{}_{\alpha}=0, \quad \omega_{\alpha\beta}u^{\beta}=\sigma_{\alpha\beta}u^{\beta}=0, \quad \theta=u^{\alpha}{}_{;\alpha}. \tag{2.44}$$

Thus, the infinitesimal transformation that the displacement vector $\delta_{\perp}x$ undergoes during proper time δs consists of the rotation $\omega^{\alpha}{}_{\beta}\delta_{\perp}x^{\beta}$, the deformation $\sigma^{\alpha}_{\beta}\delta_{\perp}x^{\beta}$ (without rotation or change in volume) and the similarity transformation $\frac{1}{3}\theta\delta_{\perp}x^{\beta}$.

Let us examine the change with s of the distance between two neighbouring particles

$$\delta l^2=-g_{\alpha\beta}\delta_{\perp}x^{\alpha}\delta_{\perp}x^{\beta}=-p_{\alpha\beta}\delta x^{\alpha}\delta x^{\beta}. \tag{2.45}$$

We have

$$\frac{(\delta l)^{\cdot}}{\delta l}=\tfrac{1}{3}\theta-\sigma_{\alpha\beta}e^{\alpha}e^{\beta}, \tag{2.46}$$

where

$$e^\alpha = \frac{\delta_\perp x^\alpha}{\delta l}, \quad e^\alpha e_\alpha = -1, \tag{2.47}$$

while

$$p^\alpha{}_\beta \dot{e}^\beta = (\omega^\alpha{}_\beta + \sigma^\alpha{}_\beta + \sigma_{\rho\kappa} e^\rho e^\kappa \delta^\alpha{}_\beta) e^\beta . \tag{2.48}$$

Since $\sigma^\alpha_\alpha = 0$, if we average $\dfrac{(\delta l)^{\cdot}}{\delta l}$ over all directions, we shall obtain

$$\theta = 3\left\langle \frac{(\delta l)^{\cdot}}{\delta l} \right\rangle = \frac{(\delta V)^{\cdot}}{\delta V}; \tag{2.49}$$

here V is the volume of the element considered. The parameter θ describes the rate of change of small volumes of the medium. It follows from equation (2.48) that when e^α is a principal direction of the shear tensor $\sigma_{\alpha\beta}$, then

$$p^\alpha{}_\beta \dot{e}^\beta = \omega^\alpha{}_\beta e^\beta . \tag{2.50}$$

If we choose the principal directions of the shear tensor as the basis vectors of a local frame in a plane orthogonal to u^α, the motion of this basis will be determined by the tensor of infinitesimal rotation $\omega_{\alpha\beta}$.

The vorticity tensor $\omega_{\alpha\beta}$, the shear tensor $\sigma_{\alpha\beta}$ and the expansion scalar θ are given by

$$\omega_{\alpha\beta} = u_{[\alpha;\beta]} - \dot{u}_{[\alpha} u_{\beta]}, \tag{2.51}$$

$$\sigma_{\alpha\beta} = u_{(\alpha;\beta)} - \dot{u}_{(\alpha} u_{\beta)} - \tfrac{1}{3}\theta p_{\alpha\beta}, \tag{2.52}$$

$$\theta = u^\alpha{}_{;\alpha} . \tag{2.53}$$

From the vorticity tensor we can form the vorticity vector

$$\omega^\alpha = \tfrac{1}{2}\eta^{\alpha\beta\gamma\delta} u_\beta \omega_{\gamma\delta} = \tfrac{1}{2}\eta^{\alpha\beta\gamma\delta} u_\beta u_{\gamma,\delta}, \tag{2.54}$$

$$\omega^\alpha u_\alpha = 0, \tag{2.55}$$

where $\eta^{\alpha\beta\gamma\delta} = \dfrac{1}{\sqrt{-g}} \epsilon^{\alpha\beta\gamma\delta}$ and $\epsilon^{\alpha\beta\gamma\delta}$ is the antisymmetric Levi-Civita symbol. The vector ω^α is spacelike and, together with u^α, it is dual to $\omega_{\alpha\beta}$ in the sense that

$$\omega_{\alpha\beta} = \eta_{\alpha\beta\gamma\delta} \omega^\gamma u^\delta, \tag{2.56}$$

whence it is readily seen that

$$\omega_{\alpha\beta} \omega^\beta = 0 . \tag{2.57}$$

In general, a timelike world-line has nine second-order differential invariants, i.e. quantities independent of the choice of the coordinate system and formed as algebraic combinations of x^α, $\dot{x}^\alpha = u^\alpha$ and derivatives of u^α. They are: six independent components of the vectors $\dot{u}^\alpha$ and ω^α, two independent eigenvalues of the shear tensor $\sigma_{\alpha\beta}$, and the expansion θ. If we know the values of these invariants at some point of space-time, we can reconstruct the curve, up to homogeneous Lorentz transformations, in the infinitesimal neighbourhood of this point. Using the definitions of $\omega_{\alpha\beta}$, $\sigma_{\alpha\beta}$ and θ, we can write the covariant derivative of the velocity as

$$u_{\alpha;\beta}=\omega_{\alpha\beta}+\sigma_{\alpha\beta}+\tfrac{1}{3}\theta p_{\alpha\beta}+\dot{u}_\alpha u_\beta\,. \tag{2.58}$$

In addition to θ, the following scalar quantities characterizing the motion of the fluid can be introduced:

$$\dot{u}^2=-\dot{u}^\alpha\dot{u}_\alpha\,,\qquad \omega^2=-\omega^\alpha\omega_\alpha=-\tfrac{1}{2}\omega^{\alpha\beta}\omega_{\alpha\beta}\,,\qquad \sigma^2=\tfrac{1}{2}\sigma^{\alpha\beta}\sigma_{\alpha\beta}\,. \tag{2.59}$$

Using these invariants, we can classify motions of the fluid. A motion is said to be free or inertial when $\dot{u}=0$, irrotational when $\omega=0$, and volume preserving when $\theta=0$. If the fluid moves so that $\sigma=0$, the motion is called shear-free; the body may then only contract or expand isotropically. If the separation of particles is unchanged during motion, we call the motion rigid. It is characterized by the vanishing of θ and σ.

If the space-time in which the motion of the fluid takes place has a group of symmetries G, the motion is called symmetric if it proceeds along the trajectories of the group. If ξ^α is the tangent vector to a trajectory of a one-parameter subgroup G_1 of G and u^α is the tangent vector to the particle world-line, the motion will be symmetric if

$$\underset{\xi}{\pounds}u^\alpha=u^\alpha{}_{;\beta}\,\xi^\beta-\xi^\alpha{}_{;\beta}\,u^\beta=0\,. \tag{2.60}$$

If the trajectories of the group G_1 are timelike, a motion invariant under G_1 is called stationary. If the trajectories of G_1 are orthogonal to a family of spacelike hypersurfaces, a motion along these trajectories is called static. A free, or inertial, motion is described by the equation

$$u^\alpha{}_{;\beta}\,u^\beta=0\,. \tag{2.61}$$

A motion is irrotational when the world-lines of the fluid are orthogonal to a family of hypersurfaces:

$$u_{[\alpha;\beta}\,u_{\gamma]}=0\Leftrightarrow\omega_{\alpha\beta}=0\,, \tag{2.62}$$

i.e. when there is a scalar function t such that

$$u_\alpha=At_{,\alpha};\qquad p^\alpha{}_\beta t_{,\alpha}=0\,. \tag{2.63}$$

where A is an arbitrary scalar function. The function t defines a time scale along the world-lines of the fluid; if it is chosen to be the new time coordinate, the resulting coordinate system is called the proper frame. In the case of a free motion, along a geodesic, the time coordinate so chosen coincides with the proper time. For irrotational free motions, $u_\alpha=t_{,\alpha}$, where t is a solution of the Hamilton–Jacobi equation

$$g_{\alpha\beta}\,t^{,\alpha}t^{,\beta}=1\,. \tag{2.64}$$

Irrotational motions preserve volume if the world-lines of the fluid are orthogonal to a family of spacelike hypersurfaces and are minimal, i.e.

$$u^\alpha{}_{;\alpha}=0\Leftrightarrow\theta=0\,. \tag{2.65}$$

An irrotational motion is rigid when the hypersurfaces to which the world-lines of the particles are orthogonal are completely geodesic. We recall that a hypersurface is completely geodesic if geodesics on this surface are at the same time geodesics in the space-time. Now since for an irrotational motion $\omega=0$, if $x^\alpha(\lambda)$ is a geodesic with the tangent vector $t^\alpha=\frac{dx^\alpha}{d\lambda}$ then

$$\frac{d}{d\lambda}(u_\alpha t^\alpha)=u_{\alpha;\beta}t^\alpha t^\beta=(\sigma_{\alpha\beta}+\tfrac{1}{3}\theta p_{\alpha\beta}+\dot{u}_\alpha u_\beta)t^\alpha t^\beta. \tag{2.66}$$

For $u_\alpha t^\alpha=0$ along $x^\alpha(\lambda)$, it is enough that $\sigma=0$ and $\theta=0$. A motion is conformal when

$$\sigma=0, \quad (\dot{u}_{[\alpha}-\tfrac{1}{3}\theta u_{[\alpha})_{;\beta]}=0, \tag{2.67}$$

hence

$$\dot{u}_\alpha-\tfrac{1}{3}\theta u_\alpha=U_{,\alpha}, \quad \dot{U}=-\tfrac{1}{3}\theta. \tag{2.68}$$

It is easy to check that $\xi^\alpha=e^{-U}u^\alpha$ satisfies the equation $\xi_{(\alpha;\beta)}=(e^{-U})^{\cdot}g_{\alpha\beta}$ and therefore ξ^α is a conformal Killing vector.

A motion is isometric when

$$\sigma=\theta=0, \quad \dot{u}_{[\alpha;\beta]}=0, \tag{2.69}$$

so that

$$\dot{u}_\alpha=U_{,\alpha}, \quad \dot{U}=0,$$

and the vector $\xi^\alpha=e^{-U}u^\alpha$ is a Killing vector.

Let us now consider monochromatic electromagnetic radiation. As is well known, in the approximation of geometrical optics, a light ray is described by a null geodesic $x^\alpha(v)$

$$k^\alpha=\frac{dx^\alpha}{dv}, \quad k_\alpha k^\alpha=0, \quad k_{\alpha;\beta}k^\beta=0, \tag{2.71}$$

where v is the affine parameter.

For a monochromatic wave with the propagation vector k^α, an observer moving with velocity u^α will find that $k_\alpha u^\alpha=\frac{2\pi}{\lambda}$, where λ is the wavelength. It follows that along the ray we have

$$\frac{d}{dv}(k^\alpha u_\alpha)=u^\alpha{}_{;\beta}k_\alpha k^\beta=(\sigma_{\alpha\beta}+\tfrac{1}{3}\theta p_{\alpha\beta}+\dot{u}_\alpha u_\beta)k^\alpha k^\beta. \tag{2.72}$$

In addition

$$\delta l^2=-p_{\alpha\beta}dx^\alpha dx^\beta=(k^\alpha u_\alpha)^2 dv^2=\left(\frac{2\pi}{\lambda}\right)^2 dv^2, \tag{2.73}$$

and so

$$\frac{d\lambda}{\lambda}=-\frac{\lambda}{2\pi}d\left(\frac{2\pi}{\lambda}\right)=(\tfrac{1}{3}\theta-\sigma_{\alpha\beta}e^\alpha e^\beta)\delta l-\dot{u}_\alpha\delta_\perp x^\alpha. \tag{2.74}$$

Using (2.46) we get

$$\frac{d\lambda}{\lambda}=(\delta l)^{\cdot}-\dot{u}_\alpha\delta_\perp x^\alpha. \tag{2.75}$$

When the observer moves shear-free on a geodesic, then $\frac{d\lambda}{\lambda}=\frac{1}{3}\theta\,\delta l$, and so

$$\frac{d\lambda}{dl}=\tfrac{1}{3}\theta\lambda. \tag{2.76}$$

Now return to the general case. The kinematic quantities describing the motion of a medium are not independent but satisfy several differential identities involving the metric of the space-time and the Riemann tensor.

We define the relative acceleration of two neighbouring particles to be

$$a^{\alpha}=p^{\alpha}_{\beta}\dot{v}^{\beta}. \tag{2.77}$$

Since $v^{\alpha}=u^{\alpha}{}_{;\beta}\delta_{\perp}x^{\beta}$ and $u_{\alpha;[\beta;\delta]}=\frac{1}{2}u_{\rho}R^{\rho}{}_{\alpha\beta\delta}$, we obtain

$$a^{\alpha}=(-R^{\alpha}{}_{\beta\gamma\delta}u^{\beta}u^{\delta}+p^{\alpha}_{\beta}\dot{u}^{\beta}{}_{;\gamma}-\dot{u}^{\alpha}\dot{u}_{\gamma})\,\delta_{\perp}x^{\gamma}. \tag{2.78}$$

The above relationship is a generalization of the equation of geodesic deviation to the case where particles do not necessarily move along geodesics. The first term on the right-hand side contains the Riemann tensor and in this way it connects the relative motion of the particles with the geometrical properties of the space-time.

Writing out the equality $u_{\alpha;[\beta;\delta]}=\frac{1}{2}u_{\rho}R^{\rho}{}_{\alpha\beta\gamma}$, we find

$$\begin{aligned}\tfrac{1}{2}R_{\alpha\beta\gamma\delta}u^{\delta}=&-\omega_{\gamma[\alpha;\beta]}-\sigma_{\gamma[\alpha;\beta]}-\tfrac{1}{3}p_{\gamma[\alpha}\theta_{,\beta]}+\tfrac{1}{3}\theta\,(u_{\gamma}\omega_{\alpha\beta}+u_{\gamma}\dot{u}_{[\alpha}u_{\beta]}+\omega_{\gamma[\beta}u_{\alpha]}+\sigma_{\gamma[\beta}u_{\alpha]}\\&+\tfrac{1}{3}\theta g_{\gamma[\beta}u_{\alpha]})-\dot{u}_{\gamma}(\omega_{\alpha\beta}+\dot{u}_{[\alpha}u_{\beta]})-\dot{u}_{\gamma;[\beta}u_{\alpha]}.\end{aligned} \tag{2.79}$$

Using this relationship, we can calculate $R_{\alpha\beta}u^{\beta}$, $R_{\alpha\beta}u^{\alpha}u^{\beta}$ and $p^{\alpha\beta}R_{\beta\delta}u^{\delta}$. The results are

$$R_{\alpha\beta}u^{\beta}=\omega^{\beta}_{\alpha;\beta}+\sigma^{\beta}_{\alpha;\beta}-\theta_{,\alpha}+\tfrac{1}{3}p^{\beta}_{\alpha}\,\theta_{,\beta}-\dot{u}^{\beta}\omega_{\beta\alpha}+\dot{u}^{\beta}\sigma_{\beta\alpha}-\tfrac{1}{3}\theta^{2}u_{\alpha}+\dot{u}^{\beta}{}_{;\beta}u_{\alpha}, \tag{2.80}$$

$$R_{\alpha\beta}u^{\alpha}u^{\beta}=-\dot{\theta}-\tfrac{1}{3}\theta^{2}+\dot{u}^{\alpha}{}_{;\alpha}+2(\omega^{2}-\sigma^{2}), \tag{2.81}$$

$$p^{\alpha\beta}R_{\beta\delta}u^{\delta}=\omega^{\beta\alpha}{}_{;\beta}+2u^{\alpha}(\omega^{2}+\sigma^{2})+\sigma^{\beta\alpha}{}_{;\beta}-\tfrac{2}{3}p^{\alpha\beta}\theta_{,\beta}-\dot{u}^{\beta}\omega^{\alpha}_{\beta}+\dot{u}^{\beta}\sigma^{\alpha}_{\beta}. \tag{2.82}$$

Equation (2.81), first obtained by Raychaudhuri, will play an important part in our further considerations.

In a similar way, one can find

$$\begin{aligned}R_{\alpha\delta\beta\rho}u^{\delta}u^{\rho}=&-\dot{\sigma}_{\alpha\beta}+\omega_{\alpha}\omega_{\beta}+p_{\alpha\beta}\omega^{2}-\sigma_{\alpha\rho}\sigma^{\rho}{}_{\beta}-\tfrac{2}{3}\theta\sigma_{\alpha\beta}-\tfrac{1}{3}(\dot{\theta}+\tfrac{1}{3}\theta^{2})\,p_{\alpha\beta}-\dot{u}_{\alpha}\dot{u}_{\beta}+p_{\alpha\rho}\,p_{\beta\sigma}\dot{u}^{\rho;\sigma}\\&-\sigma^{\rho}_{\alpha}\dot{u}_{\rho}u_{\beta}-\sigma^{\rho}_{\beta}\dot{u}_{\rho}u_{\alpha}.\end{aligned} \tag{2.83}$$

As follows from (2.49), if l denotes the average distance between the neighbouring particles, then

$$\frac{\dot{l}}{l}=\tfrac{1}{3}\theta,\quad \frac{\ddot{l}}{l}=\tfrac{1}{3}(\dot{\theta}+\tfrac{1}{3}\theta^{2}). \tag{2.84}$$

Equations (2.81) and (2.83) can be regarded as the equations of propagation of the quantities θ and $\sigma_{\alpha\beta}$. Let us find the propagation laws for the vorticity vector ω^{α}. By tedious, though basically simple, calculations it can be shown that

$$p^{\alpha}{}_{\beta}\dot{\omega}^{\beta}=\sigma^{\alpha}{}_{\beta}\omega^{\beta}+\tfrac{1}{2}\eta^{\alpha\beta\gamma\delta}u_{\beta}\dot{u}_{\gamma;\delta}-\tfrac{2}{3}\theta\omega^{\alpha} \tag{2.85}$$

or, equivalently,

$$p^{\alpha}_{\cdot\beta}\dot{\omega}^{\beta}=u^{\alpha}_{;\beta}\omega^{\beta}-\theta\omega^{\alpha}+\tfrac{1}{2}\eta^{\alpha\beta\gamma\delta}u_{\beta}\dot{u}_{\gamma;\delta} \tag{2.86}$$

The vorticity scalar ω^2 satisfies the following propagation equation:

$$(\omega^2)^{\cdot}=2\sigma_{\alpha\beta}\omega^{\alpha}\omega^{\beta}+\omega_{\alpha\beta}\dot{u}^{\alpha;\beta}-\tfrac{4}{3}\theta\omega^2. \tag{2.87}$$

For completeness, here is one more formula satisfied by the vorticity vector:

$$\omega^{\alpha}_{;\alpha}=2\dot{u}_{\alpha}\omega^{\alpha}, \tag{2.88}$$

analogous to the known formula of classical hydrodynamics.

To close this section, let us consider in some detail the properties of some of the classes of motions. Besides particle world-lines, also vortex lines will be of interest to us, i.e. space-like curves to which the vorticity vector is tangent. They are orthogonal to the world-lines of the medium.

It can be shown that the world-lines of particles moving rigidly are geodesics in a space conformal to the given one, with $\tilde{g}_{\alpha\beta}=w^2 g_{\alpha\beta}$, if

$$\dot{u}_{\alpha}=-p^{\beta}_{\alpha}(\ln w)_{,\beta}. \tag{2.89}$$

This result can be obtained, for example, from the variational principle

$$\delta\int w\,ds=0. \tag{2.90}$$

Such a motion is a particular case of the conformal motion (cf. (2.68)).

Let us recall that two vector fields ξ^{α} and η^{α} span a surface if their commutator (Lie bracket) can locally be expressed as their linear combination. The particle world-lines and the vortex lines span a surface if

$$\omega^{[\alpha}\epsilon^{\beta]\gamma\delta\kappa}u_{\gamma}\dot{u}_{\kappa;\delta}=0. \tag{2.91}$$

This is equivalent to saying that $p^{\rho}_{\alpha}p^{\kappa}_{\beta}\dot{u}_{[\kappa;\rho]}\approx\omega_{\alpha\beta}$. This condition is automatically satisfied for free motions, and for motions for which (2.89) holds because then

$$\epsilon^{\alpha\beta\gamma\delta}u_{\beta}\dot{u}_{\gamma;\delta}=-2\frac{\dot{\omega}}{\omega}\omega^{\alpha}. \tag{2.92}$$

For arbitrary motions such that $\dot{u}_{\alpha}=-p^{\beta}_{\alpha}(\ln w)_{,\beta}$ the following propagation law holds

$$p^{\alpha}_{\beta}(wu^{\beta})^{\cdot}=\sigma^{\alpha}_{\beta}w\omega^{\beta}-\tfrac{2}{3}\theta w\omega^{\alpha}. \tag{2.93}$$

Thus, no vortex can appear or vanish along a world-line. The local direction of the axis of rotation is fixed along the world-line of a particle (in Fermi–Walker transport) when it coincides with the principal direction of the shear tensor $\sigma_{\alpha\beta}$.

2.4. Relativistic Equations of Hydrodynamics

As mentioned before, the equations of motion are contained in the equations

$$T^{\alpha\beta}_{;\beta}=0. \tag{2.94}$$

We shall supplement these relations with the law of particle number conservation (the continuity equation). If we denote the particle number density by n, the continuity equation takes the form

$$(nu^\alpha)_{;\alpha}=0, \tag{2.95}$$

where u^α is the particle velocity four-vector.

Assuming that

$$T_{\alpha\beta}=(\varepsilon+p)\,u_\alpha u_\beta-pg_{\alpha\beta}, \tag{2.96}$$

we can write out equation (2.94) as

$$T^\beta_{\alpha;\beta}=[(\varepsilon+p)\,u^\beta]_{;\beta}\,u_\alpha+(\varepsilon+p)\,u^\beta u_{\alpha;\beta}-p_{,\alpha}=0. \tag{2.97}$$

Projecting these equations on the particle velocity and using the equalities $u^\alpha u_\alpha=1$ and $u^\alpha u^\beta u_{\alpha;\beta}=0$, we get

$$u^\alpha T^\beta_{\alpha;\beta}=[(\varepsilon+p)\,u^\beta]_{;\beta}-p_{,\beta}\,u^\beta=0. \tag{2.98}$$

Equation (2.98) can be written in the equivalent form

$$\left(\frac{\varepsilon+p}{n}\,nu^\alpha\right)_{;\alpha}-\frac{1}{n}\,p_{,\alpha}\,nu^\alpha=0. \tag{2.99}$$

Applying the continuity equation gives

$$nu^\alpha\left[\left(\frac{\varepsilon+p}{n}\right)_{,\alpha}-\frac{1}{n}\,p_{,\alpha}\right]=0. \tag{2.100}$$

Let us recall that $\epsilon+p$ is the enthalpy density and n the particle number density, so that $\dfrac{\varepsilon+p}{n}$ is the enthalpy per particle. These quantities are connected with the temperature T and the entropy per particle s by the thermodynamic identity

$$\mathrm{d}\frac{\varepsilon+p}{n}-\frac{1}{n}\mathrm{d}p=T\mathrm{d}s. \tag{2.101}$$

Using (2.101), we can reduce equation (2.100) to

$$nTu^\alpha s_{,\alpha}=0. \tag{2.102}$$

Applying the continuity equation (2.95) again, we obtain ($T\neq0$) the equation of continuity of the entropy flow

$$(nsu^\alpha)_{;\alpha}=0. \tag{2.103}$$

Equation (2.103) tells us that the motion is adiabatic ,i.e. the entropy is constant along the particle world-lines. This result is no surprise because the energy-momentum tensor used does not allow for the processes of internal friction and thermal conduction: it describes the motion of a perfect fluid. Projecting the equations $T^\beta_{\alpha;\beta}=0$ on the direction perpendicular to u^α, we get

$$p^\rho_\alpha\,T^\beta_{\rho;\beta}=(\varepsilon+p)\,u^\beta u_{\alpha;\beta}-p^\beta_\alpha\,p_{,\beta}=0, \tag{2.104}$$

or

$$(\varepsilon+p)\,\dot{u}_\alpha=p^\beta_\alpha\,p_{,\beta}\,. \tag{2.105}$$

In the general case a particle does not travel on a geodesic, and its acceleration is connected with the pressure gradient by relation (2.105). The motion is free only when the pressure vanishes or when the pressure gradient has the same direction as the particle velocity.

Using the thermodynamic identity (2.101), we can write the equations of motion for a perfect fluid in the form

$$\dot{u}_\alpha = p_\alpha{}^\beta (\ln F)_{,\beta} - \frac{T}{F} p_\alpha{}^\beta s_{,\beta}, \tag{2.106}$$

where $F = \frac{\varepsilon + p}{n}$ is the enthalpy per particle. Now introduce the enthalpy flow vector $v_\alpha = F u_\alpha$ and examine the expression

$$\Omega_{\mu\nu} = v_{\mu;\nu} - v_{\nu;\mu}. \tag{2.107}$$

It is easy to verify that

$$\Omega_{\mu\nu} = 2F\omega_{\mu\nu} + T(u_\mu s_{,\nu} - u_\nu s_{,\mu}), \tag{2.108}$$

where $\omega_{\mu\nu}$ is the vorticity tensor. The enthalpy flow vector will be potential if $\Omega_{\mu\nu}=0$, which together with the adiabaticity of the motion ((2.103)) leads to $\omega_{\mu\nu}=0$, $s=\text{const}$. Thus the motion of the fluid is in this case irrotational and isentropic. The entropy is then constant in spacelike cross-sections perpendicular to the particle world-lines. It can be shown that the converse also holds: if the enthalpy flow vector is potential then the motion is irrotational and isentropic (Taub, 1959).

Let us now look at certain integral quantities connected with the motion of a fluid. In classical hydrodynamics one considers tubes formed in the following way. A closed curve is chosen; every point of this curve belongs to a streamline of some particle; the set of all these lines forms the surface of the tube. To give a space-time description of this situation, suppose $z^\alpha(x)$ is a continuous vector field. Solutions of the equation

$$\frac{dx^\alpha}{d\lambda} = z^\alpha(x) \tag{2.109}$$

define a three-parameter family of curves intersecting certain initial hypersurface Σ. We can write these curves as

$$x^\alpha = x^\alpha(y^i, \lambda), \tag{2.110}$$

where y^i represents the point of intersection with the surface Σ. If a closed curve Γ described parametrically as

$$y^i = y^i(\tau), \qquad \tau_1 \leqslant \tau \leqslant \tau_2, \tag{2.111}$$

with $y^i(\tau_1) = y^i(\tau_2)$, is given on Σ, the equations

$$x^\alpha = x^\alpha(y^i(\tau), \lambda) = x^\alpha(\tau, \lambda) \tag{2.112}$$

define a two-surface in the space-time. We shall call it a tube.

The vector tangent to a τ-curve on the tube (2.112) is given by

$$t^\alpha = \left(\frac{\partial x^\alpha}{\partial \tau}\right)_\lambda, \tag{2.113}$$

and therefore

$$\frac{\partial t^\alpha}{\partial \lambda}=\frac{\partial^2 x^\alpha}{\partial \lambda \partial \tau}=\frac{\partial z^\alpha}{\partial \tau}. \tag{2.114}$$

We are interested in the integral

$$C(\lambda)=\int_{\tau_1}^{\tau_2} v_\alpha(x(\tau,\lambda))\,t^\alpha \mathrm{d}\tau, \tag{2.115}$$

which describes the circulation of the enthalpy flow vector along a closed curve. The circulation depends on λ, and changes when the original contour is carried along the particle world-lines. Let us find how the circulation changes along the tube generated by the curve Γ. To this end, calculate

$$\frac{\mathrm{d}C(\lambda)}{\mathrm{d}\lambda}=\int_{\tau_1}^{\tau_2}\left(\frac{\partial v_\alpha}{\partial x^\sigma}u^\sigma t^\alpha+v_\alpha\frac{\partial t^\alpha}{\partial \lambda}\right)\mathrm{d}\tau=\int_{\tau_1}^{\tau_2}\left(\Omega_{\alpha\sigma}u^\sigma t^\alpha+\frac{\partial}{\partial\tau}(v_\alpha u^\alpha)\right)d\tau. \tag{2.116}$$

Since we integrate over a closed curve, we finally get

$$\frac{\mathrm{d}C(\lambda)}{\mathrm{d}\lambda}=\int_{\tau_1}^{\tau_2}\Omega_{\alpha\sigma}u^\sigma t^\alpha \mathrm{d}\tau. \tag{2.117}$$

The circulation vanishes along an arbitrary tube generated by the particle velocity vector field u^α if

$$\Omega_{\mu\nu}u^\nu=0. \tag{2.118}$$

The circulation is constant along an arbitrary tube if and only if

$$(\Omega_{\mu\nu}u^\nu)_{;\alpha}-(\Omega_{\alpha\nu}u^\nu)_{;\mu}=0. \tag{2.119}$$

Using the fact that the fluid is perfect and, consequently, that the motion is adiabatic, we can write

$$\Omega_{\mu\nu}u^\nu=-Ts_{,\mu}. \tag{2.120}$$

Substituting this in (2.119), we find

$$T_{,\mu}s_{,\nu}-T_{,\nu}s_{,\mu}=0, \tag{2.121}$$

and therefore the circulation is independent of the proper time when the surfaces of constant temperature coincide with the surfaces of constant entropy per particle. Using (2.101), we can write this condition in a different form:

$$n_{,\mu}p_{,\nu}-n_{,\nu}p_{,\mu}=0. \tag{2.122}$$

The circulation does not depend on the proper time if there is a functional relationship between pressure and the particle number density, i.e. if $p=p(n)$. Fluids having this property are called barotropic.

Finally, let us look again at the continuity equation (2.95) which can be written

$$\dot{n}\quad n\theta=0; \tag{2.123}$$

hence

$$n = n_0 e^{-\int \theta ds}. \tag{2.124}$$

The dependence of the entropy per particle on θ can be written in a similar way. As one would expect, the particle number density and the entropy per particle decrease with the increasing average separation of particles, i.e. when $\theta > 0$.

2.5. Relativistic Equations of Dissipative Processes

To allow for dissipative processes: viscosity and thermal conduction, one introduces additional terms into the energy-momentum tensor. The energy flow vector, previously proportional to the particle velocity vector, will also be changed. In general, we shall have

$$T_{\alpha\beta} = (\varepsilon + p) u_\alpha u_\beta - p g_{\alpha\beta} + \tau_{\alpha\beta}, \tag{2.125}$$

$$n_\alpha = n u_\alpha + \nu_\alpha, \tag{2.126}$$

$$T^{\alpha\beta}{}_{;\beta} = 0, \quad n^\alpha{}_{;\alpha} = 0. \tag{2.127}$$

Here $\tau_{\alpha\beta}$ represents the dissipative processes, and n_α is the particle density flux vector. The question arises how to define the velocity vector u^α if energy exchange in the form of heat is allowed. In the proper frame comoving with an element of matter the momentum of that element is zero and its energy is expressed in terms of other thermodynamic quantities by the same relations as those applying when dissipation is absent. This means that the components τ_{00} and $\tau_{0\alpha}$ should be zero in the reference frame considered, and so in general we have

$$\tau_{\alpha\beta} u^\beta = 0. \tag{2.128}$$

In the proper frame, by definition, the component n^0 of the particle density flux four-vector should coincide with the particle number density n, so

$$u^\alpha \nu_\alpha = 0. \tag{2.129}$$

The tensor $\tau_{\alpha\beta}$ and the vector ν_α can be determined using the requirement that the entropy should grow. This should be implied by the equations of motion. The equation $u^\alpha T^\beta_{\alpha;\beta} = 0$ can be written

$$T(nsu^\alpha)_{;\alpha} - \mu \nu^\alpha{}_{;\alpha} + u^\alpha \tau_\alpha{}^\beta{}_{;\beta} = 0, \tag{2.130}$$

where $\mu = \dfrac{\varepsilon + p}{n} - Ts$ is the chemical potential of the substance. We can rewrite this in the equivalent form

$$\left(nsu^\alpha - \frac{\mu}{T}\nu^\alpha\right)_{;\alpha} = -\nu^\alpha \left(\frac{\mu}{T}\right)_{,\alpha} + \frac{1}{T}\tau_\alpha{}^\beta u^\alpha{}_{;\beta}. \tag{2.131}$$

The expression on the left-hand side describes changes in the entropy flux, and that on the right-hand side the entropy growth due to dissipation. In this way, the entropy density flux vector is found to be

$$s^\alpha = nsu^\alpha - \frac{\mu}{T}\nu^\alpha; \tag{2.132}$$

near equilibrium, $\tau_{\alpha\beta}$ and v^α should be linear combinations of the gradients of velocity and thermodynamic variables, such that the right-hand side of equation (2.131) is positive. This condition, together with the conditions $\tau_{\alpha\beta}u^\beta=0$ and $v^\alpha u_\alpha=0$, uniquely determines the form of the tensor $\tau_{\alpha\beta}$ and the four-vector v_α:

$$\tau_{\alpha\beta}=\eta(u_{\alpha;\beta}+u_{\beta;\alpha}-u_\alpha\dot{u}_\beta-u_\beta\dot{u}_\alpha)+(\zeta-\tfrac{2}{3}\eta)\,\theta p_{\alpha\beta}, \tag{2.133}$$

$$v_\alpha=-\frac{\kappa}{c}\left(\frac{nT}{\varepsilon+p}\right)^2\left[\left(\frac{\mu}{T}\right)_{,\alpha}-u_\alpha u^\beta\left(\frac{\mu}{T}\right)_{,\beta}\right]. \tag{2.134}$$

Here η and ζ may be identified as the shear and the bulk viscosities and κ as the thermal conductivity. Using the notation introduced before, we can represent $\tau_{\alpha\beta}$ and v_α more simply as

$$\tau_{\alpha\beta}=2\eta\sigma_{\alpha\beta}+\zeta\theta p_{\alpha\beta}, \tag{2.135}$$

$$v_\alpha=\frac{\kappa}{c}\left(\frac{nT}{\varepsilon+p}\right)^2 p_\alpha^{\ \beta}\left(\frac{\mu}{T}\right)_{,\beta}. \tag{2.136}$$

The equations of motion now take the form

$$\left(nsu^\alpha-\frac{\mu}{T}v^\alpha\right)_{;\alpha}=-\frac{\kappa}{c}\left(\frac{nT}{\varepsilon+p}\right)^2 p^{\alpha\beta}\left(\frac{\mu}{T}\right)_{,\alpha}\left(\frac{\mu}{T}\right)_{,\beta}+\frac{4\eta}{T}\sigma^2+\frac{\zeta}{T}\theta^2, \tag{2.137}$$

$$(\varepsilon+p)\dot{u}_\alpha-p_\alpha^\beta p_{,\beta}+p_{\alpha\rho}\tau^{\rho\beta}{}_{;\beta}=0. \tag{2.138}$$

Using the thermodynamic identity

$$\mathrm{d}\frac{\mu}{T}=-\frac{\varepsilon+p}{nT^2}\mathrm{d}T+\mathrm{d}p\frac{1}{nT}, \tag{2.139}$$

we can rewrite the formula for v_α as

$$v_\alpha=\frac{\kappa}{c}\frac{n}{\varepsilon+p}p_\alpha^{\ \beta}\left(T\frac{p_{,\beta}}{\varepsilon+p}-T_{,\beta}\right). \tag{2.140}$$

As we can see, the convection flux vector v_α depends on temperature and pressure gradients.

The proper frame which we have used in the above considerations had the property that the energy vanished in it; such a reference frame is called dynamic. A frame bound to a chosen element of the fluid, such that the particle flow vanishes in it, is also of physical interest. A frame having this property is called kinematic. For adiabatic processes, these two frames coincide.

In a dynamic frame we always have

$$T^{\alpha\beta}u_\beta=\varepsilon u^\alpha; \tag{2.141}$$

this relationship does not hold, in general, in a kinematic frame.

In a kinematic frame, the energy-momentum tensor will take the form

$$T^{\alpha\beta}=(\varepsilon+p)u^\alpha u^\beta-pg^{\alpha\beta}-2u^{(\alpha}q^{\beta)}+\tilde{\tau}^{\alpha\beta} \tag{2.142}$$

with

$$q_\alpha u^\alpha=0 \quad \text{and} \quad \tilde{\tau}^{\alpha\beta}u_\beta=0. \tag{2.143}$$

The equations of motion then read

$$\dot{\varepsilon}+(\varepsilon+p)\theta+q^{\beta}{}_{;\beta}+u_{\alpha}\dot{q}^{\alpha}+u_{\alpha}\tilde{\tau}^{\alpha\beta}{}_{;\beta}=0 \tag{2.144}$$

and

$$(\varepsilon+p)\dot{u}^{\alpha}+p^{\alpha\rho}(\dot{q}_{\rho}-p_{,\rho}+\tilde{\tau}_{\rho}{}^{\beta}{}_{;\beta})+(\omega^{\alpha}{}_{\beta}+\sigma^{\alpha}{}_{\beta})q^{\beta}+\tfrac{4}{3}\theta q^{\alpha}=0. \tag{2.145}$$

Equation (2.144) can be written in the equivalent form

$$\left(nsu^{\alpha}+\frac{1}{T}q_{\alpha}\right)_{;\alpha}=\frac{1}{T}q^{\alpha}(\dot{u}_{\alpha}-(\ln T)_{,\alpha})+\frac{1}{T}u_{\alpha;\beta}\tilde{\tau}^{\alpha\beta}. \tag{2.146}$$

In a kinematic frame, $nsu^{\alpha}+\frac{1}{T}q^{\alpha}$ is the density of the entropy flux.

The vector q^{α} and the tensor $\tilde{\tau}^{\alpha\beta}$ can be found from the requirement that the entropy of the element under consideration does not decrease, or, in other words, that the right-hand side of equation (2.146) be non-negative. In addition, we require that $\tilde{\tau}^{\alpha\beta}$ be a linear combination of the derivatives of the velocity four-vector. It is straightforward to check that

$$q_{\alpha}=\frac{\kappa}{c}p_{\alpha}{}^{\beta}(T_{,\beta}-T\dot{u}_{\beta}), \tag{2.147}$$

and

$$\tilde{\tau}^{\alpha\beta}=2\eta\sigma^{\alpha\beta}+\zeta\theta p^{\alpha\beta} \tag{2.148}$$

satisfy all the restrictions imposed. As before, η and ζ are shear and bulk viscosities and κ is thermal conductivity. In the general case, the coefficients of viscosity and heat conduction depend on the pressure p and the particle number density n.

The weakness of this approach of describing the dissipative processes lies in the fact that the coefficients η, ζ and κ are phenomenological parameters and must be given as functions of n and s. A more satisfactory description of these processes can be obtained by means of the Boltzmann transport equation. This, however, would require developing a relativistic kinetic theory, a task which is outside the scope of this book. For basic information on this method the reader is referred to the works of Anderson (1970), Ehlers (1971), Stewart (1971) or Israel and Stewart (1980).

Of course, if we set η, ζ and κ equal to zero, we would obtain the equations of motion of a perfect fluid.

2.6. Equations of Hydrodynamics in the Post-Newtonian Approximation

In the preceding section, we have formulated the relativistic equations of hydrodynamics. Knowing the gravitational field described in a chosen coordinate system by means of the metric tensor $g_{\alpha\beta}$, we can solve these equations and find the motion of matter. The presence of matter affects the properties of the space-time; it is therefore necessary to solve the field equations and the equations of motion simultaneously. When the distribution of matter is bounded and the gravitational field is weak, i.e. $\varepsilon=\frac{GM}{c^2R}\ll 1$, where M is the mass of the system and R represents its size, and when the relative velocities of motions of matter are small compared with the speed of light, i.e. $\lambda = v/c \ll 1$, simultaneous solution of motion and field equations can be sought by the E.I.H. method, worked out by Einstein, Infeld and Hoffman in 1938. It consists in expanding the metric and the components of the energy-momentum tensor into power series in λ. The field equations and equations of motion are then replaced by an infinite system of equations having the property that the solution of the equations of motion to order k permits determining the metric to order $k+1$, which in turn makes it possible to solve the equations of motion of order $k+1$, and so on. Of course, in order to solve field equations of order k one has to impose boundary conditions of appropriate order. This becomes complicated for high order equations, in which the radiative terms begin to play an important part.

Our aim here will be to find the metric and the equations of hydrodynamics in the first post-Newtonian approximation. This comparatively simple example will permit us to see how the equations of motion follow from the field equations and to examine the influence of the post-Newtonian corrections on the motion of matter. The problem was investigated by Bażański and Plebański; a new approach has recently been proposed by Chandrasekhar.

In the Newtonian approximation, we shall assume that the system to be examined is described by the hydrodynamic equations of an incompressible fluid

$$\frac{\partial \rho}{\partial t} + (\rho v_b)_{,b} = 0, \tag{2.149}$$

$$\rho \frac{\partial v_a}{\partial t} + \rho v_b v_{a,b} = -p_{,a} - \rho \phi_{,a}, \tag{2.150}$$

where v_a are the components of the three-dimensional velocity, ρ is the density of the fluid, p the pressure and ϕ the gravitational potential defined by the Poisson equation

$$\Delta\phi = 4\pi G\rho. \tag{2.151}$$

The distribution of matter in space-time will be described by the energy-momentum tensor

$$T_{\alpha\beta} = (\varepsilon + p) u_\alpha u_\beta - p g_{\alpha\beta}. \tag{2.152}$$

For our further purposes, it will be convenient to represent the energy density ε as a sum of the rest-mass energy density ρc^2 and the internal energy density:

$$\varepsilon = \rho c^2 (1 + \pi/c^2). \tag{2.153}$$

As we remember, in the first approximation the components of the metric tensor are

$$\begin{aligned} g_{00} &= 1 + \frac{2\phi}{c^2} + o(c^{-4}), \\ g_{ab} &= -\left(1 - \frac{2\phi}{c^2}\right)\delta_{ab} + o(c^{-4}), \\ g_{0a} &= o(c^{-3}). \end{aligned} \tag{2.154}$$

By solving, in this approximation, the equations of motion $T^{\alpha\beta}{}_{;\beta} = 0$ we obtain the equations of hydrodynamics.

If we know g_{00} and g_{ab} to the accuracy of c^{-2}, we can find those components of the energy-momentum tensor which we need to determine g_{00} to order c^{-4} and g_{0a} to order c^{-3} from the field equations. The knowledge of these components is then enough for the equations of motion to be written in the post-Newtonian approximation.

We represent the metric in the form

$$g_{\alpha\beta}=\eta_{\alpha\beta}+h_{\alpha\beta}, \tag{2.155}$$

where

$$h_{00}=\frac{2\phi}{c^2}+o(c^{-4}), \quad h_{0a}=o(c^{-3}), \quad h_{ab}=\frac{2\phi}{c^2}\delta_{ab}+o(c^{-4}), \tag{2.156}$$

and $\eta_{\alpha\beta}$ is the metric tensor of Minkowski space-time. In the first approximation the indices in $h_{\alpha\beta}$ can be raised and lowered by means of $\eta_{\alpha\beta}$; for higher orders the relationship $g_{\alpha\rho}g^{\rho\beta}=\delta^{\beta}_{\alpha}$ must be used.

The velocity four-vector of an element of the fluid, $u^{\alpha}=\frac{dx^{\alpha}}{ds}$, is in this approximation given by

$$\begin{aligned} u^0&=1+\frac{1}{c^2}(\tfrac{1}{2}v^2-\phi)+o(c^{-4}),\\ u^a&=\left[1+\frac{1}{c^2}(\tfrac{1}{2}v^2-\phi)\right]\frac{v^a}{c}+o(c^{-4}). \end{aligned} \tag{2.157}$$

Knowing the velocity, we can write the energy-momentum tensor as

$$\begin{aligned} T_{00}&=\rho c^2\left[1+\frac{1}{c^2}(v^2+2\phi+\pi)\right]+o(c^{-2}),\\ T_{0a}&=-\rho c v_a+o(c^{-1}),\\ T_{ab}&=\rho v_a v_b+\delta_{ab}\,p+o(c^{-2}). \end{aligned} \tag{2.158}$$

The field equations we are to solve can be written in the equivalent form

$$R_{\alpha\beta}=\frac{8\pi G}{c^4}(T_{\alpha\beta}-\tfrac{1}{2}g_{\alpha\beta}T), \tag{2.159}$$

where $T=T^{\alpha}_{\alpha}$.

Calculating the Ricci tensor for the metric (2.155), to second order in $h_{\alpha\beta}$ we get for R_{00} the expression

$$R_{00}=(h_0{}^a{}_{,a}-\tfrac{1}{2}h_a{}^a{}_{,0})_{,0}+\tfrac{1}{2}\Delta h_{00}+\tfrac{1}{2}h^{ab}h_{00,ab}-\tfrac{1}{4}h_{00,a}h_{00,}{}^a-\tfrac{1}{4}h_{00,b}(2h^{ab}{}_{,a}-h_a{}^{a,b}). \tag{2.160}$$

If we perform a coordinate transformation which in the zero approximation leaves the form of the tensor $\eta_{\alpha\beta}$ unchanged, i.e. a transformation which can be written $x^{\mu}\to L^{\mu}_{\nu}x^{\nu}+\xi^{\mu}(x^{\nu})$, where L^{μ}_{ν} represents a Lorentz transformation and $\xi^{\mu}(x^{\nu})$ a small coordinate transformation, then $h_{\alpha\beta}\to h_{\alpha\beta}-\xi_{(\alpha;\beta)}$. By a suitable choice of ξ^{μ} four of the ten components of the metric $h_{\mu\nu}$ can be eliminated. We shall now use this possibility; by imposing on $h_{\mu\nu}$ the condition

$$h_0{}^a{}_{,a}-\tfrac{1}{2}h_a{}^a{}_{,0}=0, \tag{2.161}$$

we eliminate h_{0a} from R_{00}. Substituting the known expression for $h_{\alpha\beta}$ in (2.160), we get in the post-Newtonian approximation

$$R_{00}=\tfrac{1}{2}\Delta h_{00}-\frac{1}{c^4}\Delta\phi^2+\frac{4}{c^4}\phi\Delta\phi, \tag{2.162}$$

where, for simplicity, h_{00} denotes the correction to order $1/c^4$ of g_{00}.

Similarly, using the gauge condition (2.161), we find

$$R_{0a}=\tfrac{1}{2}\Delta h_{0a}+\frac{1}{2c^2}\phi_{,0a}, \tag{2.163}$$

R_{ab} need not be considered since, as we shall see later, the knowledge of h_{00} and h_{0a} in the post-Newtonian approximation is enough for us to write the hydrodynamic equations in this approximation.

We now go on to solving the field equations. The 00 component of these equations is

$$\tfrac{1}{2}\Delta h_{00}-\frac{1}{c^4}\Delta\phi^2+\frac{16\pi G}{c^4}\rho\phi=\frac{4\pi G}{c^2}\rho+\frac{8\pi G}{c^4}\rho(v^2+\phi+\tfrac{1}{2}\pi+\tfrac{3}{2}p/\rho). \tag{2.164}$$

This can be written as

$$\Delta\left(\tfrac{1}{2}h_{00}-\frac{1}{c^4}\phi^2-\frac{\phi}{c^2}\right)=\frac{8\pi G}{c^4}\rho\Sigma, \tag{2.165}$$

where

$$\Sigma=v^2-\phi+\tfrac{1}{2}\pi+\tfrac{3}{2}\frac{p}{\rho}. \tag{2.166}$$

If Γ is the solution of the equation

$$\Delta\Gamma=4\pi G\rho\Sigma, \tag{2.167}$$

then

$$h_{00}=\frac{2\phi}{c^2}+\frac{2\phi^2}{c^4}+\frac{4}{c^4}\Gamma+o(c^{-6}); \tag{2.168}$$

the component $0a$ of the field equations reduces to

$$\tfrac{1}{2}\Delta h_{0a}+\frac{1}{2c^2}\phi_{,0a}=-\frac{8\pi G}{c^3}\rho v_a. \tag{2.169}$$

In order to solve this equation, we introduce two additional functions χ and U_a defined by

$$\Delta\chi=2\phi, \tag{2.170}$$

$$\Delta U_a=-4\pi G\rho v_a; \tag{2.171}$$

then

$$h_{0a}=\frac{4}{c^3}U_a-\frac{1}{2c^2}\chi_{,0a}. \tag{2.172}$$

In solving the field equations we have restricted $h_{\alpha\beta}$ by imposing the gauge condition (2.161). Let us now check whether it is satisfied. Substituting h_{0a} found above in (2.161), we get

$$h_0{}^a{}_{,a}-\tfrac{1}{2}h_a{}^a{}_{,0}=\frac{4}{c^3}(U_{a,a}-\phi_{,0}). \tag{2.173}$$

On the other hand, using equations (2.151) and (2.171), we find that

$$\frac{1}{c^3}\Delta(U_{a,a}-\phi_{,0})=-\frac{4\pi G}{c^3}\left(\frac{\partial\rho}{\partial t}+(\rho v_a)_{,a}\right)=0 \tag{2.174}$$

in virtue of the continuity equation. In deriving this equality we used the fact that $x^0=ct$, so that $\phi_{,0}=\dfrac{1}{c}\dfrac{\partial\phi}{\partial t}$. We assume that the functions ϕ and U_a are continuous and vanish at infinity; the only solution of equation (2.174), satisfying these boundary conditions is

$$U_{a,a}-\phi_{,0}=0, \tag{2.175}$$

which is just what is necessary for the gauge condition to be satisfied.

Thus, the components of the metric tensor in the post-Newtonian approximation are

$$\begin{aligned} g_{00}&=1+\frac{2\phi}{c^2}+\frac{2\phi^2}{c^4}+\frac{4}{c^4}\Gamma+o(c^{-6}),\\ g_{0a}&=\frac{1}{c^3}(4U_a-\tfrac{1}{2}\chi_{,ta})+o(c^{-5}),\\ g_{ab}&=-\left(1-\frac{2\phi}{c^2}\right)\delta_{ab}+o(c^{-4}), \end{aligned} \tag{2.176}$$

and the corresponding contravariant components can be found from the relationships $g^{\alpha\rho}g_{\rho\beta}=\delta^\alpha_\beta$.

The form of the metric tensor depends on the coordinate system it is expressed in. Let us recall that in our case a bounded distribution of matter is considered, and the coordinate system is chosen so as to be Cartesian at infinity. More precisely, in this coordinate system

$$h_{\alpha\beta}\underset{r\to\infty}{\to}o\left(\frac{1}{r}\right),\quad h_{\alpha\beta,\gamma}\underset{r\to\infty}{\to}o\left(\frac{1}{r^2}\right). \tag{2.177}$$

The equations of hydrodynamics in the post-Newtonian approximation can be obtained from the general equations $T^{\alpha\beta}{}_{;\beta}=0$ by neglecting terms of order higher than c^{-4}. After considerable calculations, from $T^{0\alpha}{}_{;\alpha}=0$ we get

$$\frac{\partial}{\partial t}\left\{\rho\left[1+\frac{1}{c^2}(v^2-2\phi+\pi)\right]\right\}+\left\{\rho v_a\left[1+\frac{1}{c^2}(v^2-2\phi+\pi+p/\rho)\right]\right\}_{,a}-\frac{1}{c^2}\rho\frac{\partial\phi}{\partial t}=0. \tag{2.178}$$

If we put

$$\sigma=\rho\left[1+\frac{1}{c^2}(v^2-2\phi+\pi+p/\rho)\right].$$

the equation will read more simply:

$$\frac{\partial \sigma}{\partial t}+(\sigma v_a)_{,a}-\frac{1}{c^2}\left(\rho\frac{\partial \phi}{\partial t}+\frac{\partial p}{\partial t}\right)=0. \tag{2.179}$$

Using the Newtonian continuity equation, we simplify this further to

$$\frac{\partial \rho^*}{\partial t}+(\rho^* v_a)_{,a}=0, \tag{2.180}$$

where

$$\rho^*=\rho\left[1+\frac{1}{c^2}(\tfrac{1}{2}v^2-3\phi)\right]. \tag{2.181}$$

This relationship is the generalized continuity equation.

Writting out the equation $T^{a\alpha}{}_{;\alpha}=0$, we obtain

$$\sigma\frac{\partial v_a}{\partial t}+\sigma v_b v_{a,b}+\frac{v_a}{c^2}\left(\rho\frac{\partial \phi}{\partial t}+\frac{\partial p}{\partial t}\right)+\left[\left(1-\frac{2\phi}{c^2}\right)p\right]_{,a}+\rho\phi_{,a}$$
$$-\frac{4}{c^2}\rho\frac{d}{dt}(v_a\phi+U_a)+\frac{4}{c^2}\rho v_b U_{b,a}+\frac{1}{2c^2}\rho\frac{d}{dt}(U_a-U_{b,ab})$$
$$-\frac{1}{2c^2}\rho v_b(U_a-U_{c,ac})_{,b}-\frac{2}{c^2}\rho(\Gamma_{,a}-\Sigma\phi_{,a})=0. \tag{2.182}$$

This is the generalized Euler equation.

As in classical hydrodynamics, from the equations of motion one can derive the conservation laws for momentum, energy and angular momentum. Integrating the equations of motion over the region V containing matter gives

$$\frac{d}{dt}\int_V\left[\sigma v_a-\frac{4}{c^2}\rho(v_a\phi+U_a)+\frac{1}{2c^2}\rho(U_a-U_{b,ab})\right]d^3x=0; \tag{2.183}$$

therefore, if we denote

$$P_a=\int_V\left[\sigma v_a-\frac{4}{c^2}\rho(v_a\phi+U_a)+\frac{1}{2c^2}\rho(U_a-U_{b,ab})\right]d^3x, \tag{2.184}$$

the equation $\dfrac{dP_a}{dt}=0$ will express the law of conservation of the total momentum of the system. It is now natural to introduce the momentum density

$$\pi_a=\sigma v_a-\frac{4}{c^2}\rho(v_a\phi+U_a)+\frac{1}{2c^2}\rho(U_a-U_{b,ab}). \tag{2.185}$$

If we multiply the equation of motion by x_b and then antisymmetrize it with respect to the indices a and b and integrate the result over the region containing matter, we obtain

$$\frac{d}{dt}\int_V(x_b\pi_a-x_a\pi_b)\,d^3x=0; \tag{2.186}$$

this equation expresses conservation of the total angular momentum of the system.

The law of energy conservation can be obtained by multiplying equation (2.182) by v_a and integrating over the region V. After rather tedious calculations, the result can be written as

$$\frac{\mathrm{d}}{\mathrm{d}t}\int_V \Big[(\sigma-\tfrac{1}{2}\rho^*)v^2+\rho^*\pi+\tfrac{1}{2}\rho^*\phi^*+\frac{\rho}{c^2}(-\tfrac{1}{8}v^4+\tfrac{1}{2}\phi^2+\pi\phi-\tfrac{1}{2}\pi v^2-\tfrac{5}{2}v^2\phi-\tfrac{7}{4}v_a U_a - \tfrac{1}{4}v_a U_{a,ab})\Big]\mathrm{d}^3x=0. \quad (2.187)$$

This relationship means that the total energy of the system is conserved. The energy density per unit of volume can be defined as

$$\varepsilon=(\sigma-\tfrac{1}{2}\rho^*)v^2+\tfrac{1}{2}\rho^*\phi^*+\rho^*\pi+\frac{\rho}{c^2}(-\tfrac{1}{8}v^4+\tfrac{1}{2}\phi^2+\pi\phi-\tfrac{1}{2}v^2\pi-\tfrac{5}{2}v^2\phi-\tfrac{7}{4}v_a U_a-\tfrac{1}{4}v_a U_{b,ab}), \quad (2.188)$$

where ϕ^* is the solution of

$$\Delta\phi^*=4\pi G\rho^*=4\pi G\rho\left[1+\frac{1}{c^2}(\tfrac{1}{2}v^2-3\phi)\right]. \quad (2.189)$$

The equations obtained above enable us, in principle, to solve, in the post-Newtonian approximation, every problem that can be solved in classical hydrodynamics. We shall use this possibility while investigating stability of equilibrium configurations.

Literature

ANDERSON, J. L. (1970) *Relativistic Boltzmann Theory and the Grad Method of Moments*, in *Relativity*, eds.: Carmeli, M., Fickler, S. I. and Witten, L., Plenum, New York.

BERGMANN, P. G. (1951) *Phys. Rev.* **84**, 1026.

CHANDRASEKHAR, S. (1965) *Astrophys. J.* **142**, 1488.

EHLERS, J. (1961) *Akad. Wiss. Lit. Mainz, Abhandl. Math. Naturw. Kl.* Nr 11, 793.

EHLERS, J. (1971) *General Relativity and Kinetic Theory*, in *General Relativity and Cosmology*, ed.: Sachs, R. K., Academic Press, New York.

EINSTEIN, A., INFELD, L. and HOFFMAN, B. (1938) *Ann. Math.* **39**, 66.

de GROOT, S. R., VAN LEEUWEN, W. A. and VAN WEERT, C. G (1980) *Relativistic Kinetic Theory*, North-Holland, Amsterdam.

INFELD, L. and PLEBAŃSKI, J. (1958) *Motion and Relativity*, PWN, Warszawa.

ISRAEL, W. and STEWART, J. M. (1980) *Progress in Relativistic Thermodynamics and Electrodynamics of Continuous Media*, in *General Relativity and Gravitation*, Vol. 2, ed.: Held, A., Plenum Press, New York.

LANDAU, L. D. and LIFSHITZ, E. M. (1959) *Fluid Mechanics*, Pergamon Press, London.

MØLLER, C. (1967) *Kgl. Danske Videnskab. Sdskab, Mat.-Fys. Medd.* **36**, No. 1.

OTT, H. (1963) *Z. Physik* **175**, 70.

PAULI, W. (1958) *Theory of Relativity*, Pergamon Press, London.

PLEBAŃSKI, J. and BAŻAŃSKI, S. (1958) *Acta Phys. Polon.* **18**, 307.

RAYCHAUDHURI, A. (1955) *Phys. Rev.* **98**, 1123.

STEWART, J. M. (1971) *Non-Equilibrium Relativistic Kinetic Theory*, Lecture Notes in Physics, Vol. 10, Springer-Verlag, Berlin.

TAUB, A. (1959) *Arch. Rat. Mech. Anal.* **3**, 312.

CHAPTER 3

GRAVITATIONAL FIELDS IN MATTER

3.1. Gravitational Fields in Matter. The Friedman Solution

So far we have been discussing the properties of gravitational fields in empty space and learning how to describe the distribution of matter by means of the energy-momentum tensor. We are now ready to solve Einstein's equations and find gravitational fields in regions filled with matter. To begin with, let us consider the simple case where the whole space is filled with dust. The energy-momentum tensor describing such a system is of the form

$$T_{\alpha\beta}=\varepsilon u_\alpha u_\beta\,, \tag{3.1}$$

where ε is the energy density of dust particles and u_α is their velocity four-vector. In the general case, the vanishing of the divergence of the tensor $T_{\alpha\beta}$, i.e. $T^{\alpha\beta}{}_{;\beta}=0$, implies that dust particles move on the geodesics

$$u_{\alpha;\beta}u^\beta=0 \tag{3.2}$$

and that the continuity equation

$$(\varepsilon u^\alpha)_{;\alpha}=0 \tag{3.3}$$

is satisfied. As in empty space, in regions filled with matter we have full freedom of choice of the coordinate system. Two possibilities are of special interest here. The coordinate system can always be chosen so that $u^\alpha=(1,0,0,0)$. We call such a system a comoving reference frame. The equations of motion of matter take in it a particularly simple form. In cosmological problems, and mainly where the properties of homogeneous models are discussed, we usually employ a coordinate system in which the components $g_{0\alpha}$ of the metric tensor vanish. Such a system is called synchronous or Gaussian. If the motion of matter is irrotational, we can always choose a reference frame which will be both comoving and synchronous. Geometrically, a synchronous frame can be realized in the following way: choose a smooth spacelike hypersurface Σ and through each point of the hypersurface construct orthogonal timelike geodesic. Adjust the proper times on the geodesics so that the points of intersection with Σ correspond to a fixed value $\tau=\tau_0$. One thus obtains a congruence of timelike geodesics which determines the desired frame. It can be shown that the points corresponding to $\tau=\text{const}$ form a spacelike surface orthogonal to this congruence. In general, a synchronous reference frame does not cover the whole space.

After this digression about coordinate systems, let us return to our problem. We shall adopt additional simplifying assumptions, namely that the distribution of matter is isotropic and homogeneous. Isotropy and homogeneity are understood in the classical sense. The distribution of matter we are interested in distinguishes no directions and no points. These

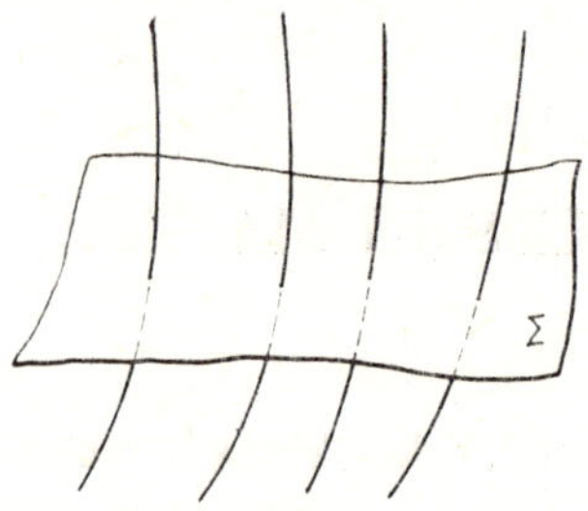

FIG. 3.1. Congruence orthogonal to the hypersurface Σ.

two assumptions strongly restrict the form of the metric. After Robertson and Walker, we take the line element in the form

$$ds^2 = c^2\,dt^2 - R^2(t)\left\{\frac{dr^2}{1-kr^2} + r^2(d\theta^2 + \sin^2\theta\, d\varphi^2)\right\}, \tag{3.4}$$

where k is a constant which may take values 1, 0 and -1, and $R(t)$ is an arbitrary function of time.

With metric (3.4), the field equations can be written

$$\frac{2\ddot{R}}{R} + \frac{\dot{R}^2}{R^2} = -\frac{kc^2}{R^2}, \tag{3.5}$$

$$\frac{\dot{R}^2}{R^2} - \frac{8\pi G}{3c^2}\varepsilon = -\frac{kc^2}{R^2}. \tag{3.6}$$

The continuity equation (3.3) takes the form

$$\varepsilon R^3 = \text{const}, \tag{3.7}$$

and the equations of motion are satisfied automatically.

The solutions of the field equations (3.5) and (3.6) can be represented parametrically. A detailed discussion of these solutions, which were first found by Friedman, will be given in Chapter 10. For the present, let us have a look at the solution for $k = +1$. It reads

$$R = \tfrac{1}{2}R_0(1 - \cos\omega_F), \tag{3.8}$$

$$ct = \tfrac{1}{2}R_0(\omega_F - \sin\omega_F). \tag{3.9}$$

At the initial moment, when $\omega_F = 0$, we have $t = 0$ and $R = 0$. As follows from equation (3.7), it is a singular state. The energy density is then infinite. Subsequently, $R(t)$ increases up to the maximum value $R_{max} = R_0$ when $ct = \frac{1}{2}R_0\pi$, and again falls to zero. At the moment of maximum expansion the first time-derivative of R vanishes. The evolution of the system is fully determined by one parameter, namely the energy density. The point of maximum expansion is of special interest; we then have $\frac{8\pi G}{3c^2}\varepsilon_0=\frac{c^2}{R_0^2}$, and $R_0^2=\frac{3c^4}{8\pi G\varepsilon_0}$.

Note that the point about which the line element (3.4) is spherically symmetric can be chosen arbitrarily.

3.2. Complete Solution for a Spherically Symmetric Dust Cloud. Matching Conditions

We saw in Chapter I that the gravitational field outside a spherically symmetric but not necessarily static distribution of mass is described by the Schwarzschild metric. In order to find the gravitational field in the whole space, we now have to connect the field inside the matter with the external field. Let us recall how an analogous problem is solved in electrostatics. At the surface of a uniformly charged sphere where there is no charged surface layer the electrostatic potential and its radial derivative must be continuous. It is only if these conditions are satisfied that the Maxwell equations hold in the whole space.

The problem of matching an internal and an external solution in general relativity is more complex; guided by the example of electrostatics, however, we can move straight in the right direction. We assume that the boundary surface Σ is regular, and spacelike or timelike. Its first fundamental form is $p_{\mu\nu}=g_{\mu\nu}\pm n_\mu n_\nu$, where n_μ is the unit normal vector; we take "$-$" if Σ is spacelike ($n_\mu n^\mu=1$), and "$+$" if it is timelike ($n_\mu n^\mu=-1$). We assume furthermore that the boundary is not a surface layer of energy and momentum.

One of the matching conditions will be satisfied if the coordinate system can be chosen so that the line elements of the two regions coincide at the boundary. More precisely, we can formulate this condition in the following way: there exists a coordinate system in which the first fundamental form of the boundary is identical in both metrics, i.e.

$$_{\mathrm{I}}p_{\mu\nu}\big|_\Sigma={}_{\mathrm{II}}p_{\mu\nu}\big|_\Sigma. \tag{3.10a}$$

The second matching condition is satisfied if

$$n^\rho{}_{\mathrm{I}}p_{\mu\nu;\rho|\Sigma}=n^\rho{}_{\mathrm{II}}\,p_{\mu\nu;\rho|\Sigma}; \tag{3.10b}$$

here $n^\rho p_{\mu\nu;\,\rho|\Sigma}$ is the second fundamental form (extrinsic curvature) of Σ.

If we can satisfy both (3.10a) and (3.10b), the two solutions are regularly matched along Σ.

To sum up, two solutions of Einstein's equations can be joined to make a complete solution if and only if the neighbourhood of the surface along which these solutions are matched admits a coordinate system in which the first and the second fundamental forms of the embedding of the boundary into one space-time are the same as the corresponding forms in the other space-time.

We shall now find a complete solution for a spherically symmetric homogeneous dust cloud. The internal solution will be described by the Friedman metric and the external one by the Schwarzschild metric. The procedure will be as follows. We start from the Friedman model with $k=1$. At the moment the expansion is maximum we cut out a sphere of a given radius. To the external boundary we then join the Schwarzschild solution, thereby obtaining a new solution of Einstein's equations.

The internal solution is given by the Robertson–Walker line element

$$ds^2 = c^2 dt^2 - R^2(ct)\{d\chi^2 + \sin^2\chi(d\theta^2 + \sin^2\theta\, d\varphi^2)\}, \tag{3.11}$$

where the new variable χ is defined by $r = \sin\chi$; we restrict the range of χ to $0 \leqslant \chi \leqslant \chi_0 < \pi/2$. The first fundamental form of the boundary surface $ct = \frac{1}{2}R_0\pi$, $R = R_0$, $\chi = \chi_0$ is therefore

$$ds^2|_\Sigma = -R_0^2 \sin^2\chi_0(d\theta^2 + \sin^2\theta\, d\varphi^2). \tag{3.12}$$

The external solution is given by the Schwarzschild metric

$$ds^2 = \left(1 - \frac{2GM}{c^2 r}\right)c^2 dt^2 - \frac{dr^2}{1 - 2GM/c^2 r} - r^2(d\theta^2 + \sin^2\theta\, d\varphi^2), \tag{3.13}$$

where r is greater than the dust cloud radius. At the initial moment, the first fundamental form of the boundary surface in this metric is

$$ds^2|_\Sigma = -r_i^2(d\theta^2 + \sin^2\theta\, d\varphi^2). \tag{3.14}$$

Comparing (3.12) with (3.14), we obtain the first matching condition

$$r_i = R_0 \sin\chi_0. \tag{3.15}$$

In the Friedman space-time, radial timelike geodesics are curves along which χ, θ and φ are constant. In the Schwarzschild space-time, radial timelike geodesics are described by the equation

$$\frac{dr}{dt} = c(1 - r_g/r)\sqrt{1 - \frac{1 - r_g/r}{1 - r_g/r_i}}. \tag{3.16}$$

The solution of this equation, given for the first time by Oppenheimer and Snyder, can be written implicitly as

$$\begin{aligned} r &= \frac{r_i}{2}(1 + \cos\eta), \\ ct &= r_g \ln\left[\frac{(r_i/r_g - 1)^{\frac{1}{2}} + \tan\eta/2}{(r_i/r_g - 1)^{\frac{1}{2}} - \tan\eta/2}\right] + r_g(r_i/r_g - 1)^{\frac{1}{2}}\left[\eta + \frac{r_i}{2r_g}(\eta + \sin\eta)\right]. \end{aligned} \tag{3.17}$$

In the proper frame of an observer falling freely in the radial direction, the Schwarzschild line element will take the form

$$ds^2 = c^2 d\tau^2 - e^\lambda dR^2 - r^2(R, \tau)(d\theta^2 + \sin^2\theta\, d\varphi^2), \tag{3.18}$$

where $\lambda(R, \tau)$ and $r(R, \tau)$ are known functions. In our case, we get for $r(R, \tau)$ the parametric relationships

$$r = \frac{R}{2}(1 - \cos\eta), \qquad c\tau = \frac{R}{2}\left(\frac{R}{r_g}\right)^{\frac{1}{2}}(\eta - \sin\eta - \pi), \tag{3.19}$$

whereas e^λ is given by

$$e^\lambda = \frac{(1 - \cos\eta)^2}{4(1 - r_g/R)}. \tag{3.20}$$

According to (3.15), at the initial moment $\eta=\pi$, $r_i=R=R_0 \sin \chi_0$. We now have to find the relation that should hold between r_g, R_0 and χ_0 for the second fundamental form of the hypersurface $\chi=\chi_0$, $R=R_0 \sin \chi_0$ to be identical in the two metrics at any time. To this end, in view of the spherical symmetry of the problem, it is enough to compare the components g_{11} of the two metrics. It turns out that they are identical if

$$r_g=R_0 \sin^3 \chi_0 . \tag{3.21}$$

In addition, as follows from a comparison of formulae (3.8) and (3.19), the parametrization of timelike geodesics has been chosen so that $\eta=\omega_F$.

The solution obtained above will be discussed in detail in the chapter devoted to gravitational collapse. For the present, let us consider the evolution of the dust cloud only in its qualitative aspect, as seen by an observer moving with the boundary surface and by a distant observer.

The equations of motion of the boundary in a comoving frame are given by (3.19). With the lapse of time, when $\eta>\pi$ the radius of the dust cloud diminishes, and when $\eta=2\pi$ it contracts to zero. Since the mass of the cloud is constant, its density increases to infinity.

From the point of view of a distant observer the motion of the boundary surface will look differently. When $\tan \eta/2$ increases from zero to $(r_i/r_g-1)^{1/2}$, $ct\to\infty$, whereas $r\to r_g$. This discordance should be no surprise. We came across a similar effect when discussing the motion of test particles in the Schwarzschild space-time.

The sum of the rest masses of the particles contained in the sphere of radius $r_i=R_0 \sin \chi_0$ can be calculated by multiplying the volume of the sphere by the particle rest mass density

$$M_F=\pi R_0^3 \varepsilon_0 c^{-2}(2\chi_0-\sin 2\chi_0) . \tag{3.22}$$

On the other hand, according to (3.21), a distant observer will assign to this sphere the mass

$$M=\frac{c^2}{2G} R_0 \sin^3 \chi_0 , \tag{3.23}$$

and hence the relationship

$$M=\frac{4}{3} \frac{\sin^3 \chi_0}{(2\chi_0-\sin 2\chi_0)} M_F . \tag{3.24}$$

The mass measured by the distant observer is less than the sum of the rest masses of the dust particles. We shall call the difference between M_F and M, multiplied by c^2, the binding energy. For small χ_0, $M=M_F$, and in the next-order approximation $M=(1-\frac{3}{10}\chi_0^2)M_F$. Using (3.22) and (3.15), one can readily check that $\chi_0^2=\dfrac{2GM_F}{c^2 r_i}$, so

$$M=M_F-\frac{3}{5}\frac{GM_F^2}{c^2 r_i} . \tag{3.25}$$

Thus, in the approximation considered, the total mass-energy M is equal to the sum M_F of the rest masses less the energy of gravitational interaction calculated according to Newton's theory.

3.3. Spherically Symmetric, Static Distribution of Matter

The theory of stellar evolution tells us that the structure of stars undergoes slow changes. In investigating the structure of a star at a given stage of its development, we can assume that the star is a static sphere. Of course, this applies to stars which do not rotate. In Newton's theory of gravitation, the equation of hydrodynamic equilibrium for a static, spherically symmetric configuration is given by the well-known relationship between the gradient of pressure and the gravitational force acting on a unit volume of matter:†

$$\frac{dp}{dr} = -\frac{Gm(r)}{r^2}\,\rho(r), \tag{3.26}$$

where $m(r) = 4\pi \int_0^r \rho(r) r^2 dr$. Let us generalize this condition to allow for the effects of general relativity.

We take the line element of the space-time in the form

$$ds^2 = e^{\nu} c^2 dt^2 - e^{\lambda} dr^2 - r^2(d\theta^2 + \sin^2\theta\, d\varphi^2), \tag{3.27}$$

where ν and λ are arbitrary functions of r only. The distribution of matter is represented by the hydrodynamic energy-momentum tensor

$$T_{\mu\nu} = (\rho c^2 + p)\, u_\mu u_\nu - p g_{\mu\nu}, \tag{3.28}$$

with

$$g^{\mu\nu} u_\mu u_\nu = 1, \tag{3.29}$$

where ρ is the density of matter and p the pressure.

In the static case, the velocity four-vector has only one non-zero component and from (3.29) we find $u_t = e^{\nu/2}$, $u_r = u_\theta = u_\varphi = 0$.

Einstein's equations can be written

$$e^{-\lambda}\left(\frac{\nu'}{r} + \frac{1}{r^2}\right) - \frac{1}{r^2} = \frac{8\pi G}{c^4}\, p, \tag{3.30}$$

$$\tfrac{1}{2} e^{-\lambda}\left(\nu'' + \frac{\nu'^2}{2} + \frac{\nu' - \lambda'}{r} - \frac{\nu'\lambda'}{2}\right) = \frac{8\pi G}{c^4}\, p, \tag{3.31}$$

$$e^{-\lambda}\left(\frac{1}{r^2} - \frac{\lambda'}{r}\right) - \frac{1}{r^2} = -\frac{8\pi G}{c^2}\, \rho, \tag{3.32}$$

where a prime denotes differentiation with respect to r. The equations of motion (2.105) are now reduced to

$$\nu' = -\frac{2p'}{\rho c^2 + p}. \tag{3.33}$$

† A formal derivation of this relationship will be given in the next chapter.

Integration of equation (3.32) yields

$$e^{-\lambda}=1-\frac{8\pi G}{c^2 r}\int_0^r \rho(r')r'^2\,dr'. \tag{3.34}$$

If we denote $m(r)=4\pi\int_0^r \rho(r')r'^2\,dr'$, we can write eqn. (3.34) as

$$e^{-\lambda}=1-\frac{2Gm(r)}{c^2 r}. \tag{3.35}$$

If matter is confined to the region $r\leqslant a$, then for $r>a$, where the space-time is described by the Schwarzschild metric, we have $M=4\pi\int_0^a \rho(r')r'^2 dr'$. Note again that the total effective mass M contained in the region $r\leqslant a$ is different from the sum M_F of the particle rest masses, which is given by

$$M_F=4\pi\int_0^a mn(r')r'^2 e^{\frac{1}{2}\lambda(r')}dr', \tag{3.36}$$

where $mn(r')$ is the rest-mass density. Using (3.33) and (3.35), we can write equation (3.30) in the form

$$\frac{dp}{dr}=-\left(\rho+\frac{p}{c^2}\right)\frac{G\left(m(r)+\frac{4\pi}{c^2}pr^3\right)}{r^2\left(1-\frac{2Gm(r)}{c^2 r}\right)}. \tag{3.37}$$

This is the desired condition for a spherically symmetric distribution of matter to be in hydrostatic equilibrium. As $c\to\infty$, it becomes (3.26). The condition was found by Tolman, Oppenheimer and Volkoff.

Note that unlike the Newtonian condition, the relativistic one involves pressure also on the right-hand side of the equality. At fixed r, $\rho(r)$ and $m(r)$, the pressure gradient is greater in the relativistic case than in Newton's theory. The expression which occurs in the denominator decreases as the gravitational field becomes stronger. The effects of general relativity strnogly influence the matter density distribution in very dense stars.

In the first post-Newtonian approximation equation (3.37) takes the form

$$\frac{dp}{dr}=-\frac{Gm(r)}{r^2}\rho(r)-\frac{G}{c^2 r^2}\left[p(m(r)+4\pi p(r)r^3)+\frac{2Gm^2}{r}\rho\right], \tag{3.38}$$

which implies that the pressure gradient depends not only on the pressure but also on the gravitational energy.

The field equation (3.33) can be integrated and written as

$$\nu=-2\int^p \frac{dp}{\rho c^2+p}. \tag{3.39}$$

In order to calculate this integral we need to know the equation of state, i.e. the relationship $f(\rho, p)=0$, which can also be presented in a parametric form

$$p=p(n, s), \quad \rho=\rho(n, s), \tag{3.40}$$

where n is the number density of particles (baryons) and s the entropy per particle.† The equation of state should locally satisfy the restrictions imposed by special relativity: the energy density ρc^2 should be positive and the speed of sound

$$v_s^2=\left(\frac{\partial p}{\partial \rho}\right)_s \tag{3.41}$$

should not exceed the speed of light.

The energy-momentum tensor which we have used to describe the distribution of matter has no terms representing transport or dissipation of energy. As follows from the equations of motion (2.103), entropy is constant along the particle world-lines. Such a description is of course an idealization. However, even the equation of state of matter at the absolute zero of temperature provides an adequate description in many real situations.

To construct a model of a static relativistic star, we begin by specifying the equation of state, the central density ρ_c and the value ν_c of the function ν at the centre of the star, and putting $m(r=0)=0$. Next, using equations (3.37), (3.39) and the equation of state, we integrate the field equations going up step by step until $p(R)=0$. The radius at which the pressure vanishes defines the boundary of the distribution of matter. We then match the internal solution with the Schwarzschild solution, putting $M=m(R)$ for $r>R$. We choose the constant ν_c so that, at $r=R$, $e^{\nu(R)}=1-2Gm(R)/c^2R$, ensuring the continuity of the metric in the whole space. The vanishing of the pressure on the surface limitting the distribution of matter, $p(R)=0$, ensures that the remaining matching conditions are satisfied.

In this way we obtain a complete, regular solution of Einstein's equations, describing the gravitational field in the whole space. Given the equation of state, both the total mass and radius of the configuration depend on the assumed value of the initial central pressure or density. The properties of this one-parameter family of configurations will allow us to predict the course of the last phases of the evolution of stars.

3.4. Weak Gravitational Fields

We usually observe gravitational fields far from their sources. Sufficiently far away, any gravitational field is weak and we can approximate it by assuming that the metric tensor in the Cartesian coordinate system does not differ much from the metric in flat space. Now, from measurements of a gravitational field we want to find out as much as possible about the structure of the field source. That this is not easy can be seen from the example of a spherically symmetric static configuration. By observing the motion of test particles in such a field we are able to determine the mass of the configuration but cannot say whether it is a small and very dense star or a vast spherical cloud of dust. Let us adopt the

† We are considering a chemically homogeneous configuration.

restrictive assumption that the gravitational field under consideration is weak everywhere, even in the region where matter is present. This means that the density of matter is small and the particle motions are slow, $v/c \ll 1$. Then, we can apply the equation derived in Chapter 1,

$$\psi_{\mu\nu,\alpha}{}^{,\alpha} = -\frac{16\pi G}{c^4} T_{\mu\nu}, \tag{3.42}$$

recalling that

$$\psi_{\mu\nu} = h_{\mu\nu} - \tfrac{1}{2}\eta_{\mu\nu} h, \tag{3.43}$$

where $h_{\mu\nu}$ are small corrections to the metric tensor of flat space. Using the procedure of solving the wave equation by means of the retarded Green function, known from electrodynamics, we obtain

$$\psi_{\mu\nu}(t, x) = \frac{4G}{c^4} \int \frac{T_{\mu\nu}\left(t - \frac{|x - x'|}{c}, x'\right)}{|x - x'|} d^3x', \tag{3.44}$$

where the integration extends over the region filled with matter. To (3.44) we can add any solution of the homogeneous equation so as to satisfy given boundary conditions. The problem of choice of suitable boundary conditions will be discussed later, when we shall consider gravitational waves. For the present, let us only remark that for a bounded distribution of matter $\psi_{\mu\nu}(t, x)$ vanishes at infinity.

Far from the source, for $|x| \gg R$, where R characterizes the dimensions of the matter-filled region, the integral in formula (3.44) can be expanded with respect to the small parameter $R/|x|$. In the comoving, centre-of-mass coordinate system we have

$$p^j = \int T^{0j} d^3x' = 0, \tag{3.45}$$

$$\int x'^j T^{00} d^3x' = 0, \tag{3.46}$$

and

$$h_{00} = -\frac{2G}{c^2 r} \int T^{00} d^2x' = -\frac{2GM}{c^2 r}, \tag{3.47}$$

$$h_{0j} = 4G\varepsilon_{jkl} \frac{x^l}{c^3 r^3} \int \varepsilon_{krs} x'^r T^{s0} d^3x' = 4G\varepsilon_{jkl} \frac{S^k x^l}{c^3 r^3}, \tag{3.48}$$

$$h_{ij} = -\frac{2G}{c^2 r} \delta_{ij} \int T^{00} d^3x' = -\frac{2GM}{c^2 r} \delta_{ij}. \tag{3.49}$$

Here we have used relationship (3.43) connecting $\psi_{\mu\nu}$ with $h_{\mu\nu}$. Thus, the line element takes the form

$$ds^2 = \left(1 - \frac{2GM}{c^2 r}\right) c^2 dt^2 + 4G\varepsilon_{jkl} \frac{S^k x^l}{c^3 r^3} c\, dt\, dx^j - \left(1 + \frac{2GM}{c^2 r}\right) \delta_{ij}\, dx^i dx^j. \tag{3.50}$$

The above result is important. It tells us that *asymptotically, the gravitational field generated by a weak source is fully characterized by the mass M and the total angular momentum of the source S^k.* We can measure both quantities by observing the motions of test particles and gyroscopes.

Literature

FRIEDMAN, A. (1922) *Z. Physik* **10**, 377.

FRIEDMAN, A. (1924) *Z. Physik* **21**, 326.

HARRISON, B. K., THORNE, K. S., WAKANO, M. and WHEELER, J. A. (1965) *Gravitation Theory and Gravitational Collapse*, Chicago University Press, Chicago.

MISNER, C. W., THORNE, K. S. and WHEELER, J. A. (1973) *Gravitation*, Freeman and Comp., San Francisco.

OPPENHEIMER, J. R. and SNYDER, H. (1939) *Phys. Rev.* **56**, 455.

OPPENHEIMER, J. R. and VOLKOFF, G. (1939) *Phys. Rev.* **55**, 374.

TOLMAN, J. R. C. (1939) *Phys. Rev.* **55**, 364.

CHAPTER 4

STELLAR EQUILIBRIUM AND STABILITY IN GENERAL RELATIVITY

4.1. Conditions of Equilibrium

In order to exist, a complex system, such as a star, must be in a state of equilibrium. Three factors are decisive for the equilibrium of a star: gravitational forces, thermal processes and the processes of generation and transport of energy. Gravitational forces are balanced by the pressure of gas. Loss of this equilibrium leads to violent changes occurring on a very short time scale $t \sim (\rho G)^{-1/2}$ comparable to the time a particle would take to fall freely from the stellar surface to the centre if the whole mass of the star were concentrated there. Thus the star must be in a state of hydrostatic equilibrium. Thermal and energy generation processes occur on a much longer time scale and, in the first approximation,

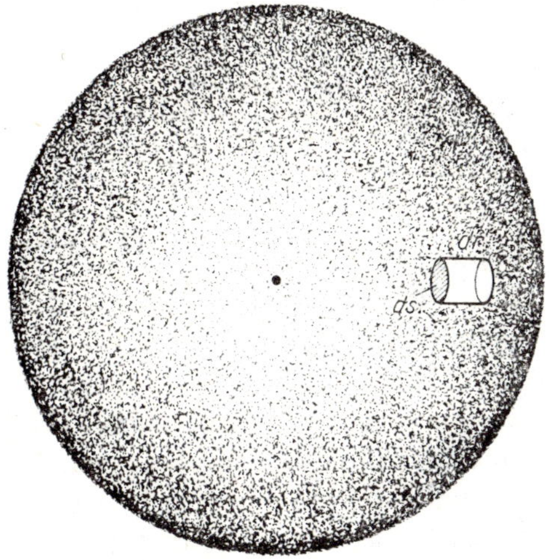

FIG. 4.1. An infinitesimal cylinder in the interior of a star. The cylinder axis coincides with the radial direction. Its height is dr; the base area dS can be taken as unity.

can be neglected. Of course, they play an essential part in the evolution of stars, and the question of conditions of thermal equilibrium is important on an astronomical time scale.

In Newton's theory of gravitation the conditions of hydrostatic equilibrium of a star can be easily derived. Consider a cylinder of unit base area and height dr, whose axis of symmetry passes through the centre of the star. The pressure difference between the bases will be balanced by the gravitational force if

$$p(r)-p(r+dr)-\rho(r)g(r)dr=0, \tag{4.1}$$

where $g(r)$ is the acceleration of gravity; hence

$$\frac{dp}{dr}=-\rho(r)\frac{Gm(r)}{r^2}; \tag{4.2}$$

as before, $m(r)$ denotes the mass contained in the sphere of radius r.

Once a configuration satisfies the equilibrium equation (4.2), we can say a good deal more about it. Namely, equation (4.2) turns out to be a necessary and sufficient condition for a given configuration to extremize the total energy of the system. If we denote the internal energy per unit of mass by $\varepsilon(\rho, s)$ and assume that there are no macroscopic motions, then the total energy of the system is given by

$$E=\int_0^M \varepsilon\, dm-G\int_0^M \frac{m(r)\,dm(r)}{r}, \tag{4.3}$$

where the second term on the right-hand side represents the gravitational energy of the system.

We shall prove that the equilibrium state described by equation (4.2) corresponds to an extremum of the total energy. Calculating the first variation of E for radial perturbations, we obtain

$$\delta E=\int_0^M \left(\frac{1}{\rho}\frac{dp}{dr}+\frac{Gm(r)}{r^2}\right)\delta r\, dm, \tag{4.4}$$

we have used the fact that $\left(\frac{\partial\varepsilon}{\partial\rho}\right)_s=\frac{p}{\rho^2}$ and restricted ourselves to adiabatic perturbations. In virtue of the hydrostatic equilibrium equation, $\delta E=0$; and conversely, if we demand that $\delta E=0$ for arbitrary adiatabic radial perturbations, we shall obtain the equation of hydrostatic equilibrium. Thus, in Newton's theory, the problem of finding equilibrium configurations can be replaced by that of finding an extremum of the total energy.

As in Newton's theory, the conditions of hydrostatic equilibrium in general relativity can be obtained from the equations of hydrodynamics by assuming that matter is at rest. In Newton's theory it can be shown that every bounded static fluid system in equilibrium is spherically symmetric. It is believed that the same is true in general relativity, but no formal proof has yet been found. In the following we shall restrict ourselves to spherically symmetric systems. We shall assume that the line element of the space-time under consideration is of the form

$$ds^2=e^{\nu}c^2dt^2-e^{\lambda}dr^2-r^2(d\theta^2+\sin^2\theta\, d\varphi^2), \tag{4.5}$$

where ν and λ are functions of r only. The four-velocity u^{α} of a chosen element of matter has only one non-zero component:

$$u^{\alpha}=(e^{-\nu/2}, 0, 0, 0). \tag{4.6}$$

The equations of motion (2.105) reduce in this case to

$$\frac{dp}{dr}=-\tfrac{1}{2}(\rho c^2+p)\,\frac{d\nu}{dr}. \tag{4.7}$$

A simple analysis of the field equations for a spherically symmetric static system shows that condition (4.7) is the integrability condition for these equations. Using the field equations we can write the condition of hydrostatic equilibrium in the form

$$\frac{dp}{dr}=-\left(\rho+\frac{p}{c^2}\right)\frac{G(m(r)+4\pi pr^3/c^2)}{r^2\left(1-\dfrac{2Gm(r)}{c^2r}\right)}. \tag{4.8}$$

We already obtained this equation before, when analysing a static spherically symmetric distribution of matter.

Two important conclusions can be drawn from the equation of hydrostatic equilibrium. For normal stars, for which typically $r>\dfrac{2Gm(r)}{c^2}$, $\dfrac{dp}{dr}$ is negative, and therefore the pressure is maximum at the centre of the star and decreases monotonically to zero at the surface. Among configurations with a given barotropic equation of state $p=p(\rho)$ there exists exactly one equilibrium configuration for every central density ρ_c. Indeed, the equation of hydrodynamic equilibrium (4.8) and the equation defining $m(r)$,

$$\frac{dm(r)}{dr}=4\pi\rho(r)r^2, \tag{4.9}$$

constitute a system of two simulataneous equations for $p(r)$ and $m(r)$, which for given initial conditions $p_c=p(\rho_c)$ and $m(0)=0$ has a unique solution. Thus, once the equation of state is fixed, the central density becomes a parameter determining the set of all possible equilibrium configurations. In order to be able to compare different configurations, we introduce one more quantity. It is known that in all physical processes known so far the number of baryons is conserved. As an invariant characteristic of a configuration we shall therefore take the total number of baryons A. If we denote the baryon number density by $n(r)$, then

$$A=4\pi\int_0^R n(r)\left(1-\frac{2Gm(r)}{c^2r}\right)^{-\frac{1}{2}} r^2 dr, \tag{4.10}$$

where R is the configuration radius. For a given total baryon number there may exist several equilibrium configurations.

In Newton's theory, the equation of hydrostatic equilibrium can be obtained, as we have seen, as an energy extremum condition. In general relativity the situation is similar. Namely, among all static, spherically symmetric configurations of cold matter which contain a specified number of baryons A, that configuration for which the total mass $M=m(R)$

is an extremum satisfies the equation of hydrostatic equilibrium. Thus also in general relativity the problem of finding an equilibrium configuration is equivalent to that of finding an extremum of the total mass, given the number of baryons, A. An important question is now how to choose among all possible equilibrium configurations those which are stable. Before taking up this problem, however, let us look at some general properties of equilibrium configurations.

4.2. Polytropic Stars in Newton's Theory

If the equation of state is given, the equation of hydrostatic equilibrium permits us to find the dependence of pressure and density on the radius r. The simplest, physically sensible, perfect-gas stellar model is that in which pressure and density are related by the polytropic equation†

$$p = K\rho^{\frac{n+1}{n}}, \tag{4.11}$$

where K is a constant and n the polytropic index. For convenience, we introduce a dimensionless function θ by putting

$$\rho = \rho_c \theta^n, \tag{4.12}$$

where ρ_c is the central density. The pressure is then given by

$$p = p_c \theta^{n+1}; \tag{4.13}$$

here $p_c = K\rho_c^{\frac{n+1}{n}}$. If the material of the star is an ideal gas, $p = \frac{R}{\mu}\rho T$, where R is the gas constant and μ the mean molecular weight, and

$$T = T_c \theta, \tag{4.14}$$

where T_c is the central temperature, $T_c = \frac{\mu p_c}{R\rho_c} = \frac{\mu}{R} K\rho_c^{1/n}$.

As we can see, the function $\theta(r)$ provides a complete description of the stellar structure. To obtain an equation for $\theta(r)$ we can proceed as follows. Using the equation of equilibrium, we write the Poisson equation as

$$\frac{1}{r^2}\frac{\mathrm{d}}{\mathrm{d}r}\left(\frac{r^2}{\rho}\frac{\mathrm{d}p}{\mathrm{d}r}\right) = -4\pi G\rho. \tag{4.15}$$

Substituting (4.12) and (4.13), and assuming that the star is homogeneous, so that n=const, we get

$$\frac{(n+1)K}{4\pi G}\rho_c^{\frac{1}{n}-1}\frac{1}{r^2}\frac{\mathrm{d}}{\mathrm{d}r}\left(r^2\frac{\mathrm{d}\theta}{\mathrm{d}r}\right) = -\theta^n. \tag{4.16}$$

† In general, a polytropic process is a quasi-static process in which specific heat remains constant: $\mathrm{d}Q/\mathrm{d}T$=const.

Introducing a new dimensionless variable ξ defined by

$$r = r_n \xi, \quad r_n^2 = \frac{(n+1)K}{4\pi G}\rho_c^{1/n-1}, \tag{4.17}$$

we can simplify this equation to the form

$$\frac{1}{\xi^2}\frac{\mathrm{d}}{\mathrm{d}\xi}\left(\xi^2 \frac{\mathrm{d}\theta}{\mathrm{d}\xi}\right) = -\theta^n. \tag{4.18}$$

This is known as the Lane–Emden equation. It is a second-order differential equation, and therefore to obtain a unique solution we have to specify, for example, the values $\theta(0)$ and $\theta'(0)$. It follows from the definition of the function θ that if $\theta(0)=1$, then ρ_c, p_c and T_c have the meanings of the central density, central pressure and central temperature, respectively. The equilibrium equation (4.2) then implies that $\theta'(0)=0$. The solution of equation (4.18) with the initial conditions $\theta(0)=1$, $\theta'(0)=0$ is called the n-th Emden function and is denoted by θ_n.

The Lane–Emden equation can be solved analytically only for a few special values of n. For $n=0$, 1 and 5 we have

$$\begin{aligned} \theta_0 &= 1 - \tfrac{1}{6}\xi^2, \\ \theta_1 &= \frac{\sin\xi}{\xi}, \\ \theta_5 &= \frac{1}{(1+\frac{1}{3}\xi^2)^{\frac{1}{2}}}. \end{aligned} \tag{4.19}$$

Knowing the function θ_n, we can easily construct a one-parameter family of equilibrium configurations. Let us calculate the most important quantities characterizing such configurations. Let ξ_1 be the first zero of the function θ_n. The radius of the star R is then given by

$$R = r_n \xi_1 = \left(\frac{(n+1)K}{4\pi G}\right)^{\frac{1}{2}} \rho_c^{\frac{1-n}{2n}} \xi_1. \tag{4.20}$$

The critical value is $n=1$; then $\xi_1 = \pi$ and

$$R = \left(\frac{K}{2\pi G}\right)^{\frac{1}{2}} \pi \tag{4.21}$$

is independent of ρ_c. Thus, when the polytropic index is $n=1$, the radius of the star depends only on K, or, in other words, on the polytropic temperature defined as $TV^{1/n} = T_n = \text{const.}$ If $n=5$, $R=\infty$ for any finite ρ_c.

The total mass M of the system for arbitrary n is

$$M = m(R) = -4\pi\left[\frac{(n+1)K}{4\pi G}\right]^{\frac{3}{2}} \rho_c^{\frac{3-n}{2n}} \left(\xi^2 \frac{\mathrm{d}\theta}{\mathrm{d}\xi}\right)_{\xi=\xi_1}. \tag{4.22}$$

Note that if $n<3$, the total mass M increases with the growing central density, if $n=3$

it is independent of the central density and if $n>3$ it decreases as the central density increases. For $n=5$ the total mass is finite, even though the configuration has infinite dimensions; for θ_5 we have

$$\lim_{\xi\to\infty}\left(-\xi^2\frac{\mathrm{d}\theta}{\mathrm{d}\xi}\right)=\sqrt{3}, \tag{4.23}$$

whence

$$M=4\sqrt{3}\pi\left(\frac{3K}{2\pi G}\right)^{\frac{3}{2}}\rho_c^{-\frac{1}{5}}. \tag{4.24}$$

By eliminating ρ_c from (4.20) and (4.22) we obtain a relation between mass and radius:

$$GM^{\frac{n-1}{n}}R^{\frac{3-n}{n}}=\frac{(n+1)K}{(4\pi)^{1/n}}\left[-\xi^{\frac{n+1}{n-1}}\frac{\mathrm{d}\theta_n}{\mathrm{d}\xi}\right]^{\frac{n-1}{n}}_{\xi=\xi_1}. \tag{4.25}$$

It is easy to find that the total gravitational energy E_g of the system is given by

$$E_g=-\frac{3}{5-n}\frac{GM^2}{R}, \tag{4.26}$$

and the total internal energy E_i is expressed by the formula

$$E_i=\frac{n}{5-n}\frac{GM^2}{R}. \tag{4.27}$$

Therefore, the total energy of the system, $E=E_i+E_g$, is

$$E=\frac{n-3}{5-n}\frac{GM^2}{R}. \tag{4.28}$$

For $n=3$ the total energy is zero irrespective of the star's radius. In this case, only a small expansion of the star with no change in energy is possible, i.e. a transition from one equilibrium state to another. In the sequence of equilibrium configurations, with n changing continuously, $n=3$ is the point of transition from stable to unstable or from unstable to stable equilibrium states. But as $n\to\infty$, $E_g\to 0$ for any value of E and the configuration cannot be stable. It follows that the system is stable only when $n<3$.

As is seen from equation (4.28), for $n<3$ the total energy of the system is negative, i.e. less than the energy of an infinitely rarified system of the same mass. In the case of stable configurations, with $n<3$, the central density increases with increasing M. When $1<n<3$, the radius of the configuration decreases with growing mass, and at $n=1$ the radius of the star is independent of mass.

To end with, note that the stellar model considered above is not quite realistic; for example, the temperature at the surface of such a star would have to vanish. For a more realistic description, one has to take into account the process of energy generation in central parts of the star due to thermonuclear reactions and the complicated process of energy transport from the center outward. Detailed discussion of these problems can be found in standard books on stellar structure, the most popular of them being those of Chandrasekhar, Schwarzschild, Cox and Giuli, and Clayton.

4.3. Adiabatic Gaseous Spheres in General Relativity

We now move on to consider the relativistic equation of hydrostatic equilibrium for a spherically symmetric configuration, assuming that matter satisfies the equation of a perfect gas changing adiabatically. For a perfect gas we have

$$p=\frac{R}{\mu}\rho_g T, \quad \epsilon=\rho_g c^2+\frac{1}{\gamma-1}p, \tag{4.29}$$

where γ is the ratio of specific heats and ρ_g the rest-mass density of the gas. Using the first law of thermodynamics, we can write the equation of state in the form

$$p=K\rho_g^{\frac{n+1}{n}}, \quad \epsilon=\rho_g c^2+np, \tag{4.30}$$

where

$$n=\frac{1}{\gamma-1} \tag{4.31}$$

is the adiabatic exponent. As in the nonrelativistic case, we introduce a new function θ defined by

$$\rho_g=\rho_{gc}\theta^n \tag{4.32}$$

and a function v defined by the relation

$$m(r)=4\pi\rho_{gc}r_n^3 v, \tag{4.33}$$

where, as before, r_n is given by formula (4.17). Using the dimensionless variable ξ defined by $r=r_n\xi$, we can write the equation of hydrodynamic equilibrium and the equation describing the distribution of matter in the form

$$\frac{1-2(n+1)\dfrac{\alpha v}{\xi}}{1+(n+1)\alpha\theta}\xi^2\frac{d\theta}{d\xi}+v+\alpha\xi^3\theta^{n+1}=0, \tag{4.34}$$

$$\frac{dv}{d\xi}=\xi^2\theta^n(1+n\alpha\theta), \tag{4.35}$$

where

$$\alpha=\frac{p_c}{\rho_{gc}c^2}=\frac{K}{c^2}\rho_{gc}^{1/n}. \tag{4.36}$$

In the Newtonian approximation, when $\alpha\to 0$, the system of equations (4.34) and (4.35) reduces to the Lane–Emden equation. In the relativistic case, we solve the equilibrium equations subject to the initial conditions $\theta(0)=1$, $v(0)=0$. The radius of the configuration is determined by the first zero of the function θ, $\theta(\xi_1)=0$, and the mass by $v(\xi_1)$.

Before we go on to a discussion of the properties of equilibrium configurations, let us find the range of the adiabatic exponent n for which the speed of sound is less than the speed of light. Using the relationship between pressure and density, we get

$$\frac{v_s^2}{c^2}=\frac{dp}{d\varepsilon}=\frac{n+1}{n}\,\frac{\alpha\theta}{1+(n+1)\alpha\theta}. \tag{4.37}$$

The speed of sound will be maximum at the centre; there $\theta(0)=1$, and for large α $v_s^2/c^2\approx 1/n$, so n should be greater than or at most equal to one.

The radius and the total mass of the configuration are given by

$$R=r_n\xi_1, \qquad M=m(R)=4\pi r_n^3\rho_{gc}v(\xi_1). \tag{4.38}$$

Thus as in the Newtonian case, when $n=1$ the radius of the configuration is independent of the central density, while for $n=3$ the mass is independent of the central density.

No explicit solution of the system of equations (4.34) and (4.35) has yet been found for n different from zero. The case $n=0$ corresponds to an incompressible fluid and therefore is not interesting from the physical point of view. Nevertheless, it is worth while to present

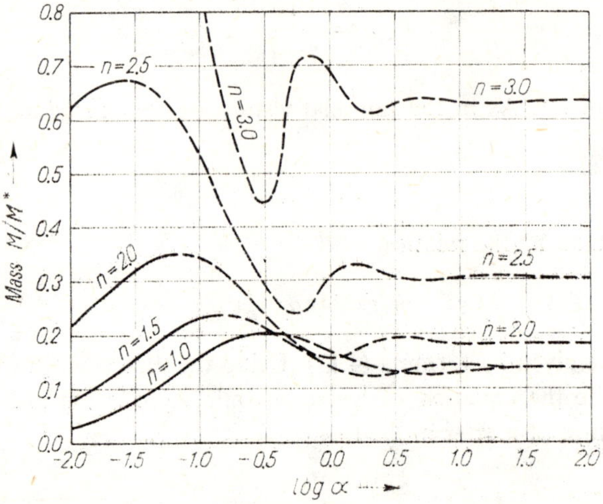

FIG. 4.2. Mass of an adiabatic gaseous sphere as a function of the parameter α for various n. The mass is measured in the units $M^*=(4\pi)^{-1/2}(n+1)^{3/2}G^{3/2}K^{n/2}c^{3-n}$. For $\log\alpha>0.5$ the mass of the system is practically independent of α (adapted from Tooper, 1965).

this solution as it is one of the few exact solutions of Einstein's equations describing a gravitational field inside matter. It is easy to check that when $n=0$,

$$v=\tfrac{1}{3}\xi^3,\ \alpha\theta=\frac{(1+3\alpha)(1-\frac{2}{3}\alpha\xi^2)^{\frac{1}{2}}-(1+\alpha)}{3(1+\alpha)-(1+3\alpha)(1-\frac{2}{3}\alpha\xi^2)^{\frac{1}{2}}} \tag{4.39}$$

satisfy equations (4.34) and (4.35). The metric components inside matter, which in the general case are given by

$$e^{-\lambda}=1-2(n+1)\alpha\frac{v}{\xi},$$

$$e^{\nu}=\frac{1-2\dfrac{Gm(r)}{c^2r}}{(1+(n+1)\alpha\theta)^2}, \tag{4.40}$$

now take the form

$$e^{-\lambda}=1-\frac{8\pi G}{3c^2}\rho_{gc}r^3,$$

$$e^{\nu}=\left[\tfrac{3}{2}\left(1-\frac{2GM}{c^2R}\right)^{\frac{1}{2}}-\tfrac{1}{2}\left(1-\frac{2GMr^2}{c^2R^3}\right)^{\frac{1}{2}}\right]^2. \tag{4.41}$$

Outside the configuration, for $r \geqslant R$, the gravitational field is described by the Schwarzschild metric. However, the internal solution cannot this time be joined regularly to the external one.

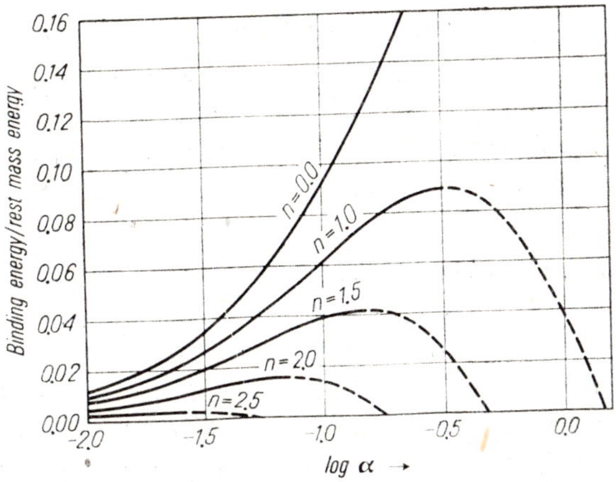

FIG. 4.3. Binding energy of an adiabatic gaseous sphere as a function of log α for various n (adapted from Tooper, 1965).

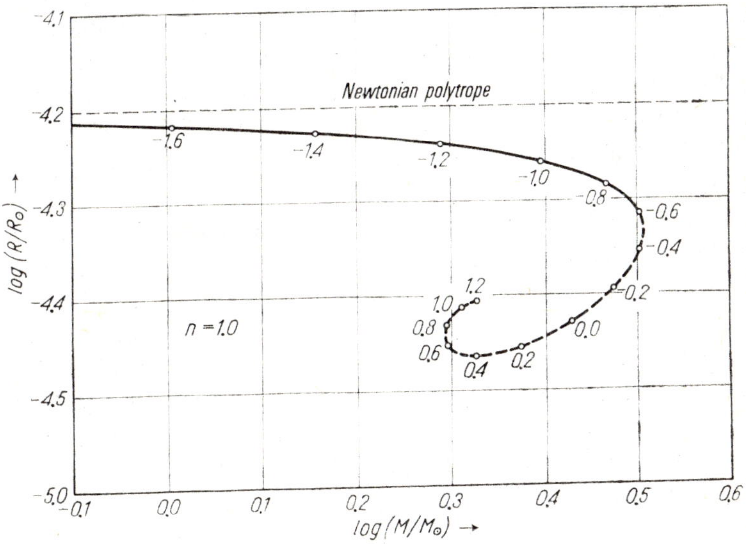

FIG. 4.4. Mass–radius relation for an adiabatic gaseous sphere with $n=1$. The same relation for a Newtonian polytrope is indicated for comparison. The values given along the curve are those of log α (adapted from Tooper, 1965).

For $n \neq 0$, equations (4.34) and (4.35) can be integrated numerically. Tooper has calculated solutions for a few values of n. Their properties are illustrated in Figures 4.2 and 4.3. As can be seen from Fig. 4.2, except the special case $n=3$ ($\gamma=4/3$), the mass, as a function of the parameter α, increases to a maximum and then oscillates with a diminishing amplitude about an asymptotic value. The binding energy behaves similarly (Fig. 4.3). Beyond the maximum point of the binding energy, for greater α, the system is no longer stable. As in the Newtonian case, there are no stable configurations for $n \geqslant 3$.

Important information about the behaviour of a system is provided by the total mass versus radius relation. A typical relationship between these two quantities is shown in Fig. 4.4, where also a Newtonian polytrope is indicated for comparison. Even when the equation of state is nonrelativistic, with $n=1.5$ (or $\gamma=5/3$), the general-relativistic effects play an essential role. The mass reaches a maximum and then the curve turns in a spiral and the mass aproaches an asymptotic value of about 0.3 M_0 while the radius of the configuration tends to about $10^{-4} R_0$. As we shall see later, such a pattern of the mass-radius relation is typical of systems in which relativistic effects become dominant.

4.4. Stellar Stability in Newton's Theory

Whether a spherically symmetric equilibrium configuration is stable or not can be decided, as we have seen, by energy considerations. In more general situations, however, the problem is not as simple. Stability of equilibrium can be investigated by at least two methods. One is to check whether the energy extremum, to which the equilibrium state corresponds, is a local minimum, in which case the equilibrium is stable, or a local maximum, in which case the equilibrium is unstable. The second method is to examine small perturbations of the system. If small perturbations, which displace the system from equilibrium, decay with time, the configuration is in a state of stable equilibrium. If, instead, the perturbation amplitude grows with time, the equilibrium is unstable. Since the equations of motion for perturbations are linear, the time-dependence of perturbations can be represented as $e^{\kappa t}$, where κ can be real or complex. Clearly, if the real part of κ is negative, the equilibrium is stable, whereas if κ has a positive real part, the configuration is in unstable equilibrium.

Dissipation of energy in a star is very small compared with the internal energy and the gravitational energy. Therefore, in the first approximation, we can regard a star as a conservative system. Since the equations of motion will then involve only second time-derivatives, κ will occur only in the second power. One obtains the condition $\kappa^2 = -w^2$; when $w^2 > 0$, the motion is oscillatory (dynamic stability) whereas if $w^2 < 0$, the perturbation grows exponentially. The characteristic period of oscillations is equal to $t_H \sim (G\rho)^{-1/2}$.

If we allow dissipative processes, in which the entropy of any element of mass can be changed owing to viscosity, heat exchange or heat generation in thermonuclear reactions, then κ is complex, and assuming dynamic stability we have

$$\kappa = iw - w'. \tag{4.42}$$

The generalized damping coefficient w' may be positive, in which case we speak of *vibrational stability*, or negative, in the absence of vibrational stability.

The characteristic damping time $\tau=1/w'$ is of the same order as the ratio of the total energy of oscillations to the dissipation rate; it is of the order of the characteristic time of thermal processes, which is defined as the ratio of the total thermal energy to luminosity.

A star emits energy continually, but its luminosity L is usually small compared with the internal energy stored; a change in the state of a star can therefore be considered a sequence of quasi-equilibrium states. The relevant question is whether the process of emission of energy is slow. If so, we speak of secular stability.

For the most part of this section we shall investigate dynamic stability. Consider the total energy of a star, given by

$$E=\int_0^M \varepsilon \mathrm{d}m - G\int_0^M \frac{m\mathrm{d}m}{r} + \tfrac{1}{2}\int_0^M v^2 \mathrm{d}m. \tag{4.43}$$

In the static case, the equilibrium position is determined by the extremum of the total energy E. For the equilibrium to be stable, it has to be a local minimum. The stability condition reduces to

$$\delta E=0, \quad \delta^2 E>0. \tag{4.44}$$

Examine the second variation of E with respect to radial perturbations. Applying the rules of variational calculus, we find

$$\delta^2 E=\tfrac{1}{2}\int \mathrm{d}V\left[\tfrac{4}{3}\frac{1}{V}\frac{\mathrm{d}p}{\mathrm{d}V}(\delta V)^2+p\Gamma_1\left(\frac{\mathrm{d}\delta V}{\mathrm{d}V}\right)^2\right], \tag{4.45}$$

where $\delta V=4\pi r^2\delta r$, and $\Gamma_1=\left(\frac{\partial \ln p}{\partial \ln \rho}\right)_s$ is the adiabatic exponent†. In the case of homogeneous contraction ($\delta V=\alpha V$, α=const) the stability condition $\delta^2E>0$ becomes

$$\int \mathrm{d}V(\Gamma_1-\tfrac{4}{3})p>0. \tag{4.46}$$

If, as assumed, the pressure p is positive inside the star, the stability condition takes the simple form

$$\Gamma_1>\tfrac{4}{3}. \tag{4.47}$$

We obtained the same conclusion while studying polytropic stars. Now it turns out to be true in general, for any stars in equilibrium.

More often, stability of stellar configurations is investigated by the perturbation method. We shall look at the working of this method in the simplest case where only radial perturbations are considered. We assume that the system under consideration can be described

† Generally, an adiabatic process is characterized by three adiabatic exponents:

$$\Gamma_1=\left(\frac{\partial \ln p}{\partial \ln \rho}\right)_s, \quad \frac{\Gamma_2}{\Gamma_2-1}=\left(\frac{\partial \ln p}{\partial \ln T}\right)_s, \quad \Gamma_3-1=\left(\frac{\partial \ln T}{\partial \ln \rho}\right)_s.$$

For a non-relativistic perfect gas $\Gamma_1=\Gamma_2=\Gamma_3=\frac{c_p}{c_v}=\text{const}$.

by the equations of classical hydrodynamics

$$\frac{\partial \rho}{\partial t}+(\rho v_a)_{,a}=0, \tag{4.48}$$

$$\frac{\partial v_a}{\partial t}+v_b v_{a,b}=-\phi_{,a}-\frac{1}{\rho}p_{,a}, \tag{4.49}$$

with

$$\Delta\phi=4\pi G\rho. \tag{4.50}$$

We shall restrict ourselves to adiabatic perturbations. To describe the motion of the system we shall use Lagrangian coordinates, i.e. the coordinates of any element of matter will be $(x(t), y(t), z(t))$.

We shall denote substantial perturbations by D and local perturbations by δ. For any quantity f they are related by

$$Df=\delta f+\delta r\cdot\nabla f. \tag{4.51}$$

We now perturb the equilibrium position defined by $v_a=0$, introducing

$$p+\delta p, \quad v=\frac{\partial}{\partial t}\delta r, \quad \rho+\delta\rho, \quad \phi+\delta\phi, \tag{4.52}$$

where p, ρ and ϕ are the equilibrium values. In the linear approximation, we obtain for the perturbations δp, $\delta\rho$, $\delta\phi$ and for v the equations

$$\frac{\partial\delta\rho}{\partial t}+\frac{1}{r^2}\frac{\partial}{\partial r}(r^2\rho v)=0, \tag{4.53}$$

$$\frac{\partial v}{\partial t}=-\frac{\partial\delta\phi}{\partial r}+\frac{\delta\rho}{\rho^2}\frac{\partial p}{\partial r}-\frac{1}{\rho}\frac{\partial\delta p}{\partial r} \tag{4.54}$$

and

$$\frac{\partial\delta p}{\partial t}+v\frac{\partial p}{\partial r}-\frac{\Gamma_1 p}{\rho}\left(\frac{\partial\delta\rho}{\partial t}+v\frac{\partial\rho}{\partial r}\right)=0; \tag{4.55}$$

this last relation ensures adiabaticity of the perturbations. There is one more relation at our disposal, namely $\delta\phi$ and $\delta\rho$ are connected by

$$\frac{1}{r^2}\frac{\mathrm{d}}{\mathrm{d}r}\left(r^2\frac{\partial\delta\phi}{\partial r}\right)=4\pi G\delta\rho. \tag{4.56}$$

Differentiating this equation with respect to time and using (4.53), we get

$$\frac{\partial}{\partial r}\left(\frac{\partial\delta\phi}{\partial t}\right)=-4\pi G\rho v. \tag{4.57}$$

Elimination of $\frac{\partial\delta\rho}{\partial t}$, $\frac{\partial\delta p}{\partial t}$ and $\frac{\partial\delta\phi}{\partial t}$ from equation (4.54) differentiated with respect to time and equations (4.53), (4.55) and (4.57) yields

$$\frac{\partial^3 \delta r}{\partial t^3}=\frac{1}{\rho}\frac{\partial}{\partial r}\left[\frac{\Gamma_1 p}{r^2}\frac{\partial}{\partial r}\left(r^2\frac{\partial \delta r}{\partial t}\right)\right]-\frac{4}{\rho r}\frac{\partial p}{\partial r}\frac{\partial \delta r}{\partial t}. \tag{4.58}$$

As this equation is linear in δr, we can assume that

$$\frac{\delta r}{r}=\xi(r)\,\mathrm{e}^{-i\omega t}, \tag{4.59}$$

where ω is the frequency. Substituting this in (4.58), we finally find the relationship

$$-\omega^2 r\xi=\frac{1}{\rho r^3}\left\{\frac{\mathrm{d}}{\mathrm{d}r}\left(\Gamma_1 p r^4\frac{\mathrm{d}\xi}{\mathrm{d}r}\right)+r^3\xi\frac{\mathrm{d}}{\mathrm{d}r}[(3\Gamma_1-4)p]\right\}. \tag{4.60}$$

The perturbations δr and δp cannot be arbitrary; they must satisfy the boundary conditions

$$\delta r=0 \quad \text{at} \quad r=0, \tag{4.61}$$

$$Dp=-\Gamma_1 p\left(3\xi+r\frac{\mathrm{d}\xi}{\mathrm{d}r}\right)=0 \quad \text{at} \quad r=R. \tag{4.62}$$

The meaning of these conditions is that neither the centre of the star nor its surface are displaced by the perturbations.

If we multiply equation (4.60) by $4\pi\rho r^3\xi$ and integrate the result over the whole volume of the star, we shall obtain the averaged equation for ω

$$\omega^2-(A-B)=0, \tag{4.63}$$

where

$$A=Q^{-1}\int_0^R 4\pi\Gamma_1 p r^4\left(\frac{d\xi}{\mathrm{d}r}\right)^2\mathrm{d}r, \tag{4.64}$$

$$B=Q^{-1}\int_0^R \xi^2\frac{\mathrm{d}}{\mathrm{d}r}[(3\Gamma_1-4)p]\,4\pi r^2\mathrm{d}r, \tag{4.65}$$

with

$$Q=\int_0^M \xi^2 r^2\mathrm{d}m. \tag{4.66}$$

For the fundamental frequency mode, we can assume to a good approximation that ξ is constant. Then

$$\omega^2=-\langle(3\Gamma_1-4)\rangle\frac{\int_0^M \frac{Gm\mathrm{d}m}{r}}{\int_0^M r^2\mathrm{d}m}=\langle(3\Gamma_1-4)\rangle\frac{E_g}{I}, \tag{4.67}$$

where $\langle\ \rangle$ denotes the average over the whole volume and I is the moment of inertia of the system with respect to the centre. The system is dynamically stable when $\omega^2>0$, i.e. when

$$\langle\Gamma_1\rangle>\tfrac{4}{3}. \tag{4.68}$$

It is worthwhile to note that equation (4.60), which can be written in an equivalent form as

$$\frac{d}{dr}\left(\Gamma_1 pr^4\frac{d\xi}{dr}\right)+\left\{r^3\frac{d}{dr}[(3\Gamma_1-4)p]+\omega^2\rho r^4\right\}\xi=0 \tag{4.69}$$

is a self-adjoint equation of the Sturm–Liouville type (S–L equation)

$$(P\xi')'+(Q+\omega^2W)\xi=0, \tag{4.70}$$

where

$$P=\Gamma_1 pr^4,\quad Q=r^3\frac{d}{dr}[(3\Gamma_1-4)p],\quad W=\rho r^4. \tag{4.71}$$

At the centre of the star the pressure is finite and therefore $P(0)=0$. On the surface of the star, on the other hand, $p(R)=0$, which implies the vanishing of P. Thus the centre and surface of the star are regular singular points of the equation.

The properties of the Sturm–Liouville equation are well known. We shall mention only the most important ones.

The spectrum of an S–L equation is discrete and has a lower bound, i.e. there exists a least eigenvalue $\omega^2=\omega_0^2$. All the eigenvalues can therefore be arranged in a series

$$\omega_0^2\leqslant\omega_1^2\leqslant\ldots\leqslant\omega_n^2\leqslant\ldots \tag{4.72}$$

The spectrum has no upper bound and extends to infinity. The above arrangement is possible because all the eigenvalues ω_n^2 are real.

The solution ξ_n of an S–L equation, corresponding to the eigenvalue ω_n^2 and satisfying the boundary conditions, is referred to as the n-th mode of radial pulsations. The mode corresponding to the least eigenvalue ω_0^2 is called the fundamental mode. Usually, the fundamental mode has no nodes (zeros) in the open interval $(0, R)$ but it may vanish at the ends. If ξ_n and ξ_{n+1} are the modes corresponding to the eigenvalues ω_n^2 and ω_{n+1}^2, the number of nodes of the function ξ_{n+1} is greater by one than that of the function ξ_n. Thus the number of nodes of the eigenfunctions of an S–L equation increases monotonically as we pass to larger eigenvalues.

An S–L equation can be obtained from the variational principle

$$\delta I[\xi]=\delta\int_a^b(P\xi'^2-Q\xi^2)\,dr \tag{4.73}$$

with the additional condition

$$\int_a^b W\xi^2 dr=\text{const}. \tag{4.74}$$

Assume that the boundary conditions are chosen so that

$$P\xi\xi'=0 \quad\text{for}\quad r=0 \text{ and } r=R, \tag{4.75}$$

and that the function $\xi(r)$ is normalized,

$$\int_0^R W\xi^2 dr=1. \tag{4.76}$$

Consider the functional $I[\xi]$, where the trial function ξ may change arbitrarily as long as it remains normalized and satisfies the boundary conditions (4.75). *The trial function which minimizes the functional $I[\xi]$ is the eigenfunction of the* S–L *equation corresponding to the least eigenvalue* ω_0. Indeed,

$$I[\xi_0]=\int_0^R (P\xi_0'^2-Q\xi_0^2)\,dr=\omega_0^2\int_0^R W\xi_0^2\,dr=\omega_0^2. \tag{4.77}$$

The problem of finding the least eigenvalue ω_0^2 is thus reduced to that of finding the minimum of the functional $I[\xi]$. Similarly, we can find any eigenvalue ω_n^2 by choosing trial functions having the corresponding number of nodes.

So far in investigating the stability of stars we have neglected the influence of thermal processes. Clearly a star which is in a state of hydrostatic equilibrium does not have to be in thermal equilibrium. Energy which is constantly released in the centre of the star has to be transported to the surface and radiated away. In most stars energy is transported by radiation and/or convection.

Radiative energy transport occurs whenever there is a temperature gradient. The distribution of temperature inside the star adjusts itself in such a way as to allow all the energy released in the centre to be transported outward. Sometimes the opacity of the material is so high that the temperature gradient necessary to drive all the energy is too steep to be maintained and then convection begins. Conditions for the occurrence of convection were derived by Schwarzschild.

Consider a small element which somehow becomes hotter than its surroundings. It expands and its density decreases. The element moves outward under the influence of the buoyant force of the surrounding material. We assume that the pressure inside the element is equal to the outside pressure and that the element is displaced adiabatically. If the density of the displaced element turns out to be smaller than the density of the surrounding matter the element tends to move further outward and convection sets in. Thus, convection occurs if

$$\left(\frac{d\rho}{dr}\right)_s<\left(\frac{d\rho}{dr}\right)_{\text{star}}; \tag{4.78}$$

here $\left(\frac{d\rho}{dr}\right)_s$ denotes the adiabatic density gradient and $\left(\frac{d\rho}{dr}\right)_{\text{star}}$ the actual density gradient in the star. For a star with uniform chemical composition this condition can be rewritten in the equivalent form

$$\left|\frac{1}{T}\left(\frac{dT}{dr}\right)_{\text{star}}\right|>\left|\frac{1}{T}\left(\frac{dT}{dr}\right)_s\right| \tag{4.79}$$

or

$$\left(\frac{dp}{dr}\right)_{\text{star}}-\Gamma_1\frac{p}{\rho}\left(\frac{d\rho}{dr}\right)_{\text{star}}<0. \tag{4.80}$$

For a perfect fluid configuration the Schwarzschild condition can be written as

$$\frac{1}{\rho^2}\left(\frac{\partial\rho}{\partial s}\right)_p\nabla p\cdot\nabla s<0. \tag{4.81}$$

4.5. Stellar Stability in the Post-Newtonian Approximation

Before we take up the question of stability in the relativistic case, let us first look at the effect of relativity on the stability of equilibrium in the post-Newtonian approximation. This restriction simplifies the stability considerations in an essential way. For arbitrary perturbations, general-relativistic considerations would have to allow for the possibility of generation of gravitational waves under perturbations changing the quadrupole or higher moments of the distribution of matter, which would complicate the problem considerably.

Using the equations of hydrodynamics in the post-Newtonian approximation, we shall examine the stability of equilibrium by analysing small perturbations. To find the equation of equilibrium, we start from the general equations of motion in this approximation, assuming that the velocity of matter is zero and neglecting all time-derivatives. The result is

$$\left[\left(1-\frac{2\phi}{c^2}\right)p\right]'=-\rho\phi'+\frac{2}{c^2}\rho(\Gamma'-\Sigma\phi'), \tag{4.82}$$

$$\frac{1}{r^2}(r^2\phi')'=4\pi G\rho, \tag{4.83}$$

$$\frac{1}{r^2}(r^2\Gamma')'=-4\pi G\rho\Sigma, \tag{4.84}$$

where

$$\Sigma=-\phi+\tfrac{1}{2}\pi+\tfrac{3}{2}p/\rho, \tag{4.85}$$

and a prime denotes differentiation with respect to r; ρ is the density of rest mass, p the pressure and π the internal energy per unit of mass, so that the total energy density is given by

$$\varepsilon=\rho c^2+\pi\rho. \tag{4.86}$$

Now let us perturb the equilibrium position by introducing local changes of density $\delta\rho$, pressure δp, gravitational potential $\delta\phi$ and position $\delta\mathbf{r}$, so that $\mathbf{v}=\frac{\mathrm{d}}{\mathrm{d}t}\,\delta\mathbf{r}$. The linearized equation of motion for the perturbed quantities is

$$\frac{\partial}{\partial t}(\sigma v_a)+\frac{1}{2c^2}\rho\frac{\partial}{\partial t}(U_a-U_{b,ab})-\frac{4}{c^2}\rho\frac{\partial}{\partial t}(v_a\phi+U_a)=-\rho\delta\phi_{,a}$$
$$-\left[\left(1-\frac{2\phi}{c^2}\right)\delta p-\frac{2}{c^2}p\delta\phi\right]_{,a}+\frac{\delta\rho}{\rho}\left[\left(1-\frac{2\phi}{c^2}\right)p\right]_{,a}-\frac{2}{c^2}\rho\,[\delta\Sigma\phi_{,a}+\Sigma\delta\phi_{,a}-\delta\Gamma_{,a}], \tag{4.87}$$

where

$$\sigma=\rho\left[1+\frac{1}{c^2}(-2\phi+\pi+p/\rho)\right], \tag{4.88}$$

$$U_a=G\int_V\frac{\rho(\mathbf{x}')\,v_a(\mathbf{x}')}{|\mathbf{x}-\mathbf{x}'|}\,\mathrm{d}^3\mathbf{x}', \tag{4.89}$$

and

$$\Delta\delta\phi=4\pi G\delta\rho, \tag{4.90}$$

$$\Delta\delta\Gamma=-4\pi G\delta(\rho\Sigma). \tag{4.91}$$

From the generalized continuity equation (2.180) it follows that

$$\frac{\partial}{\partial t}\delta\rho^{*}+(\rho^{*}v_{a})_{,a}=0, \tag{4.92}$$

where

$$\rho^{*}=\rho\left(1-\frac{3\phi}{c^{2}}\right). \tag{4.93}$$

Assume that $v_a=\frac{\mathrm{d}}{\mathrm{d}t}\delta x_a$ and $\delta x_a=\xi_a e^{i\omega t}$; equation (4.87) then takes the form

$$\omega^{2}\left[\sigma\xi_{a}+\frac{\rho}{2c^{2}}(U_{a}-U_{b,ab})-\frac{4\rho}{c^{2}}(\xi_{a}\phi+U_{a})\right]=\rho\delta\phi_{,a}+\left[\left(1-\frac{2\phi}{c^{2}}\right)\delta p-\frac{2}{c^{2}}p\delta\phi\right]_{,a}$$
$$-\frac{\delta\rho}{\rho}\left[\left(1-\frac{2\phi}{c^{2}}\right)p\right]_{,a}+\frac{2\rho}{c^{2}}[\delta\Sigma\phi_{,a}+\Sigma\delta\phi_{,a}-\delta\Gamma_{,a}]; \tag{4.94}$$

here $\delta\rho$, δp and $\delta\phi$ denote the changes in the density, pressure and gravitational potential respectively, caused by the displacement δr, and U_a is given by formula (4.89) with v_a, replaced by ξ_a. The continuity equation reduces to

$$\delta\rho^{*}=-(\rho^{*}\xi_{a})_{,a}, \tag{4.95}$$

where $\delta\rho^*$ depends on spatial coordinates only.

As in the classical case, it is convenient to pass from differentials at fixed coordinate locations δ to the material differentials D. The equations of motion and the continuity equation can then be written

$$\omega^{2}\left[\sigma\xi_{a}+\frac{\rho}{2c^{2}}(U_{a}-U_{b,ab})-\frac{4\rho}{c^{2}}(\xi_{a}\phi+U_{a})\right]=\rho(D\phi)_{,a}+\left[\left(1-\frac{2\phi}{c^{2}}\right)Dp-\frac{2p}{c^{2}}D\phi\right]_{,a}$$
$$-\frac{D\rho}{\rho}\left[\left(1-\frac{2\phi}{c^{2}}\right)p\right]_{,a}+\frac{2\rho}{c^{2}}[D\Sigma\phi_{,a}+\Sigma(D\phi)_{,a}-(D\Gamma)_{,a}], \tag{4.96}$$

$$D\rho^{*}=-\rho^{*}\xi_{a,a}. \tag{4.97}$$

We now need to express the material changes of the quantities of interest in terms of the relative displacement ξ. Following a procedure similar to that used in the Newtonian case, and integrating over the volume of the star, we obtain the equation

$$\omega^{2}\left\{\int_{V}\sigma|\xi|^{2}\mathrm{d}^{3}x+\frac{1}{2c^{2}}\int_{V}\rho(U_{a}-U_{b,ab})\xi_{a}\mathrm{d}^{3}x-\frac{4}{c^{2}}\int_{V}\rho(\phi|\xi|^{2}+U_{a}\xi_{a})\mathrm{d}^{3}x\right\}=X, \tag{4.98}$$

where X is given by

$$X = \int_V \rho(D\phi)_{,a}\xi_a \mathrm{d}^3x + \int_V \Gamma_1 p\left(1-\frac{2\phi}{c^2}\right)(\xi_{b,b})^2\mathrm{d}^3x + \int_V \xi_{b,b}\left[\left(1-\frac{2\phi}{c^2}\right)p\right]_{,a}\xi_a \mathrm{d}^3x$$
$$-\frac{1}{c^2}\int_V (3\Gamma_1-2)\,pD\phi\xi_{b,b}\mathrm{d}^3x + \frac{2}{c^2}\int_V \rho\Sigma(D\phi)_{,a}\xi_a\mathrm{d}^3x - \frac{2}{c^2}\int_V \rho(D\Gamma)_{,a}\xi_a\mathrm{d}^3x$$
$$-\frac{1}{c^2}\int_V D\phi p_{,a}\xi_a\mathrm{d}^3x - \frac{1}{c^2}\int_V (3\Gamma_1-2)\,p\xi_{b,b}\xi_a\phi_{,a}\mathrm{d}^3x. \tag{4.99}$$

In practical applications it is very convenient to expand ξ into spherical harmonics, according to the formulae

$$\xi_r = \frac{1}{r^2}\sum_{l,m}\psi_{lm}(r)\,Y_l^m(\theta,\varphi),$$
$$\xi_\theta + i\xi_\varphi = \frac{1}{r}\sum_{l,m}\frac{1}{l(l+1)}\frac{\mathrm{d}\chi_{lm}(r)}{\mathrm{d}r}\left(\frac{\partial}{\partial\theta}+\frac{i}{\sin\theta}\frac{\partial}{\partial\varphi}\right)Y_l^m(\theta,\varphi), \tag{4.100}$$

where $\psi_{lm}(r)$ and $\chi_{lm}(r)$ are real functions of r, and $Y_l^m(\theta,\varphi)$ are spherical harmonics. Since the equilibrium configuration is spherically symmetric, such an expansion is very natural.

Substituting (4.100) in (4.99) and performing the angular integrations, we get

$$X = \int_0^R \Gamma_1 p\left(1-\frac{2\phi}{c^2}\right)(\psi_{00}-\chi_{00})'^2\frac{\mathrm{d}r}{r^2} + 2\int_0^R \left[\left(1-\frac{2\phi}{c^2}\right)p\right]'\left(2\frac{\psi_{00}^2}{r}-\psi_{00}\chi_{00}'\right)\frac{\mathrm{d}r}{r^2}$$
$$-4\pi G\int_0^R \left(1+\frac{4\Sigma}{c^2}\right)\sum_l \frac{1}{2l+1}(J_lK_l'-K_lJ_l')\,\mathrm{d}r$$
$$+\frac{1}{c^2}\int_0^R \rho(D\phi)^2r^2\mathrm{d}r - \frac{2}{c^2}\int_0^R (3\Gamma_1-2)\,pD\phi(\psi_{00}-\chi_{00})'\mathrm{d}r, \tag{4.101}$$

where

$$D\phi = -4\pi G\sum_l \frac{1}{2l+1}\left(\frac{J_l(r)}{r^{l+1}} - r^lK_l(r)\right) + \frac{\psi_{00}}{r^2}\phi',$$
$$J_l(r) = \int_0^r \rho(r)\,r^l\left[l\frac{\psi_{lm}(r)}{r}+\frac{\mathrm{d}\chi_{lm}(r)}{\mathrm{d}r}\right]\mathrm{d}r, \tag{4.102}$$
$$K_l(r) = \int_r^R \frac{\rho(r)}{r^{l+1}}\left[(l+1)\frac{\psi_{lm}(r)}{r}-\frac{\mathrm{d}\chi_{lm}}{\mathrm{d}r}\right]\mathrm{d}r.$$

We have thus obtained a general equation which allows us to investigate the stability of equilibrium against arbitrary perturbations in the post-Newtonian approximation.

For radial perturbations, $l=0$ and $\chi_{lm}=0$. In order to compare the results with the Newtonian case, we assume $\psi_{00}=r^2\xi$. For the fundamental frequency, ξ can be taken as constant and then the equilibrium condition reduces to

$$\Gamma_1>\tfrac{4}{3}+\alpha\frac{2GM}{c^2R}, \tag{4.103}$$

where α is a constant dependent on the parameters of the equilibrium configuration. The relativistic correction is important in the analysis of stability against radial perturbations. It causes an earlier appearance of dynamic instability than is predicted in Newton's theory, a fact of particular relevance for systems near the stability limit.

Now consider an example of non-radial perturbations, and specifically the influence of relativistic effects on the occurrence of convective processes. In a classical star, convection occurs only when the temperature gradient is superadiabatic. An analysis of equation (4.98) in the case where $\omega=0$ and $l\geqslant 1$ leads to the conclusion that convection may occur only in the region in which

$$p'-\Gamma_1\frac{p}{\rho}\rho'=\frac{1}{c^2}\frac{p}{\Gamma-1}\phi'\left(\Gamma-\Gamma_1+\frac{1}{\Gamma-1}\frac{\Gamma'\rho}{\rho'}\right). \tag{4.104}$$

Γ is defined by the relation

$$\pi=\frac{1}{\Gamma-1}\frac{p}{\rho}. \tag{4.105}$$

In the classical case, convection processes occur when $p'-\Gamma_1\dfrac{p}{\rho}\rho'<0$. In the post-Newtonian approximation the condition will read

$$p'-\Gamma_1\frac{p}{\rho}\rho'+\frac{\pi}{c^2}p'\left(\Gamma-\Gamma_1+\frac{1}{\Gamma-1}\frac{\Gamma'\rho}{\rho'}\right)<0. \tag{4.106}$$

Using thermodynamic identities, we can write this in the equivalent form

$$\left(p'-\Gamma_1\frac{p}{\rho}\rho'\right)\left[1+\frac{\pi}{c^2}\frac{\Gamma_3-\Gamma}{\Gamma_3-1}\frac{(\ln p)'}{(\ln\rho)'}\right]<0, \tag{4.107}$$

where

$$\Gamma_3=1+\left(\frac{\partial\ln T}{\partial\ln\rho}\right)_s. \tag{4.108}$$

It is now seen that the post-Newtonian corrections have no effect on the occurrence of convection. Thorne's general-relativistic analysis of convective stability has shown that convection occurs always when the temperature gradient is superadiabatic, i.e. when

$$\left|\frac{1}{T}\left(\frac{\partial T}{\partial r}\right)_{\text{star}}\right|>\left|\frac{1}{T}\left(\frac{\partial T}{\partial r}\right)_s\right|.$$

4.6. Stability of Spherically Symmetric Equilibrium Configurations with Respect to Radial Perturbations

Thus far we have found criteria of stability against radial perturbations in Newton' theory of gravity and in the first post-Newtonian approximation. It has turned out that relativistic corrections cause an earlier appearance of instability than is predicted in the classical case. We now proceed to derive a stability criterion based on the equations of general relativity.

Consider a spherically symmetric equilibrium configuration which we subject to small spherically symmetric radial perturbations. Restricting our attention to radial perturbations greatly simplifies the problem. The gravitational field outside matter is described by the Schwarzschild metric and that inside matter by the metric

$$ds^2 = e^{\nu} c^2 dt^2 - e^{\lambda} dr^2 - r^2(d\theta^2 + \sin^2\theta\, d\varphi^2), \tag{4.109}$$

where ν and λ are arbitrary functions of r and t. To describe the distribution of matter we shall use the hydrodynamic energy-momentum tensor, which implies that only adiabatic perturbations will be considered.

We prescribe the initial configuration by choosing $\nu_0(r)$ and $\lambda_0(r)$ satisfying the equilibrium conditions (4.8) and (4.9). The field equations for the metric (4.109) are

$$\begin{aligned}
\frac{8\pi G}{c^4} T_0^0 &= -e^{-\lambda}\left(\frac{1}{r^2} - \frac{\lambda'}{r}\right) + \frac{1}{r^2},\\
\frac{8\pi G}{c^4} T_1^1 &= -e^{-\lambda}\left(\frac{\nu'}{r} + \frac{1}{r^2}\right) + \frac{1}{r^2},\\
\frac{8\pi G}{c^4} T_2^2 = \frac{8\pi G}{c^4} T_3^3 &= -\tfrac{1}{2} e^{-\lambda}\left(\nu'' + \frac{\nu'^2}{2} + \frac{\nu' - \lambda'}{r} - \frac{\nu'\lambda'}{2}\right) + \tfrac{1}{2} e^{-\nu}\left(\ddot{\lambda} + \frac{\dot{\lambda}^2}{2} - \frac{\dot{\lambda}\dot{\nu}}{2}\right),\\
\frac{8\pi G}{c^4} T_0^1 &= -e^{-\lambda}\frac{\dot{\lambda}}{r};
\end{aligned} \tag{4.110}$$

a prime denotes differentiation with respect to r, and a dot differentiation with respect to ct. We perturb these equations by adding small displacements to the parameters of the configuration: $\rho_0(r) + \delta\rho(r, t)$, $p_0(r) + \delta p(r, t)$, $\nu_0(r) + \delta\nu(r, t)$, $\lambda_0(r) + \delta\lambda(r, t)$. The velocity four-vector of particles in equilibrium had only one non-zero component: $u^\alpha = (e^{-\frac{\nu_0}{2}}, 0, 0, 0)$. Under perturbation, the particles can be displaced in the radial direction, so now

$$u^\alpha = (e^{-\frac{\nu_0}{2}}, e^{-\frac{\nu_0}{2}} v, 0, 0), \tag{4.111}$$

where $v = \frac{1}{c}\frac{d\delta r}{dt}$ is the radial velocity measured in units of the speed of light.

Note that the field equations are not independent because $T^{\alpha\beta}{}_{;\beta} = 0$, equivalent to the equations of motion

$$(\varepsilon + p) u_{\alpha;\beta} u^\beta - (\delta_\alpha^\beta - u_\alpha u^\beta) p_{,\beta} = 0. \tag{4.112}$$

To first order in the perturbations, the field equations and the equations of motion yield four independent relations

$$(re^{-\lambda_0}\delta\lambda)' = \frac{8\pi G}{c^4} r^2 \delta\varepsilon, \tag{4.113}$$

$$\frac{e^{-\lambda_0}}{r}\left(\delta\nu' - \nu_0'\,\delta\lambda - \frac{\delta\lambda}{r}\right) = \frac{8\pi G}{c^4}\,\delta p, \tag{4.114}$$

$$-\frac{e^{-\lambda_0}}{r}\,\delta\dot{\lambda} = \frac{8\pi G}{c^4}(\varepsilon_0 + p_0)\,v, \tag{4.115}$$

$$e^{\lambda_0-\nu_0}(\varepsilon_0+p_0)\,\dot{v} + \delta p' + \tfrac{1}{2}(\varepsilon_0+p_0)\,\delta\nu' + \tfrac{1}{2}(\delta\varepsilon + \delta p)\,\nu_0' = 0. \tag{4.116}$$

It is now convenient to introduce the matter displacement ξ defined by

$$v = \dot{\zeta}. \tag{4.117}$$

Equation (4.115) can be integrated; the result is

$$\frac{e^{-\lambda_0}}{r}\,\delta\lambda = -\frac{8\pi G}{c^4}(\varepsilon_0+p_0)\,\zeta. \tag{4.118}$$

To the right-hand side we might add an arbitrary function of r; however, by a suitable coordinate transformation $x^\mu \to x^\mu + \delta x^\mu$, it can always be made equal to zero. Using the field equations for the unperturbed configuration, we finally get

$$\delta\lambda = -\zeta(\lambda_0 + \nu_0)'. \tag{4.119}$$

Inserting (4.118) into (4.113), we can express $\delta\varepsilon$ in terms of ζ and $\varepsilon_0 + p_0$:

$$\delta\varepsilon = -\frac{1}{r^2}\,[r^2(\varepsilon_0+p_0)\zeta]'. \tag{4.120}$$

Similarly, equation (4.115) can be simplified to

$$(\varepsilon_0+p_0)\,\delta\nu' = \left[\delta p - (\varepsilon_0+p_0)\left(\nu_0' + \frac{1}{r}\right)\zeta\right](\lambda_0+\nu_0)'. \tag{4.121}$$

In this way, three of the four equations for perturbations involve no time-derivatives, hence if they are satisfied at some fixed point of time, they are satisfied always. The dynamics of the process is described by equation (4.116), which contains the second time-derivative of the displacement ζ. To use this equation, we need to express δp in terms of ζ. To this end, we introduce an additional assumption, which in the classical case amounts to dividing the total energy density ε into parts corresponding to the rest mass and to the internal energy of the system. It can be formulated in various ways; the most convenient one in our case will be to assume that the total number of baryons is unchanged, i.e.

$$(nu^\alpha)_{;\alpha} = 0, \tag{4.122}$$

where n is the baryon number density. Explicitly, this reads

$$(nu^\alpha)_{;\alpha} = nu^\alpha(\ln\sqrt{-g})_{,\alpha} = 0. \tag{4.123}$$

Linearizing this equation in the neighbourhood of the equilibrium position and using (4.118), we get

$$\delta n = -\frac{e^{\frac{\nu_0}{2}}}{r^2}\left(n_0 r^2 \zeta e^{-\frac{\nu_0}{2}}\right)'. \tag{4.124}$$

Now if

$$p = p(n), \quad \varepsilon = \varepsilon(n) \tag{4.125}$$

is the equation of state, then from (4.120) and (4.124) we can find the desired relationship between δp, ζ and the quantities characterizing the unperturbed configuration

$$\delta p = -\zeta p_0' - \Gamma_1 p_0 \frac{e^{\frac{\nu_0}{2}}}{r^2}\left(r^2 e^{-\frac{\nu_0}{2}}\zeta\right)', \tag{4.126}$$

where, as before,

$$\Gamma_1 = \left(\frac{\partial \ln p}{\partial \ln n}\right)_s \tag{4.127}$$

is the adiabatic exponent.†

Inserting (4.124) and (4.126) into (4.121) gives an equation for ζ:

$$-e^{\lambda_0-\nu_0}(\varepsilon_0+p_0)\ddot{\zeta} = -e^{-\frac{\lambda_0+2\nu_0}{2}}\left[e^{\frac{\lambda_0+3\nu_0}{2}}\frac{\Gamma_1 p_0}{r^2}\left(r^2 e^{-\frac{\nu_0}{2}}\zeta\right)'\right]'$$

$$+\left[\frac{4p_0'}{r} + \frac{8\pi G}{c^4}e^{\lambda_0}p_0(\varepsilon_0+p_0) - \frac{p_0'^2}{\varepsilon_0+p_0}\right]\zeta. \tag{4.128}$$

If we assume that

$$\zeta = \frac{e^{\frac{\nu_0}{2}}}{r^2}\xi(r)e^{i\omega t}, \tag{4.129}$$

we obtain the following equation for ω^2:

$$\left[e^{\frac{\lambda_0+3\nu_0}{2}}\frac{\Gamma_1 p_0}{r^2}\xi'\right]' + \left\{\left(\frac{p_0'^2}{\varepsilon_0+p_0} - \frac{4p_0'}{r} - \frac{8\pi G}{c^4}e^{\lambda_0}p_0(\varepsilon_0+p_0)\right)e^{\frac{\lambda_0+3\nu_0}{2}}\right.$$

$$\left. + \omega^2(\varepsilon_0+p_0)e^{\frac{3\lambda_0+\nu_0}{2}}\right\}\frac{\xi(r)}{r^2} = 0. \tag{4.130}$$

This is the fundamental equation describing *radial pulsations* in the relativistic case. As in the classical case, it is a Sturm–Liouville equation, and to examine its eigenvalues and eigenfunctions one can use the same methods as those discussed in the section on stellar

† The three relativistic counterparts of the non-relativistic adiabatic exponents (cf. footnote p. 79) are

$$\Gamma_1 = \left(\frac{\partial \ln p}{\partial \ln n}\right)_s = \frac{\varepsilon+p}{\varepsilon}\left(\frac{\partial \ln p}{\partial \ln \varepsilon}\right)_s,$$

$$\frac{\Gamma_2}{\Gamma_2-1} = \left(\frac{\partial \ln p}{\partial \ln T}\right)_s, \quad \Gamma_3 - 1 = \left(\frac{\partial \ln T}{\partial \ln n}\right)_s = \frac{\varepsilon+p}{\varepsilon}\left(\frac{\partial \ln T}{\partial \ln \varepsilon}\right)_s.$$

stability in Newton's theory (§ 4).

The solutions which are of interest to us should satisfy the boundary conditions

$$\xi=0 \text{ at } r=0 \quad \text{and} \quad Dp=0 \text{ at } r=R, \tag{4.131}$$

where Dp is a substantial perturbation.

As in the classical case, $Dp=0$ at $r=R$ is an implicit condition for $\xi(R)$ and $\xi'(R)$. The specific form of $\xi(R)$ and $\xi'(R)$ depends on the physical conditions at the surface of the star.

We shall now find a sufficient condition for dynamic instability of the system. To this end, multiply equation (4.130) by ξ and integrate both sides with respect to r. The result is

$$\omega^2 \int_0^R e^{\frac{3\lambda_0+\nu_0}{2}}(\varepsilon_0+p_0)\frac{\xi^2}{r^2}\,dr = 4\int_0^R e^{\frac{\lambda_0+3\nu_0}{2}} p_0' \frac{\xi^2}{r^3}\,dr + \int_0^R e^{\frac{\lambda_0+3\nu_0}{2}} \frac{\Gamma_1 p_0}{r^2}\xi'^2\,dr$$

$$+\frac{8\pi G}{c^4}\int_0^R e^{\frac{3(\lambda_0+\nu_0)}{2}} p_0(\varepsilon_0+p_0)\frac{\xi^2}{r^2}\,dr - \int_0^R \frac{p_0'^2}{\varepsilon_0+p_0} e^{\frac{\lambda_0+3\nu_0}{2}} \frac{\xi^2}{r^2}\,dr. \tag{4.132}$$

A sufficient condition for the system to be dynamically unstable is the vanishing of the right-hand side of this equation for some trial function ξ satisfying the given boundary conditions.

To pass to the Newtonian approximation, one has to put $\lambda_0=\nu_0=0$, replace ξ by ξr^3 and neglect pressure as small compared with the rest-mass density $\rho_0=\varepsilon_0/c^2$.

The conditions of stability of spherical configurations against radial perturbations in general relativity were first derived by Chandrasekhar in 1964. Studies of stability of relativistic objects require numerical methods. We shall discuss some of those methods while investigating the stability of cold spherical configurations of dense matter and models of neutron stars.

4.7. Spherical Configurations of Cold, Catalyzed Matter. Critical Mass

Having presented the general conditions of equilibrium of spherical objects in the general theory of relativity, the equations describing radial pulsations of such objects and the ways of investigating their stability, we can now move on to considering some specific examples and tracing the influence of the relativistic effects on the stability of such systems. Earlier, we considered relativistic adiabatic gaseous spheres; now we shall look at another extreme case, namely that of a spherical system of cold, catalyzed matter. By "catalyzed matter" we understand matter in which all nuclear reactions, admissible at the given thermodynamic state of the system, are already completed. The example is not merely academic, for we have in mind the late stages of the evolution of a star which has reached the end of its thermonuclear cycle and has its store of thermal and gravitational energy as its sole energy source.

The state of cold, catalyzed matter is described by the dependence of its density and pressure on the baryon number density n:

$$\rho=\rho(n), \quad p=p(n). \tag{4.133}$$

The temperature of cold, catalyzed matter need not be the absolute zero. Let ε_T be the density of thermal energy, defined as the difference between the total energy density $\varepsilon(n, T, X_i)$, which depends on the baryon number density n, temperature T and fractional abundances X_i, and the rest-mass energy density $c^2\rho_0(n, X_i)$, i.e.

$$\varepsilon_T=\varepsilon(n, T, X_i)-c^2\rho_0(n, X_i). \tag{4.134}$$

We can regard matter as cold and catalyzed when $\varepsilon_T \ll c^2\rho$. This may hold even when the temperature T is quite high.

For adiabatic changes of state of cold matter, the first law of thermodynamics takes the form

$$\frac{d\rho}{dn}=\frac{\rho+p/c^2}{n}. \tag{4.135}$$

The first law of thermodynamics and the equation of state for catalyzed matter allow us to choose one of the three variables n, ρ, p as independent. In the following we shall assume that we know the dependence of pressure on density,

$$p=p(\rho), \tag{4.136}$$

and we shall call this relationship the equation of state for cold catalyzed matter.

Over a fairly wide range of density, the dependence of matter density on baryon number density can be represented in the form

$$\rho=\rho_0\left(\frac{n}{n_0}\right)^{\Gamma_4}, \tag{4.137}$$

where

$$\Gamma_4=\left(\frac{d\ln\rho}{d\ln n}\right)_s=\text{const}. \tag{4.138}$$

Using (4.135), we can write the equation of state as

$$p=(\Gamma_4-1)\rho c^2. \tag{4.139}$$

Local stability of matter requires that the pressure be positive, whence $\Gamma_4 \geqslant 1$; on the other hand, the speed of sound cannot exceed the speed of light, which gives the upper limit $\Gamma_4 \leqslant 2$, and thus

$$1 \leqslant \Gamma_4 \leqslant 2. \tag{4.140}$$

The equation of state of real catalyzed matter depends on micro-scale physical processes. It connects phenomena occurring in the microworld with the macroscopic properties of matter. In order to derive the equation of state one must know precisely the forces of interaction between the particles which form the system. At the present time, the equation of state of catalyzed matter is known for densities up to the order of nuclear density, i.e. up to $\rho \sim 10^{14}$ g/cm^3. At higher densities, besides electrons, protons and neutrons, other particles appear, for example hyperons, and then we find ourselves skating on very thin ice because their interactions with one another and with the electrons, protons or neutrons, are very little known. The equation of state for the density range 10^{14}–10^{16} g/cm^3 should be regarded as merely a first approximation.

After these general remarks concerning the equation of state, we pass to examples.

The relation between Γ_4 and density for the Harrison–Wheeler equation of state and that for the Cameron–Cohen–Langer–Rosen equation of state are shown in Fig. 4.5. It is

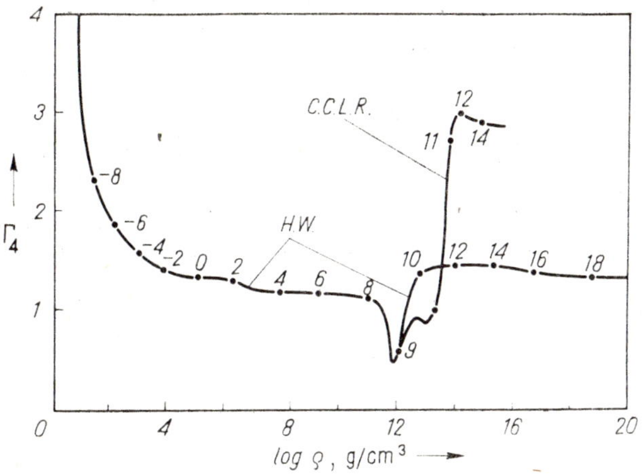

FIG. 4.5. Γ_4 vs. density for the Harrison–Wheeler and the Cameron–Cohen–Langer–Rosen equations of state. The two equations agree in the region of densities lower than 10^{12} g/cm³. The values of $\log p$ are indicated along the curves (adapted from Rees, Ruffini and Wheeler, 1974).

seen from the graph that the two equations agree in the region $\rho<10^{12}$ g/cm³; at higher densities the curves differ considerably. The reason is that different authors take into account different physical processes.

Meltzer and Thorne were first to carry out a comprehensive analysis of the properties of spherical equilibrium configurations of cold, catalyzed matter, and to examine their stability. They adopted the Harrison–Wheeler equation of state and integrated numerically the equilibrium configuration equations

$$\frac{dm}{dr}=4\pi r^2\rho, \tag{4.141}$$

$$\frac{dp}{dr}=-\left(\rho+\frac{p}{c^2}\right)\frac{G\left(m(r)+4\pi r^3\dfrac{p}{c^2}\right)}{r^2\left(1-\dfrac{2Gm(r)}{c^2r}\right)}, \tag{4.142}$$

subject to the initial conditions

$$m(0)=0, \quad \rho(0)=\rho_c. \tag{4.143}$$

As a result, they obtained a one-parameter family of solutions. For a given ρ_c, the radius R of the star was obtained from the condition $p(R)=0$, and the total mass M from the condition $M=m(R)$. If we solve the first of these conditions for R, we find the radius as a function of the central density ρ_c, $R=R(\rho_c)$, and thereby the dependence of the total mass M on ρ_c, $M=M(\rho_c)$.

In order to find the relation $M=M(R)$ at densities exceeding 10^{15} g/cm³, Meltzer and Thorne used the equation of state of relativistic gas,

$$p = \tfrac{1}{3}\rho c^2 . \tag{4.144}$$

Results of their calculations are illustrated in Fig. 4.6 and Fig. 4.7. The parameter on the curve $M = M(R)$ is the central density.

The properties of the curve $M = M(R)$ for densities lower than 10^{15} g/cm³ have long been known. In the early 1930's, starting from the Newtonian equations of equilibrium and the equation of state of degenerate electron gas, Chandrasekhar came to the conclusion that there exists a maximum mass above which the pressure of the electron gas is no longer able to balance the forces of gravitational attraction. White dwarfs are stars in which the electron gas is the main source of pressure. Chandrasekhar calculated the limit for the maximum mass of white dwarfs. About the same time, Landau furnished a very simple argument supporting Chandrasekhar's results and predicted the existence of a similar limit for the maximum mass of stars whose pressure is produced by degenerate neutron gas. His pre-

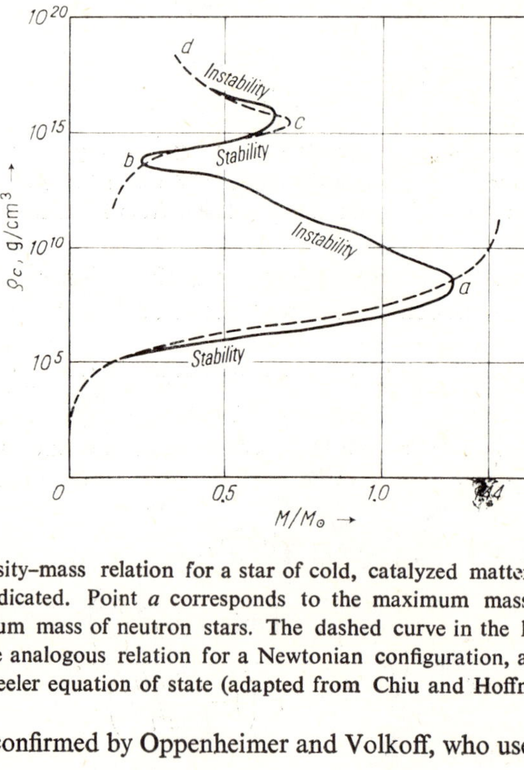

FIG. 4.6. Central density–mass relation for a star of cold, catalyzed matter. Stable and unstable regions are indicated. Point a corresponds to the maximum mass of white dwarfs, point c to the maximum mass of neutron stars. The dashed curve in the lower part of the diagram represents the analogous relation for a Newtonian configuration, and that at the top for the Harrison–Wheeler equation of state (adapted from Chiu and Hoffman, 1964).

dictions were later confirmed by Oppenheimer and Volkoff, who used relativistic equations of equilibrium.

Consider, after Landau, a spherical system of cold matter in which pressure is produced by a degenerate neutron gas, so that

$$p = \tfrac{1}{4}(3\pi^2)^{1/3}\hbar c n^{4/3} . \tag{4.145}$$

Using the Newtonian equation of equilibrium, we find

$$\frac{\langle p \rangle}{R} = \frac{GM}{R^2}\langle \rho \rangle ; \tag{4.146}$$

but since

$$n\mu = \langle \rho \rangle = \frac{3M}{4\pi R^3}, \tag{4.147}$$

where μ is the mass per electron, we get

$$\langle p \rangle = \frac{G(4\pi)^{1/3}}{3^{1/3}} M^{2/3} n^{4/3} \mu^{4/3}. \tag{4.148}$$

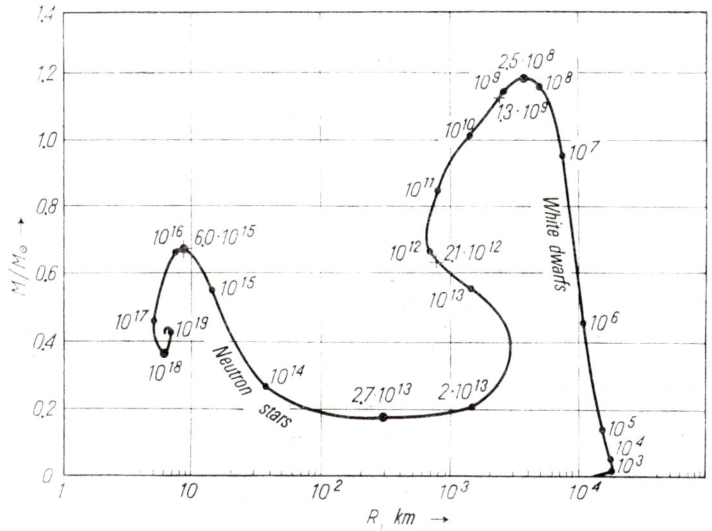

FIG. 4.7. Mass–radius relation for relativistic, spherically symmetric equilibrium configurations satisfying the Harrison–Wakano–Wheeler equation of state. Beginning from the large R end, the first maximum corresponds to the maximum mass of white dwarfs; in passing through this point the system loses stability. At the next extremal point, a minimum, the system becomes stable again; there follows the region of central densities of neutron stars. The next maximum point corresponds to the maximum mass of neutron stars. Beyond this point there are no more stable configurations. Higher and higher modes become unstable. The same general picture is obtained for other realistic equations of state satisfying the restrictions imposed by the theory of relativity (adapted from Meltzer and Thorne 1966).

Comparing (4.145) with (4.148), we obtain

$$M = \frac{3\pi^{1/2}}{16} \frac{1}{\mu^2} \left(\frac{\hbar c}{G}\right)^{3/2}. \tag{4.149}$$

Assuming that the mass per electron is equal to the mass of two protons, i.e. $\mu = 2m_p$, and substituting numerical data, we find $M = 0.2\, M_\odot$. By integrating the equation of equilibrium, Chandrasekhar obtained

$$M_{Ch} = 1{\cdot}46 M_\odot. \tag{4.150}$$

An inspection of the curve $M = M(R)$ shows that in the relativistic case the mass of any equilibrium configuration of cold matter obeying the Harrison–Wakano–Wheeler equation of state is less than the maximum mass of white dwarfs. For more realistic equations of state of dense matter the mass of a neutron star may exceed the Chandrasekhar limit (see Chapter 6).

Misner and Zapolsky have shown that for large densities the curve $M(R)$ forms a spiral described parametrically by the equations

$$M-M_{\infty}=C_M \rho_c^{-\alpha/2} \cos\left[\tfrac{1}{2}\beta \ln \rho_c+\delta_M\right],$$
$$R-R_{\infty}=C_R \rho_c^{-\alpha/2} \cos\left[\tfrac{1}{2}\beta \ln \rho_c+\delta_R\right], \tag{4.151}$$

where C_M, C_R, $\delta_M \neq \delta_R$, α and β are constants, with α and β related to Γ_4 by

$$\alpha=\frac{3}{2}-\frac{1}{\Gamma_4}, \quad \beta=\left(\frac{11}{\Gamma_4}-\frac{9}{\Gamma_4^2}-\frac{1}{4}\right)^{1/2}. \tag{4.152}$$

It is to be stressed that the above considerations concern stars formed from cold, catalyzed matter, and cannot be applied to hot stars, in which thermonuclear processes are still under way.

Certain general conclusions concerning the stability of a relativistic spherical configuration can be drawn from the fact that the gravitational field outside the system is described by the Schwarzschild metric.

As we remember from the discussion of the maximal extension of this metric, no particle can be at rest in the region $r<r_g$; it is a dynamic region where all particles fall towards the centre. It follows that the radius of an equilibium configuration must be greater than the gravitational radius, i.e.

$$R>\frac{2Gm(R)}{c^2}. \tag{4.153}$$

More restrictive inequalities of this type can be obtained by a closer analysis of the structure of the star. Using the post-Newtonian approximation, Chandrasekhar showed that

$$R>\frac{K}{\Gamma_4-\frac{4}{3}}\frac{2GM}{c^2}, \tag{4.154}$$

where K depends on the density distribution. For a few specific density distributions Bondi obtained the limit

$$R>1.6\frac{2GM}{c^2}. \tag{4.155}$$

Harrison, Wakano, Wheeler and Thorne have given general criteria which allow us to investigate the stability of a given configuration if the relation $M=M(R)$ is known.

First, at each extremal point of the curve $M(R)$ one of the modes of radial pulsations changes stability. To prove this statement, let us add δN baryons to the system and calculate the resulting change in the total mass. Measured at infinity, the change in the mass δM will be equal to the local energy increment $\mu\delta N=\left(\frac{\partial\rho}{\partial n}\right)_s \delta N$, where μ is the chemical potential, multiplied by the factor by which it diminishes when seen by a distant observer, i.e. by $\sqrt{g_{00}}$; thus

$$\delta M=\left(\frac{\partial \rho}{\partial n}\right)_s \sqrt{g_{00}}\,\delta N. \tag{4.156}$$

It follows that to every extremal point of the curve $M=M(R)$ there corresponds an extremum on the curve $N=N(R)$. This implies that the configuration corresponding to an extremum on the curve $M=M(R)$ can be carried to a nearby equilibrium configuration without any change in the total number of particles. The density change corresponding to such a transition,

$$\rho_0(r)-\rho(r)=\delta\rho\,, \tag{4.157}$$

can be regarded as a small perturbation of the equilibrium state. Being a radial perturbation, it can be represented in the form

$$\delta\rho=F(r)\,\xi_n(r)\,e^{i\omega_n t}\,, \tag{4.158}$$

where $F(r)$ is a function of r. In the equilibrium state, ρ cannot be a function of time and therefore

$$\omega_n=0\,. \tag{4.159}$$

Thus, if ω_n are continuous functions of ρ_c, changes in stability occur only at the extremal points of the curve $M=M(R)$.

Second, at every extremal point of the curve $M=M(R)$ in the neighbourhood of which the radius of the configuration decreases with increasing central density, the stability character of the even modes ($n=0, 2, 4, \ldots$) changes. At other extremal points the odd modes ($n=1, 3, 5, \ldots$) change stability. As we have seen above, the configuration corresponding to an extremal point of the curve $M=M(R)$ can be carried to a nearby equilibrium configuration without any change in the total number of particles. For a simultaneous growth of the central density and a decrease in the configuration radius, δr should be negative near the centre (compression) and near the surface. The sign of δr is determined by the sign of ξ_n. ξ_n can be negative near the centre and near the surface, in accordance with the general character of oscillations, if between the centre and the surface it becomes zero an even number of times. Only even modes have this property.

Third, at an extremal point of the curve $M=M(R)$ one mode becomes unstable as the central density rises if and only if the curve winds anti-clockwise in the neighbourhood of this point; otherwise, one mode becomes stable. This follows easily from the previous criterion; it is enough to represent the curve $M(R)$ in a diagram where ordinate is the radius and abscissa is the mass. Additionally, it is to be noted that configurations having low central density, for example planets, are stable.

A glimpse at the curve $M=M(R)$ is now enough to decide which equilibrium configurations are stable. In the region of central densities from zero to $2.5\cdot10^8$ g/cm^3 all configurations are stable, with the extremal point corresponding to white dwarfs of maximum possible mass. In the neighbourhood of this extremum, the radius decreases as the central density rises and therefore an even mode changes stability. The third criterion implies that here the fundamental mode becomes unstable. All the subsequent configurations up to the point where $\rho_c=2.7\cdot10^{13}$ g/cm^3 are unstable, the only unstable mode being the fundamental one. At the minimum at $\rho_c=2.7\cdot10^{13}$ g/cm^3 the configuration radius is again decreasing as the central density rises and so again an even mode will change its stability. According to the third criterion, the fundamental mode regains its stability at this point. The next extremum, a maximum, occurs when $\rho_c=6\cdot10^{15}$ g/cm^3, and here the fundamental mode loses its stability again. Going further along the curve, we shall find no stable configurations. In this region, as the central density grows, modes of higher and higher order become unstable.

The analysis of Meltzer and Thorne leads to very important conclusions. There exist only two different, disjoint regions of central densities where configurations are stable. One is the region of low densities from zero to $2.5 \cdot 10^8$ g/cm^3. Its upper part corresponds to white dwarfs. The other region extends from the density $2.7 \cdot 10^{13}$ g/cm^3 to $6 \cdot 10^{15}$ g/cm^3 and corresponds to neutron stars. Above the central density of $6 \cdot 10^{15}$ there are no stable configurations. Thus, if a star of cold, catalyzed matter contracts so that its central density exceeds the value $6 \cdot 10^{15}$ g/cm^3, it will no longer be able to assume a stable configuration. A further, catastrophic contraction will be inevitable.

The question may arise to what extent the conclusions obtained by Meltzer and Thorne depend on the equation of state used. Calculations carried out for other equations of state satisfying the restrictions imposed by general relativity lead to conclusions which are qualitatively the same. The maximum values of the mass of a white dwarf or a neutron star may vary, but there always exists a maximum density, above which no stable configuration may occur.

Literature

BARDEEN, J. M., THORNE, K. S. and MELTZER, D. W. (1966) *Astrophys. J.* **145**, 505.

CHANDRASEKHAR, S. (1957) *An Introduction to the Study of Stellar Structure*, Dover Publications, N. Y.

CHANDRASEKHAR, S. (1965) *Astrophys. J.* **142**, 1519.

CHIU H. Y. and HOFFMAN W. F., eds. (1964) *Gravitation and Relativity*, W. A. Benjamin, New York.

CLAYTON, D. D. (1968) *Principles of Stellar Evolution and Nucleosynthesis*, McGraw-Hill, New York.

COX, J. P. and GIULI, R. T. (1968) *Principles of Stellar Structure*, Gordon and Breach, New York.

HARRISON, B. K., THORNE, K. S., WAKANO, M. and WHEELER, J. A. (1965) *Gravitation Theory and Gravitational Collapse*, The University of Chicago Press, Chicago.

IPSER, J. R. (1970) *Astrophys. Space Sci.*, **7**, 361.

LEDOUX, P. (1963) *Stellar Stability and Stellar Evolution*, Proc. Int. Sch. Phys. "E. Fermi" Course 28, Academic Press, N. Y.

LEDOUX, P. (1965) *Stellar Stability*, in *Stellar Structure — Stars and Stellar systems*, VIII, ed.: Aller L. H., McLaughlin, D. B., The University of Chicago Press, Chicago.

LEDOUX, P. (1978) *Stellar Stability*, in *Theoretical Principles of Astrophysics and Relativity*, ed.: Lebovitz, N. R., Reid, W. H. and Vandervoort, P., The University of Chicago Press, Chicago.

LEDOUX, P. and WALRAVEN, T. (1958) *Variable Stars*, in *Handbuch der Physik*, vol. LI, ed.: Flügge, Springer, Berlin.

MELTZER, D. and THORNE, K. S. (1966) *Astrophys. J.* 145, 514.

MISNER, C. W., THORNE, K. S. and WHEELER, J. A. (1973) *Gravitation*, W. Freeman, San Francisco.

REES, M., RUFFINI, R., WHEELER, J. A. (1974) *Black Holes, Gravitational Waves and Cosmology*, Gordon and Breach, New York.

SAHAKYAN, G. S. (1974) *Equilibrium Configurations of Degenerate Gaseous Masses*, J. Wiley and Sons, New York.

SCHWARZSCHILD, M. (1965) *Structure and Evolution of the Stars*, Dover–New York.

THORNE, K. S. (1966) *Astrophys. J.* **144**, 201.

THORNE, K. S. (1967) *Relativistic Stellar Structure and Dynamics*, in *High Energy Astrophysics*, Vol. III, ed. DeWitt, C., Schatzman, E. and Veron, P., Gordon and Breach, N. Y.

TOOPER, R. (1965) *Astrophys. J.* **142**, 1541.

WHEELER, J. A. (1964) in *Gravitation and Relativity*, eds.: Chiu, H. Y., Hoffmann, W. F., Benjamin, New York.

ZELDOVICH, YA. B. and NOVIKOV, I. D. (1971) *Relativistic Astrophysics*, Vol. 1: *Stars and Relativity*, The University of Chicago Press, Chicago.

CHAPTER 5

ROTATING STARS

5.1. Critical Mass of Rotating Stars

The relativistic theory of rotating stars is of great astrophysical interest for several reasons. Perhaps the most important one is the existence of pulsars, which, beyond any doubt, are rotating neutron stars; we shall discuss them in greater detail in the next chapter. Typical densities of neutron stars are of the order of 10^{15} g/cm^3, and their dimensions of the order of 10 km. This means that the number $2GM/Rc^2$, which characterizes the influence of the relativistic effects, is not small compared with unity, which is the reason why these effects play an important part in the structure of neutron stars. Typical linear velocities (at the equator) connected with the rotation of pulsars are $\sim$1000 km/s. The theory of rotating bodies is also relevant to the important problem of determining the critical masses of dense stars, i.e. the upper limits for the masses of white dwarfs and neutron stars. It is a very difficult problem, the reason being that the calculated value of the critical mass strongly depends on the equation of state used for the calculations, and for densities greater than 10^{15} g/cm^3 our knowledge of that equation is very imperfect.

As we saw in the previous chapter, the problem of the critical mass of nonrotating white dwarfs and neutron stars in general relativity has been dealt with successfully enough. The critical mass of white dwarfs is equal to the classical Chandrasekhar limit (1.46 $M_\odot$), the critical mass of neutron stars cannot be greater than 3.3 $M_\odot$ (Rhodes and Ruffini, 1971) and is probably not greater than $3M_\odot$ (Hartle, 1978). A rapid differential rotation can boost the critical mass of white dwarfs to much higher values (theoretically up to infinity; Ostriker and Bodenheimer, 1968). Very massive white dwarfs, however, cannot be stable. Gravitational radiation becomes a source of secular instability, which causes a rapid loss of angular momentum and a contraction (Friedman, Schutz, 1978). A white dwarf of mass $4M_\odot$ would lose its angular momentum within two weeks.

If the general theory of relativity is not the correct theory of gravitation, the problem of determining the critical mass of neutron stars appears quite hopeless. The masses of neutron stars calculated according to different possible gravitation theories differ drastically (Fig. 5.1). The effect of rotation on the critical mass of neutron stars is not known well enough. Numerical investigations of Hartle and Thorne (1968) have shown that rigid rotation may increase the critical mass by about 10%.

The problem of stability of rotating neutron stars is nowhere near a final solution either.

It may be that in the not too distant future we shall be able to determine the critical mass of neutron stars directly from observation. One such possibility is connected with the fact that the moment of inertia of a neutron star depends much more strongly on the equation of state of the stellar matter than on the mass of the star (Carter and Quintana,

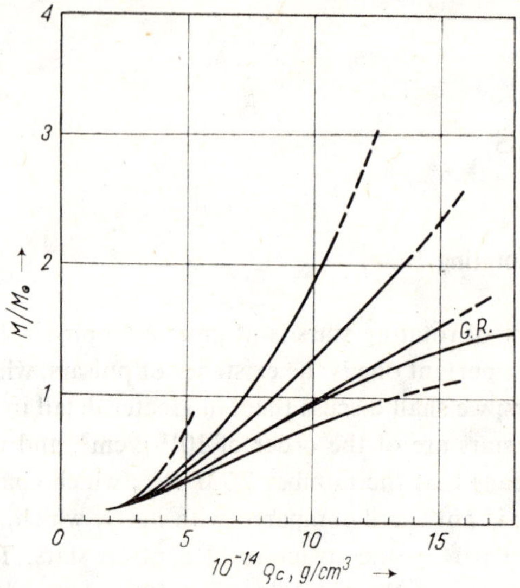

FIG. 5.1. Mass of a neutron star as a function of central density according to different gravitation theories which have Newtonian limit.

1973). If we know the period P of a pulsar, its rate of slowing down $\dot{P}$ and the energy loss per unit of time $\dot{E}$, we can find the moment of inertia from the formula

$$I = -\frac{4\pi^2 P^3}{\dot{P}} \dot{E}. \tag{5.1}$$

Despite considerable uncertainty in the determination of $\dot{E}$, a comparison of the theoretical value of I with that calculated from formula (5.1) allows us to reject some types of the equation of state (the so-called soft equations of state) as inconsistent with observation. In this way one can also come to the conclusion that the mass of the pulsar in the Crab Nebula is between $0.5M_\odot$ and $1M_\odot$. Another possibility is connected with the greatest astronomical sensation of 1974, the discovery of the binary pulsar PSR 1913+16. At the time this chapter is being written we already know of several pulsars in binary systems. For the most part they are X-ray pulsars. Knowledge of a larger number of such objects will permit observational determination of the masses of neutron stars.†

The relativistic theory of rotating bodies emerged at a time when big computers already existed, which made it possible to construct—right from the start—models of stars actually existing in nature. But, as one of the Russian writers put it, "there is a bit of

† Observations of the pulsar PSR 1913+16 enable us, for the very first time, to verify the predictions of general relativity, and not only its post-Newtonian approximations.

misfortune in every good fortune". In the Newtonian theory of rotating bodies, the vast theoretical knowledge accumulated over generations gives us a deep insight into what is important and what is incidental in the numerically constructed models, and permits us to analyse them with discrimination. Such analysis is still beyond us in the case of relativistic rotating stars.

5.2. Newtonian Theory of Rotating Bodies

The Newtonian theory of rotating bodies has a long and exciting history (Chandrasekhar, 1969). It occupied the attention of the most prominent mathematicians and physicists of the 18th, 19th and 20th centuries, to mention Newton, Maclaurin, Jacobi, Kelvin, Poincaré, Darwin, Jeans, Cartan, Liapunow, Chebyshev, Lichtenstein and Chandrasekhar. For the most part, the theory deals with figures of equilibrium, i.e. bodies formed from incompressible fluids. Even though real stars are formed from compressible gases, the theory of equilibrium figures is of great importance for investigations of rotating stars, the two main reasons being:

(1) Most problems in the theory of equilibrium figures are solved and presented in the form of strict, mathematical theorems;

(2) Computer calculations show that bodies of compressible gases have properties similar to those of classical equilibrium figures.

Let us look at the properties of axisymmetric rotating bodies from the point of view of thermodynamics. Divide a given body into a large number ($N \gg 1$) of elements, each characterized by its mass m_i, energy $\mathscr{E}_i$, entropy S_i, momentum $\boldsymbol{p}_i$ and the internal energy $\mathscr{E}_i^* = \mathscr{E}_i - \dfrac{p_i^2}{2m}$. A state of thermodynamic equilibrium is reached when the entropy is maximum for a given total angular momentum. If $\boldsymbol{\alpha}$ is an undetermined Lagrange multiplier and $\boldsymbol{r}_i \times \boldsymbol{p}_i$ the angular momentum of the i-th element, the condition for an extremum of the entropy takes the form

$$0 = \frac{\partial}{\partial \boldsymbol{p}_i} \sum_i^N \{S_i(\mathscr{E}_i^*) + \boldsymbol{\alpha}(\boldsymbol{r}_i \times \boldsymbol{p}_i)\}. \tag{5.2}$$

Using the definition of temperature, $T = \dfrac{d\mathscr{E}^*}{dS}$, we can write this condition as

$$\boldsymbol{v}_i = \boldsymbol{\Omega} \times \boldsymbol{r}_i, \tag{5.3}$$

where $\boldsymbol{v}_i$ is the velocity and $\boldsymbol{\Omega} = T\boldsymbol{\alpha}$ is a constant vector equal to the angular velocity of the rotation. Thus, *a body in thermodynamic equilibrium must rotate rigidly.*

It is known that in the absence of dissipative processes the angular momentum distribution relative to mass,

$$j = j(m), \tag{5.4}$$

where $j = |\boldsymbol{r} \times \boldsymbol{p}|$ and $m = \int_0^j dm$, does not change (Ostriker and Mark, 1968). Therefore a body rotating differentially, i.e. not as a rigid body, cannot alter its angular momentum distribution so as to attain a state of thermodynamic equilibrium, in which rotation must be rigid. Any small viscosity, however, will tend to stiffen the rotation. It will proceed on a thermal time scale, which is much longer than a dynamic time scale. Notice that in the case where the thermodynamic coupling between different parts of the body is small (for example when the pressure gradient is much less than the gradient of inertial forces – a typical situation for thin discs), the fact that rotation must be differential follows from the laws of dynamics (Kepler's laws). Generally, dissipative processes render rotation, no matter of what kind, unstable. The immediate reason why an instability occurs is the existence of an infinitely close state of lower energy (higher entropy). If this state can be attained irrespective of the presence of dissipative processes, we speak of dynamic instability; otherwise the instability is called secular. In the Newtonian theory of nonrotating bodies (without energy sources) we deal with dynamic instabilities only. In the previous chapter we saw that they can be divided into two classes: local instabilities and global instabilities. These terms are best explained by examples of criteria for the occurrence of such instabilities (Ledoux, 1958; Ledoux and Walraven, 1958):

$$\frac{1}{\rho^2}\left(\frac{\partial\rho}{\partial s}\right)_p \nabla p\cdot\nabla s<0, \tag{5.5}$$

$$\langle(\Gamma_1-\tfrac{4}{3})\rangle\frac{E_g}{I}<0. \tag{5.6}$$

The first criterion is the well-known Schwarzschild criterion for the occurrence of a convective instability. Instability of this kind, accompanying non-radial perturbations, is indeed local: the quantities which appear in formula (5.5), viz. the density of matter ρ, the entropy per particle s, the pressure p and their gradients, are all connected with a particular point of the star. The second criterion applies to global instabilities that accompany radial perturbations; its form depends on the global properties of the star: the total potential energy E_g, the moment of inertia I and $\langle\Gamma_1\rangle$. For more details about these criteria look back to Chapter 4.

The above criteria can be generalized to the case of rotating, axisymmetric stars. The following results have been obtained (Ledoux and Walraven, 1958; Chandrasekhar and Lebowitz, 1958).

The criterion for convective instability, derived on the assumption that the perturbations are axisymmetric, is

$$\begin{gathered} 2\boldsymbol{r}\cdot\nabla l+\frac{1}{\rho^2}\left(\frac{\partial\rho}{\partial s}\right)_p\nabla p\cdot\nabla s<0,\\ \left(\frac{\partial\rho}{\partial s}\right)_p(\boldsymbol{r}\times\nabla p)\cdot(\nabla s\times\nabla l)<0, \end{gathered} \tag{5.7}$$

where $l=r^2\Omega$, $r=|\boldsymbol{r}|$.

Global instability, also for axial perturbations, appears when

$$\langle(\Gamma_1-\tfrac{4}{3})\rangle\frac{|E_g|}{I}+2\langle(\tfrac{5}{3}-\Gamma_1)\rangle\frac{K}{I}<0, \tag{5.8}$$

where K is the kinetic energy, and E_g the gravitational potential energy. In the case of slow rotation of a homogeneous body, it follows from the last formula that

$$\langle \Gamma_1 \rangle - \tfrac{4}{3} + \tfrac{4}{3}\frac{\Omega^2}{2\pi G\rho} < 0, \tag{5.9}$$

i.e., a slow rotation may stabilize an unstable star (Chandrasekhar and Lebowitz, 1968).

Besides the types of instability described above, which occur both in rotating bodies and in nonrotating ones, there exist instabilities connected with more general perturbations, typical for rotating bodies only. They may be dynamic as well as secular.

Classification of different types of instabilities becomes important when we want to generalize the results of the Newtonian theory to general relativity. Since general relativity theory locally (but not globally) gives the same results as Newton's theory, and since general relativity predicts a dissipative process unknown in the classical theory, namely gravitational radiation, we can expect that

(1) Global dynamic stability criteria will be different in the two theories,

(2) Local dynamic stability criteria will be similar in these theories, and

(3) Gravitational radiation will in some cases cause a secular instability of stars which are stable in Newton's theory (Friedman, 1978; Schutz and Sorkin, 1977).

The Newtonian theory of equilibrium of rotating bodies (without energy sources) is embodied in the Poisson equation and the Euler equation with appropriate boundary conditions,

$$\Delta\phi = 4\pi G\rho,$$
$$\nabla\phi + \boldsymbol{F} = -\frac{1}{\rho}\nabla p, \tag{5.10}$$

where ϕ is the gravitational potential and $\boldsymbol{F}$ the centrifugal force. The latter equation implies that

$$\nabla\times\boldsymbol{F} = -\frac{1}{\rho^2}\nabla p\times\nabla\rho, \tag{5.11}$$

i.e. the centrifugal force is a potential force if and only if the surfaces of constant pressure coincide with the surfaces of constant density. It is easy to find that in this case the angular velocity of rotation depends only on the distance from the rotation axis

$$\Omega = \Omega(r), \tag{5.12}$$

and that the potential of the centrifugal force, defined by $\boldsymbol{F} = \nabla V$, is equal to

$$V = V(r) = \int_0^r \Omega^2(r)\, r\, dr. \tag{5.13}$$

The above statments are known as the von Zeipel theorems or Poincaré conditions (Poincaré, 1902; von Zeipel, 1924).

Any one-parameter family of solutions of equations (5.10) is called a linear sequence. The best known linear sequence is the Maclaurin sequence, discovered in the 18th century. It can be defined as follows:

Consider an axisymmetric, homogeneous, rigidly rotating body with fixed angular momentum J and mass M. Let its surface be an ellipsoid of revolution with the axes a, a, c ($a \geqslant c$). Now if we change the eccentricity of the meridional cross-section $e = (1-c^2/a^2)^{1/2}$ over the whole range from 0 to 1, we obtain a sequence of solutions of (5.10), given by the equations

$$\left(\frac{3M}{4\pi\rho}\right)^{1/3} = \frac{25}{18}\frac{J^2}{GM^3} e^3 (1-e^2)^{1/6} \left[\left(1-\tfrac{2}{3}e^2\right)\arcsin e - e(1-e^2)^{1/2}\right]^{-1}, \tag{5.14}$$

$$\Omega = \frac{162}{125}\frac{G^2 M^5}{J^3} e^{-6} \left[\left(1-\tfrac{2}{3}e^2\right)\arcsin e - e(1-e^2)^{1/2}\right]^{-1}. \tag{5.15}$$

This is the Maclaurin sequence. It includes spheroids of any density from the interval $0 \leqslant \rho < \infty$ and any mean radius $\langle R \rangle = (3M/4\pi\rho)^{1/3}$, $0 \leqslant \langle R \rangle < \infty$. The angular velocity changes between the limits 0 and $(27\pi/125)G^2M^5/I^3$ and never exceeds the critical angular velocity at which a "rotation catastrophe" might occur, i.e. a fragmentation of the star or an outflow of matter from the equatorial zone (when the centrifugal forces become greater then the gravitational pull). However, not all Maclaurin spheroids are stable. The most important data concerning the stability of the Maclaurin equilibrium figures are presented in Table 5.1.

At point B the Maclaurin sequence intersects another sequence of equilibrium figures, the so-called Jacobi sequence. The Jacobi equilibrium figures are three-axial rigidly rotating homogeneous ellipsoids. As we restrict ourselves here to the theory of axisymmetric rotating bodies, we shall leave the Jacobi and other classical linear sequences aside, and consider the Maclaurin equilibrium figures only. A comprehensive discussion of the whole of the theory of classical equilibrium figures can be found in monographs by Chandrasekhar (1969) and by Tassoul (1978), and in books by Lichtenstein (1933) and by Lyttleton (1953).

TABLE 5.1. *Characteristic points of the Maclaurin sequence*

Point	e	$\frac{\Omega^2}{2\pi G\rho}$	$\frac{c}{a}$	$\frac{K}{\lvert E_g \rvert}$
A. Nonrotating sphere. Stable configuration	0.00	0.00	1.00	0.00
B. Jacobi branch-off point. Secular loss of stability against non-axisymmetric perturbations	0.81	0.19	0.58	0.14
C. Maximum of the function $\frac{\Omega^2}{2\pi G\rho}$	0.93	0.22	0.37	0.24
D. Dynamic loss of stability against non-axisymmetric perturbations	0.95	0.22	0.30	0.27
E. Post-Newtonian corrections become singular. Secular loss of stability against axisymmetric perturbations. Instability with respect to differential rotation	0.98	0.17	0.17	0.36
F. Dynamic loss of stability against axisymmetric perturbations	0.99	0.04	0.05	0.46
G. Infinitely thin disc. Complete loss of stability	1.00	0.00	0.00	0.50

Linear sequences of configurations of compressible gases with the equation of state $p=p(\rho)$ were investigated numerically by Osteriker and his collaborators. They used the self-consistent field method: If the density distribution $\rho=\rho(x)$ is known, the gravitational potential distribution can be found by formally solving the Poisson equation:

$$\phi(x)=-G\int \frac{\rho(x')}{|x-x'|}\,\mathrm{d}^3x'. \tag{5.16}$$

On the other hand, the knowledge of the gravitational potential permits us to find the density distribution using the first integral of the Euler equation,

$$\phi+\int_0^r \Omega^2(r)\,r\,\mathrm{d}r=-\int_0^p \frac{\mathrm{d}p}{\rho}. \tag{5.17}$$

In the first step, one specifies the density distribution. Then, the potential distribution is found from formula (5.16) and subsequently the density distribution from formula (5.17). The new density distribution obtained in this way is treated as an improvement on the one guessed in the first step and the whole procedure is repeated until the difference between the succesive density distributions becomes less than a prescribed number λ determining the accuracy of the calculations. Another accuracy test is to check whether the *virial theorem*

$$E_g+2K+3\int p\,\mathrm{d}^3x=0 \tag{5.18}$$

is satisfied. The results obtained by this method are as follows:

Consider a linear sequence of configurations with a fixed, mass-dependent angular momentum distribution $j=j(m)$. For each value of the total mass M and the total angular

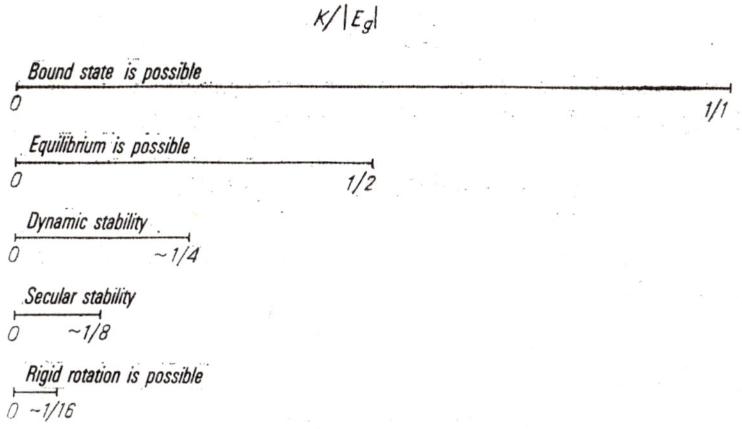

FIG. 5.2. Restrictions imposed on the ratio of the kinetic energy of rotation to the gravitational potential energy by the equilibrium and stability conditions.

momentum J, an equilibrium model can be constructed. The central density range is $0\leqslant\rho_c<\infty$; no configuration is liable to a rotational catastrophe. The dependences of a, c, Ω, $K/|E_g|$ on ρ_c are similar to the corresponding relationships for the classical Maclaurin equilibrium figures. A secular loss of stability occurs when $K/|E_g|=0.14$, and a dynamic one when $K/|E_g|=0.26$, irrespective of the equation of state $p=p(\rho)$ and the rotation law $j=j(m)$. The most important properties of rotating configurations, connected with the ratio $K/|E_g|$, are presented schematically in Fig. 5.2. The very general Ostriker criteria

$$K/|E_g|\geqslant 0.14 \text{ (secular instability)},$$

$$K/|E_g|\geqslant 0.26 \text{ (dynamic instability)}, \tag{5.19}$$

have one shortcoming: they cannot be generalized to relativistic rotating bodies, the reason being that the concepts of gravitational energy and kinetic energy of a rotating body have no invariant meanings in general relativity.

5.3. Assumptions and Definitions

Usually, by a relativistic rotating star one understands an idealized mathematical model representing an axisymmetric, stationary, bounded perfect-fluid configuration in an asymptotically flat space-time. More precisely:

(A) The gravitational field of a rotating star is asymptotically flat. This implies that in a region sufficiently distant from the matter world-tube one can introduce a spherical coordinate system (t, r, θ, φ) in which the line element is given by the formula

$$ds^2=c^2dt^2-dr^2-r^2(d\theta^2+\sin^2\theta\, d\varphi^2). \tag{5.20}$$

(B) The gravitational field of a rotating star is stationary, i.e., there exists an open-trajectory Killing vector ξ^α which asymptotically is a unit timelike vector:

$$\lim_{r\to\infty} \xi\cdot\xi=1. \tag{5.21}$$

(C) The gravitational field of a rotating star is axisymmetric, i.e., there is a closed-trajectory Killing vector η^α which is spacelike everywhere and asymptotically orthogonal to ξ^α:

$$\lim_{r\to\infty} \eta\cdot\eta=-r^2\sin^2\theta, \qquad \lim_{r\to\infty} \eta\cdot\xi=0. \tag{5.22}$$

It can be shown (Carter, 1969) that assumptions (A), (B) and (C) warrant commutation of the vectors ξ^α and η^α:

$$\underset{\xi}{\pounds}\eta^\alpha=-\underset{\eta}{\pounds}\xi^\alpha=0; \tag{5.23}$$

where £ is the Lie derivative†

$$\underset{\eta}{\pounds}\xi^\alpha=\eta^\beta\nabla_\beta\xi^\alpha-\xi^\beta\nabla_\beta\eta^\alpha. \tag{5.24}$$

(D) The energy-momentum tensor of the stellar matter is given by

$$T_{\alpha\beta}=(\varepsilon+p)u_\alpha u_\beta-pg_{\alpha\beta} \tag{5.25}$$

† Covariant derivative is often denoted by ∇_α.

and satisfies the equation $\underset{\eta}{\pounds} T_{\alpha\beta}=0=\underset{\xi}{\pounds} T_{\alpha\beta}$, i.e., the distribution of energy and momentum is axisymmetric and stationary. All thermodynamic quantities, such as energy density ε, pressure p, particle number density n, chemical potential μ, temperature T and entropy per particle s, are measured in a comoving frame. From the equation of state $\varepsilon=\varepsilon(n, s)$ and the first law of thermodynamics it follows that

$$T=\frac{1}{n}\left(\frac{\partial\varepsilon}{\partial s}\right)_n \tag{5.26}$$

and

$$p=n\left(\frac{\partial\varepsilon}{\partial n}\right)_s-\varepsilon. \tag{5.27}$$

(E) The velocity vector of the matter is given by the expression

$$u^\alpha=Z\left(\xi^\alpha+\frac{\Omega}{c}\eta^\alpha\right) \tag{5.28}$$

and satisfies the equation $\underset{\eta}{\pounds} u^\alpha=0=\underset{\xi}{\pounds} u^\alpha$, which excludes from our considerations phenomena such as meridional circulation (Carter, 1969). The quantity Ω which occurs in this formula is called the angular velocity. It is not the rotation scalar of the congruence of the world-lines of the rotating matter, ω, defined by

$$\omega^2=\omega^\alpha\omega_\alpha,\qquad \omega^\alpha=\tfrac{1}{2}\epsilon^{\alpha\beta\gamma\delta}u_{\gamma;\beta}u_\delta. \tag{5.29}$$

The name is justified by the fact that a distant observer will see the matter rotate with the angular velocity Ω.

The quantity Z is called, for reasons explained below, the red shift factor.

The rotation axis is the space-time region in which

$$\boldsymbol{\eta}\cdot\boldsymbol{\eta}=0. \tag{5.30}$$

From the relativistic equations of motion it follows that†

$$(T^\alpha{}_\beta\eta^\beta)_{;\alpha}=0=(T^\alpha{}_\beta\xi^\beta)_{;\alpha}. \tag{5.31}$$

Using the continuity equation $(nu^\alpha)_{;\alpha}=0$, one can show that in any motion of a small element of matter the energy E, the entropy per particle s and the angular momentum per particle j remain constant along the world-line of this element:

$$E=\mu(\boldsymbol{u}\cdot\boldsymbol{\xi}), \tag{5.32}$$

$$j=-(\mu/c)(\boldsymbol{u}\cdot\boldsymbol{\eta}); \tag{5.33}$$

here μ is the chemical potential per particle.

Define also the geometric angular momentum of a particle as

$$l=-\frac{c\boldsymbol{u}\cdot\boldsymbol{\eta}}{\boldsymbol{u}\cdot\boldsymbol{\xi}}=jc^2/E. \tag{5.34}$$

† This theorem holds in the general case of a gravitational field obeying only assumptions (B), (C) and (D).

On the rotation axis, of course, $l=0=j$. In the Newtonian limit

$$j=\Omega m_0 \tilde{\rho}^2, \quad l=\Omega\tilde{\rho}^2, \tag{5.35}$$

where m_0 is the mass per particle and $\tilde{\rho}$ the distance from the rotation axis.†

The energy E^* of a given element of matter, as measured by an observer comoving with this element, is related with the quantities defined above by

$$E^*=Z(E-\Omega j)=Z(1-\Omega l/c^2)E. \tag{5.36}$$

Thus, in order to find the energy E measured by a distant stationary observer, one has to divide the "proper" energy E^* of the element (with zero angular momentum, $l=0$) by Z.

For photons with zero angular momentum $l=-\frac{p\eta}{p\xi}=0$ (where p^α is the photon's 4-momentum), a similar reasoning leads to the formula

$$\frac{E_f}{E_{f\infty}}=\frac{\text{photon energy on the surface of the star}}{\text{photon energy at infinity}}=1+z=Z, \tag{5.37}$$

where z is the red shift; hence the name of Z.

Note that the condition $\boldsymbol{u}\cdot\boldsymbol{u}=1$ implies

$$\left(\frac{1}{Z}\right)^2=(\xi\cdot\xi)+2\frac{\Omega}{c}(\eta\cdot\xi)+\frac{\Omega^2}{c^2}(\eta\cdot\eta). \tag{5.38}$$

Following Thorne (1967) we define the injection energy E_I as the energy which is added to the total energy Mc^2 of the star when a small portion of matter, containing $\mathfrak{N}$ particles, is injected (in a quasi-stationary manner) into the star and put in the same thermodynamic state as that of the surrounding matter (in the limit $\mathfrak{N}\to 1$). Thorne showed that

$$E_I=\mu/Z=E-\Omega j. \tag{5.39}$$

5.4. Basic Equations

In the first chapter we showed that in a static space-time the trajectories of a timelike Killing vector are orthogonal to a family of spacelike hypersurfaces Σ. One of the important consequences of this is the possibility of introducing an invariant division of a static space-time into space and time. In a stationary but non-static space-time, the trajectories of an asymptotically timelike Killing vector are, in general, not even locally orthogonal to a family of hypersurfaces. It can be shown, however, (Carter, 1969; Bardeen, 1970) that in the space-time of a rotating star there exists an invariantly defined family of spacelike hypersurfaces $\tilde{\Sigma}$ whose orthogonal trajectories are tangent to the vector field

$$n^\alpha=e^{-\nu}\left(\xi^\alpha+\frac{\tilde{\omega}}{c}\eta^\alpha\right), \tag{5.40}$$

† $\tilde{\rho}$ denotes the distance from the rotation axis and ρ the rest-mass density.

where

$$\nu = \ln \kappa - \psi, \tag{5.41}$$

$$\psi = \tfrac{1}{2} \ln |(\eta \cdot \eta)|, \tag{5.42}$$

$$\kappa^2 = (\eta \cdot \xi)^2 - (\eta \cdot \eta)(\xi \cdot \xi), \tag{5.43}$$

$$\frac{\tilde{\omega}}{c} = e^{-2\psi}(\eta \cdot \xi). \tag{5.44}$$

The vector n^α is not a Killing vector and is not parallel to one if the space-time is not static.

Denote by ${}_\perp\nabla_\alpha$ the operator of covariant differentation and by $\tilde{\Delta}$ the Laplace operator on the hypersurface $\tilde{\Sigma}$, and define

$$v = l/\mathscr{R}, \quad \mathscr{R} = \frac{Z\kappa}{u \cdot \xi}. \tag{5.45}$$

Using these definitions, we can write some of the Einstein field equations in an elegant, invariant form:

$$R_{\alpha\beta}\eta^\alpha n^\beta = -\tfrac{1}{2} {}_\perp\nabla_\alpha \left(e^{2\psi - \nu} {}_\perp\nabla_\alpha \frac{\tilde{\omega}}{c} \right) = \frac{32\pi G}{c^4} e^\nu (\varepsilon + p) \frac{lc}{c^2 - \Omega l}, \tag{5.46}$$

$$R_{\alpha\beta}\xi^\alpha n^\beta = -{}_\perp\nabla^\alpha \left(e^\nu {}_\perp\nabla_\alpha \nu + \tfrac{1}{2} e^{-\nu + 2\psi} \frac{\tilde{\omega}}{c^2} {}_\perp\nabla_\alpha \tilde{\omega} \right) = \frac{4\pi G}{c^4} e^\nu (\varepsilon + 3p) + \frac{8\pi G}{c^4} e^\nu (\varepsilon + p) \frac{l\Omega}{c^2 - \Omega l}, \tag{5.47}$$

$$R_{\alpha\beta} n^\alpha n^\beta = -\tilde{\Delta}\nu + {}_\perp\nabla^\alpha {}_\perp\nabla_\alpha \nu - \tfrac{1}{2} e^{2\psi - \nu} c^{-2} {}_\perp\nabla_\alpha \tilde{\omega} {}_\perp\nabla^\alpha \tilde{\omega} = \frac{4\pi G}{c^4} \left(\varepsilon + 3p + \frac{2v^2(\varepsilon + p)}{c^2 - v^2} \right). \tag{5.48}$$

Carter (1969) proved that the surfaces of transitivity† of the group of motions generated by ξ^α and η^α are globally orthogonal to a family of spacelike two-surfaces $\mathscr{P}$. Denote by $*$ the Laplace operator on the surface $\mathscr{P}$; one can easily show that

$$R_{\alpha\beta}[(\xi \cdot \xi)\eta^\alpha \eta^\beta - 2(\eta \cdot \xi)\xi^\alpha \eta^\beta + (\eta \cdot \eta)\xi^\alpha \xi^\beta] = \kappa^{-1} * \kappa = -\frac{16\pi G}{c^4} p. \tag{5.49}$$

This means that if $p = 0$, κ is a harmonic function on the surface $\mathscr{P}$.

Owing to their invariance and simplicity, equations (5.46), (5.47) and (5.48) are often employed in theoretical studies of the properties of rotating stars. However, they are of little use in numerical practice, where concrete coordinate systems are necessary. One of the coordinate systems used most frequently is the system $(t, \tilde{\rho}, z, \varphi)$ defined as follows:

(1) The Killing vectors in these coordinates are given by

$$\xi^\alpha = \delta^\alpha_0, \quad \eta^\alpha = \delta^\alpha_3. \tag{5.50}$$

† A surface S is called the surface of transitivity of a group of motions G if for any two points p and q of this surface there exists an element in the group which carries p into q.

(2) The metric of the surface $\mathscr{P}$ is

$$ds^2_{\mathscr{P}} = -e^{2\mu}(d\tilde{\rho}^2 + dz^2); \tag{5.51}$$

the function μ is connected with the Gauss curvature of the surface by the equation

$$-\tfrac{1}{2}R_{\text{Gauss}} = *\mu. \tag{5.52}$$

It is straightforward to show (using Carter's theorem about the orthogonal transitivity of the group of motions) that the space-time metric is in these coordinates given by the formula

$$ds^2 = e^{2\nu}c^2dt^2 - e^{2\psi}(d\varphi - \tilde{\omega}\,dt)^2 - e^{2\mu}(d\tilde{\rho}^2 + dz^2). \tag{5.53}$$

which involves four structural functions

$$\begin{aligned} \nu &= \nu(\tilde{\rho}, z), \\ \psi &= \psi(\tilde{\rho}, z), \\ \tilde{\omega} &= \tilde{\omega}(\tilde{\rho}, z), \\ \mu &= \mu(\tilde{\rho}, z). \end{aligned} \tag{5.54}$$

The number of the structural functions can be reduced either by additional assumptions or by solving some of the field equations. For example, when $p=0$, the function κ is harmonic and we can assume $\kappa = \tilde{\rho}$. Relation (5.41) then allows reduction of the number of the unknown structural functions to three.

Since the first three of the metric functions (5.54) are formed from scalar products of Killing vectors and have invariant geometrical meaning, there should exist a method of determining them by measurements. This can be done in the following way (Bardeen, 1970):

Suppose that an observer, having placed a wave-guide on a circle $\tilde{\rho}=$const, $z=$const, sends light signals in the $+\varphi$-direction and in the $-\varphi$-direction. If the observer's four-velocity is u^{α}_{obs} and if

$$v^{\alpha}{}_{\text{obs}} = (\delta^{\alpha}{}_{\beta} - n^{\alpha}n_{\beta})\,u^{\beta}{}_{\text{obs}}(c^2 - v^2{}_{\text{obs}})^{\frac{1}{2}}, \qquad v^2{}_{\text{obs}} = -v^{\alpha}{}_{\text{obs}}v^{\beta}{}_{\text{obs}}\,g_{\alpha\beta}, \tag{5.55}$$

then in the first case the time necessary for a light signal to travel round the circle is equal to

$$c\tau_1 = 2\pi e^{\psi}\left(\frac{c + v_{\text{obs}}}{c - v_{\text{obs}}}\right)^{\frac{1}{2}}, \tag{5.56}$$

and in the second case it is equal to

$$c\tau_2 = 2\pi e^{\psi}\left(\frac{c - v_{\text{obs}}}{c + v_{\text{obs}}}\right)^{\frac{1}{2}}. \tag{5.57}$$

Notice that v^{α}_{obs} is the "physical" velocity of the observer on the hypersurface $\tilde{\Sigma}$. It can be shown to be equal to $v^{\alpha}_{\text{obs}} = (\Omega_{\text{obs}} - \tilde{\omega})e^{-\nu}\eta^{\alpha}$, i.e.

$$v_{\text{obs}} = (\Omega_{\text{obs}} - \tilde{\omega})e^{\psi - \nu}, \tag{5.58}$$

where Ω_{obs} is the observer's angular velocity. On comparing formulae (5.56) and (5.57) with the corresponding special-relativistic formulae, we see that v_{obs} can be interpreted as the locally measured velocity of the observer. Thus, for locally nonrotating observers

$$v_{\text{obs}}=0 \Rightarrow v_{\text{obs}}=v_{\text{lnr}}=0. \tag{5.59}$$

From the last few formulae it follows that

$$\Omega_{\text{lnr}}=\tilde{\omega}, \qquad c\tau_1=c\tau_2=2\pi e^{\psi}\big|_{\text{lnr}}. \tag{5.60}$$

The last equality means that $2\pi e^{\psi}$ is the "proper" length of the circle $\tilde{\rho}$=const, z=const on the hypersurface $\tilde{\Sigma}$. In the Newtonian limit,

$$e^{\psi}=\tilde{\rho}. \tag{5.61}$$

We saw before that Z is the red-shift factor. Comparing the expansion

$$z+o(c^{-2})=1-Z^{-1}=1-e^{\nu}(1-v^2/c^2)^{\frac{1}{2}}=\tfrac{1}{2}v^2/c^2-\nu+o(c^{-4}) \tag{5.62}$$

with the corresponding Newtonian limit, we find that ν is a "gravitational potential". Note furthermore that

$$u^{\alpha}_{\text{lnr}}=n^{\alpha}. \tag{5.63}$$

The angular velocity of locally nonrotating observers (equal to $\tilde{\omega}$) is called the rate of dragging of inertial frames.

For metric (5.53), the Einstein field equations take the form (Seguin, 1975)

$$e^{-\psi}\nabla^2 e^{\psi}+\nabla^2\mu+\tfrac{1}{4}e^{2(\psi-\nu)}\nabla\tilde{\omega}/c\cdot\nabla\tilde{\omega}/c=-\frac{8\pi G}{c^4}e^{2\mu}\left(\frac{\varepsilon+p}{1-v^2/c^2}-p\right), \tag{5.64}$$

$$e^{-\nu}\nabla^2 e^{\nu}+\nabla\nu\cdot\nabla\psi-\tfrac{1}{2}e^{2(\psi-\nu)}\nabla\tilde{\omega}/c\cdot\nabla\tilde{\omega}/c=\frac{4\pi G}{c^4}e^{2\mu}\left[(\varepsilon+p)\frac{c^2+v^2}{c^2-v^2}+2p\right], \tag{5.65}$$

$$\nabla[e^{3\psi-\nu}\nabla\tilde{\omega}]=-\frac{16\pi G}{c^4}e^{2(\psi+\mu)}\frac{\varepsilon+p}{1-v^2/c^2}v, \tag{5.66}$$

$$e^{-(\psi-\nu)}\nabla^2 e^{\psi+\nu}=\frac{16\pi G}{c^4}e^{2\mu}p, \tag{5.67}$$

and the equations of motion, $T^{\alpha\beta}{}_{;\beta}=0$, can be written

$$\frac{\nabla p}{\varepsilon+p}=\nabla\ln Z-\frac{l}{c^2-\Omega l}\nabla\Omega. \tag{5.68}$$

The operators ∇^2 and ∇ in the above equations are the Laplace operator and the gradient, respectively, in the flat Euclidean space $(\tilde{\rho}, z)$. A dot denotes the scalar product.

The functions ν, ψ, $\tilde{\omega}$, Ω, p and ε cannot be arbitrary. Inside the star, we have

$$\varepsilon^2+p^2\neq 0. \tag{5.69}$$

The functions ν, ψ, $\tilde{\omega}$ and μ must be regular and continuous in the whole space-time.

In particular, at the boundary of the interior, i.e. on the surface given by the equation

$$p=0, \tag{5.70}$$

they must satisfy the matching conditions which we discussed in Chapter 3.

According to the "dominant energy condition", the pressure and density of matter inside the star cannot be negative,

$$p\geqslant 0, \quad \varepsilon\geqslant 0. \tag{5.71}$$

The asymptotic behaviour of the metric functions is determined by the Papapetrou conditions (1948)

$$M=\tfrac{1}{2}\lim_{\tilde{\rho}^2+z^2\to\infty} c^2/G\{(\tilde{\rho}^2+z^2)^{\frac{1}{2}}(1-e^{2\nu}+e^{2\psi}\tilde{\omega}^2/c^2)\}=\tfrac{1}{2}\lim_{\tilde{\rho}^2+z^2\to\infty} c^2/G\{(\tilde{\rho}^2+z^2)^{\frac{1}{2}}(e^{2\mu}-1)\}, \tag{5.72}$$

$$J=\tfrac{1}{2}\lim_{\tilde{\rho}^2+z^2\to\infty} c^2/G\{(\tilde{\rho}^2+z^2)^{3/2}\tilde{\omega}\}, \tag{5.73}$$

$$Q=\lim_{\substack{z=0\\ \tilde{\rho}\to\infty}}\{[c^2/G(1-e^{2\nu}+e^{2\psi}\tilde{\omega}^2/c^2)(\tilde{\rho}^2+z^2)^{\frac{1}{2}}-2M](\tilde{\rho}^2+z^2)\}, \tag{5.74}$$

where M, J and Q are the total mass, the angular momentum and the quadrupole moment of the rotating star. As follows from formulae (5.72)–(5.74), these quantities can be determined from observation of test particles moving in the star's gravitational field. This does not apply to two other global characteristics of the rotating star, namely to its rest-mass M_0 and the binding energy E_B

$$E_B=(M_0-M)c^2. \tag{5.75}$$

For the system of five equations (5.64)–(5.68) with the seven unknown functions

$$p,\ \varepsilon,\ \Omega,\ \nu,\ \psi,\ \tilde{\omega},\ \mu \tag{5.76}$$

to yield a unique solution, two more equations must be added. One may be the rotation law

$$\Omega=\Omega(\tilde{\rho},z), \tag{5.77}$$

and the other the equation of state of the matter,

$$p=p(\varepsilon,s). \tag{5.78}$$

If the rotation proceeds according to the law

$$\Omega=\Omega_0=\text{const}, \tag{5.79}$$

then, as we shall see later, the shear of the congruence of the world-lines of the matter vanishes. It can be shown that the expansion of this congruence vanishes irrespective of the form of the rotation law. Thus, a rotation obeying law (5.79) is a rigid motion. We call it a rigid rotation. If ω and σ are the rotation and shear scalars of the congruence of the matter world-lines, then in the Newtonian limit

$$\omega=\sigma+\Omega, \tag{5.80}$$

which implies that if a rotation is rigid, $\omega=\Omega$. Since formula (5.80) is not true in general relativity, for a rigid rotation, in general,

$$\omega \neq \Omega . \tag{5.81}$$

A star formed from matter satisfying an equation of state of the form $p=p(\varepsilon)$ is called a barotropic star, and a star in which $s=s_0=\text{const}$ an isentropic star.

We shall see later that, in general, the rotation law and the equation of state cannot be postulated independently.

5.5. Preferred Observers. Dragging of Inertial Frames

In the gravitational field of a rotating star some observers are naturally distinguished by the fact that their four-velocities u^α_{obs} are combinations of the Killing vectors ξ^α and η^α, of the form

$$u^\alpha_{\text{obs}} = Z\left(\xi^\alpha + (\Omega_{\text{obs}}/c)\eta^\alpha\right) \tag{5.82}$$

and satisfy the condition $\underset{\eta}{\pounds} u^\alpha_{\text{obs}} = 0 = \underset{\xi}{\pounds} u^\alpha_{\text{obs}}$. From the normalization condition $u^\alpha_{\text{obs}} u^\beta_{\text{obs}} g_{\alpha\beta} = 1$ it follows that

$$Z^{-2}_{\text{obs}} = (\xi\cdot\xi) + 2(\Omega_{\text{obs}}/c)(\eta\cdot\xi) + (\Omega^2_{\text{obs}}/c^2)(\eta\cdot\eta), \tag{5.83}$$

which implies that different observers can be characterized by means of a single function $\Omega_{\text{obs}} = \Omega_{\text{obs}}(x^i)$. Since u^α_{obs} is a timelike vector, this function must satisfy the inequality

$$|\Omega_{\text{obs}} - \tilde{\omega}| < -\frac{c\kappa}{\eta\cdot\eta}. \tag{5.84}$$

If in some region of the space-time the function Ω_{obs} is chosen in such a way that every

TABLE 5.2. *Preferred observers*

Observer	Four-velocity	Angular velocity with respect to stationary observers
Stationary (sta)	$\mu^\alpha_{\text{sta}} = (\xi\cdot\xi)^{-\frac{1}{2}}\xi^\alpha$	$\Omega_{\text{sta}} = 0$
Locally nonrotating (lnr)	$u^\alpha_{\text{lnr}} = e^{-\nu}(\xi^\alpha + (\tilde{\omega}/c)\eta^\alpha)$	$\Omega_{\text{lnr}} = \tilde{\omega}$
Comoving with matter (com)	$u^\alpha_{\text{com}} = Z(\xi^\alpha + (\Omega/c)\eta^\alpha)$	$\Omega_{\text{com}} = \Omega$

point of this region belongs to exactly one world-line of any one of observers (5.82), the set of all world-lines of these observers will form a congruence in this region.†

Three classes of observers are of particular importance: locally nonrotating observers (lnr), stationary observers (sta) and observers comoving with matter (com). Definitions of the three types are given in Table 5.2.

† An arbitrary con ruence will be denoted by $\mathscr{C}$; a subscript will be added to denote the congruence of the world-lines of the observers of a specific type, for example $\mathscr{C}_{\text{lnr}}$ is the congruence of the world-lines of the locally nonrotating observers.

We define the angular momentum of an observer as

$$l_{\rm obs} = -\frac{c u_{\rm obs} \cdot \eta}{u_{\rm obs} \cdot \xi}. \tag{5.85}$$

This definition permits us to write the formulae for the kinematic invariants of any congruence $\mathscr{C}_{\rm obs}$ in a particularly simple form:

$$\text{acceleration} \quad \dot{u}^{\alpha}{}_{\rm obs} = \nabla^{\alpha} \ln Z_{\rm obs} - \frac{l_{\rm obs}}{c^2 - \Omega_{\rm obs} l_{\rm obs}} \nabla^{\alpha} \Omega_{\rm obs}, \tag{5.86}$$

$$\text{expansion} \quad \theta_{\rm obs} = 0, \tag{5.87}$$

$$\text{rotation} \quad \omega^2_{\rm obs} = -\tfrac{1}{4}\mathscr{R}^{-2}(1 - (\Omega_{\rm obs}/c^2)\, l_{\rm obs})^{-2} \nabla^{\alpha} l_{\rm obs} \nabla_{\alpha} l_{\rm obs}, \tag{5.88}$$

$$\text{shear} \quad \sigma^2_{\rm obs} = -\tfrac{1}{4}\mathscr{R}^{2}(1 - (\Omega_{\rm obs}/c^2)\, l_{\rm obs})^{-2} \nabla^{\alpha} \Omega_{\rm obs} \nabla_{\alpha} \Omega_{\rm obs}, \tag{5.89}$$

where

$$\mathscr{R} = \kappa Z_{\rm obs}/(u_{\rm obs}\, \xi).$$

Every observer can construct in his neighbourhood a locally Euclidean, three-dimensional space with the positive-definite metric

$$h^{\alpha\beta} = u^{\alpha}_{\rm obs} u^{\beta}_{\rm obs} - g^{\alpha\beta}. \tag{5.90}$$

Let $e^{\alpha}_{\hat{a}}$ denote an arbitrary triad of linearly independent vectors in this space. Together with the vector $u^{\alpha}_{\rm obs}$, every such triad forms a tetrad connected with the observer. Of course, an observer may define an arbitrary tetrad at any point of his world-line. However, if he wants to make sensible measurements, the tetrads he chooses at different points of his world-line must be related in some way. Mathematically, this relation will correspond to some kind of transport (displacement) of vectors along the observer's world-line (e.g. parallel transport, Fermi transport), and physically – to the measurements by which the lengths and directions of the vectors $e^{\alpha}_{\hat{a}}$ are determined. Two possibilities deserve particular attention. The first is connected with the fact that the distant galaxies establish a primary standard of a globally nonrotating reference frame (to be more precise, they *probably* do so: no rotation of the universe has been observed, though such a possibility cannot be excluded). Therefore, by setting his telescopes in the directions of $e^{\alpha}_{\hat{a}}$ and measuring the motion of the galaxies in the visual field of the telescopes, an observer can find out whether the axes of the coordinate system he erected rotate in a global sense. The other possibility is to observe the precession of free gyroscopes relative to the axes $e^{\alpha}_{\hat{a}}$; in this way the observer can find out whether the axes of his frame rotate in a local sense. Whether or not these two experiments will give the same result is not obvious *a priori*. If the universe does not rotate as a whole, then according to Newton's theory of gravitation both measurements should give the same result. The theory of relativity, on the other hand, shows that these measurements yield different results. We leave till later the calculation of the angular velocity of the rotation of an "inertial frame" determined locally, relative to the "inertial frame" determined globally. Now let us only give an estimate of its value, based on dimensional analysis. According to such an analysis,

$$\omega_{\rm prec} = \alpha \frac{GM}{c^2 R} \Omega. \tag{5.91}$$

In this formula G is the gravitational constant, c the speed of light and M, R and Ω are the mass, radius and angular velocity of the body which is the source of the gravitational field. The dimensionless constant α can be determined exactly. Lense and Thirring (1918) showed that when the velocity of the relative rotation of the two inertial frame standards is calculated at the pole of a rotating massive body, $\alpha=4/3$. For the Earth

$$\omega_{\text{prec}\,\circlearrowleft}=10^{-14}\ \text{s}^{-1}. \tag{5.92}$$

This value seems too small to have any real significance. However, as follows from estimate (5.91), for neutron stars, in extreme cases, it is of the order of the angular velocity Ω and the effect becomes very important.

If an observer decides to propagate his tetrad by Lie transport,† he will measure the angular velocity of gyroscopic precession as being exactly equal to the vorticity vector of $\mathscr{C}_{\text{obs}}$,

$$\omega^{\alpha}_{\text{obs}}=\tfrac{1}{2}\sqrt{-g}\,g^{\alpha\beta}\epsilon_{\beta\rho\sigma\tau}\nabla^{\rho}u^{\sigma}_{\text{obs}}u^{\tau}_{\text{obs}}. \tag{5.93}$$

The absolute value of this velocity is given by formula (5.88).

Stationary observers

Using formulae (5.86)–(5.89), one can easily find that for the congruence $\mathscr{C}_{\text{sta}}$

$$\theta_{\text{sta}}=0=\sigma_{\text{sta}}, \tag{5.94}$$

$$\dot{u}^{\alpha}_{\text{sta}}=\nabla^{\alpha}\gamma, \tag{5.95}$$

$$\omega^{2}_{\text{sta}}=-\tfrac{1}{4}e^{2\nu-2\psi-2\gamma}\nabla^{\alpha}(\tilde{\omega}e^{2\psi-\gamma})\nabla_{\alpha}(\tilde{\omega}e^{2\psi-\gamma}). \tag{5.96}$$

where $e^{2\gamma}=\xi\cdot\xi$. It is also easy to see that stationary observers do not rotate with respect to the distant galaxies, whose four-velocities are given by

$$u^{\alpha}_{\text{gal}}\overset{*}{=}(\xi\xi)^{-\frac{1}{2}}\xi^{\alpha}=u^{\alpha}_{\text{sta}} \tag{5.97}$$

($\overset{*}{=}$ denotes an asymptotic equality).

Formula (5.94) implies that if an orthonormal tetrad associated with a stationary observer is Lie-transported, it remains orthonormal along the entire world-line of the observer. Formula (5.97) tells us that it is bound to the distant galaxies. Therefore, the velocity of the relative precession of the global and the local "inertial frames" is

$$\omega^{\alpha}{}_{\text{prec}}=\omega^{\alpha}{}_{\text{sta}}. \tag{5.98}$$

Thorne (1971) gave an asymptotic expression for the vector $\omega^{\alpha}_{\text{sta}}$:

$$\lim_{r\to\infty}\omega^{\alpha}_{\text{sta}}=\frac{G}{c^{2}}\frac{J}{r^{3}}\{\sin\theta r^{-1}e^{\alpha}_{\hat{\theta}}+2\cos\theta e^{\alpha}_{\hat{r}}\}, \tag{5.99}$$

here the vectors $e^{\alpha}_{\hat{r}}$ and $e^{\alpha}_{\hat{\theta}}$ and the coordinates r and θ are associated with the coordinate system (5.20), and J is the total angular momentum of the rotating star, defined by equa-

† We say that a tetrad is Lie-transported along the observer's world-line if $\pounds_{e_{\hat{0}}} e^{\alpha}_{\hat{a}}=0$.

tion (5.73). It can be seen that in the vicinity of the rotation axis ($\theta=0$) the axes of the locally defined "inertial frame" corotate with the star, whereas at the equator ($\theta=\pi/2$) they rotate in the opposite direction. It is also seen that

$$\lim_{\substack{r\to\infty\\ \theta=0}} \omega^{\alpha}{}_{\mathrm{sta}}=\tilde{\omega}\delta^{\alpha}_{\theta}. \tag{5.100}$$

This is the reason why the angular velocity $\tilde{\omega}$ is called the rate of dragging of inertial frames.

Locally nonrotating observers

As we know, for locally nonrotating observers $u^{\alpha}_{\mathrm{lnr}}=n^{\alpha}$ and

$$\omega_{\mathrm{lnr}}=0=\theta_{\mathrm{lnr}}. \tag{5.101}$$

From formulae (5.86)–(5.89) we obtain for the congruence $\mathscr{C}_{\mathrm{lnr}}$

$$\dot{u}^{\alpha}{}_{\mathrm{lnr}}=\dot{n}^{\alpha}=-\nabla^{\alpha}\nu, \tag{5.102}$$

$$\sigma^{2}{}_{\mathrm{lnr}}=-\tfrac{1}{4}\kappa^{2}\nabla^{\alpha}\tilde{\omega}\nabla_{\alpha}\tilde{\omega}. \tag{5.103}$$

The fact that $\sigma^{2}_{\mathrm{lnr}}\neq 0$ implies that

$$\underset{n}{\pounds} h^{\alpha\beta}_{\mathrm{lnr}}\neq 0, \tag{5.104}$$

which means that orthonormal tetrads are not Lie-transported along the world-lines of lnr observers. Therefore, the gyroscope axes will rotate in the lnr-bound tetrad (a tetrad associated with an lnr observer) defined by

$$\begin{aligned}
e^{\alpha}_{\hat{t}}&=(e^{-\nu},0,0,\tilde{\omega}e^{-\nu}),\\
e^{\alpha}_{\hat{\rho}}&=(0,e^{-\mu},0,0),\\
e^{\alpha}_{\hat{z}}&=(0,0,e^{-\mu},0),\\
e^{\alpha}_{\hat{\varphi}}&=(0,0,0,e^{-\psi}),
\end{aligned} \tag{5.105}$$

which by definition is orthonormal on lnr world-lines. Tetrad (5.105) plays an important role in the theory of rotating stars since, as shown by Bardeen, Press and Teukolsky (1972), the physically or geometrically distinguished quantities connected with rotating stars (or black holes) take a particularly simple form in it.

Clearly, the gyroscopes will not rotate in a Lie-transported lnr-bound tetrad; this follows from relation (5.101) and the properties of the Lie transport.

Observers comoving with matter

For these observers (and also – which is obvious – for the congruence of the world-lines of the rotating matter) the following relationships hold:

$$\dot{u}^{\alpha}_{\mathrm{com}}=\dot{u}^{\alpha}=\nabla^{\alpha}\ln Z-\frac{l}{c^{2}-\Omega l}\nabla^{\alpha}\Omega, \tag{5.106}$$

$$\sigma^2_{\text{com}} = -\tfrac{1}{4}\mathscr{R}^2(1-\Omega l/c^2)^{-2}\nabla^\alpha\Omega\nabla_\alpha\Omega. \tag{5.107}$$

$$\omega^2_{\text{com}} = -\tfrac{1}{4}\mathscr{R}^{-2}(1-\Omega l/c^2)^{-2}\nabla^\alpha l\nabla_\alpha l, \tag{5.108}$$

$$\theta_{\text{com}} = 0. \tag{5.109}$$

We thus see that the rotation is rigid (Ω=const) if the configuration $\mathscr{C}$ is rigid, $\theta=0=\sigma$. If the rotation is rigid, Einstein's field equations give

$$(\mathscr{R}^{-2}l_{;\alpha})^{;\alpha}=0. \tag{5.110}$$

We can integrate this equation in the coordinate system (t, $\tilde{\rho}$, z, φ) bound to an observer comoving with matter. Knowing the solution, we can write the metric of the rotating star in the form

$$ds^2 = e^{-2Z}\{(c\,dt-\Omega\tilde{\rho}^2 d\varphi/c)^2-\tilde{\rho}^2 e^{2\gamma-4Z}d\varphi^2-e^{2\lambda}(d\tilde{\rho}^2+e^{2\gamma}dz^2)\}, \tag{5.111}$$

where only three functions, Z, λ and γ are not known *a priori*. Metric (5.111) was given by Harrison (1970); a generalization of this result to the case of differential rotation was found by Abramowicz and Muchotrzeb (1975).

If the angular velocity and angular momentum are related by an equation of the type $f(\Omega, l)=0$, then acceleration is the gradient of some scalar function,

$$\dot{u}^\alpha = \nabla^\alpha \ln W, \tag{5.112}$$

which can be called the acceleration potential. It satisfies the Raychaudhuri equation

$$\nabla^\alpha\nabla_\alpha \ln W = \frac{4\pi G}{c^4}(\varepsilon+3p)+\frac{2}{c^2}(\sigma^2-\omega^2). \tag{5.113}$$

By integrating this equation over the world-tube of the rotating star and using the fact that $p>0$ inside the star one can show that

$$\langle\omega^2\rangle \leqslant \frac{2\pi G}{c^2}(\langle\varepsilon\rangle+3\langle p\rangle)+\langle\sigma^2\rangle, \tag{5.114}$$

where $\langle\rangle$ denotes averaging over the invariantly defined volume of the star. This result is known as the Poincaré condition. If it is not satisfied, the star cannot exist in a stationary state. It must then either expand at an increasing rate, or contract at a decreasing rate

$$\langle\dot{\theta}\rangle \geqslant 0. \tag{5.115}$$

The Poincaré condition is a reminder that in general relativity, as in Newton's theory, the distribution of angular velocity in a star cannot be arbitrary.

5.6. Relativistic von Zeipel Theorems. Convective Stability

Relativistic von Zeipel's theorems, like their Newtonian counterparts discussed before, describe very general properties of some special surfaces inside stars. In order to derive them, let us write the equations of motion in two equivalent forms, of which the second is obtained with the use of the thermodynamic identities (5.26) and (5.27), the definition of the injection energy E_I (5.39) and the definition of the angular momentum density j (5.33):

$$(\varepsilon+p)^{-1}\nabla p=\nabla \ln Z-l(c^2-\Omega l)^{-1}\nabla\Omega; \tag{5.116}$$

$$-(T/Z)\nabla s+\nabla E_I+j\nabla\Omega=0. \tag{5.117}$$

the operator ∇ has the same meaning as in formulae (5.64)–(5.68). Taking the curl ($\nabla\times$) of both sides of equation (5.116), we obtain

$$(\varepsilon+p)^{-2}\nabla\varepsilon\times\nabla p=c^2(c^2-\Omega l)^{-2}\nabla l\times\nabla\Omega. \tag{5.118}$$

The implication is that the surfaces l=const coincide with the surfaces Ω=const if and only if the surfaces ϵ=const coincide with the surfaces p=const. It can be shown that in this case the surfaces l=const and Ω=const are topologically cylinders. We shall call them von Zeipel cylinders. The surfaces p=const are usually called equipotential surfaces. For an isentropic star, equation (5.117) gives

$$\nabla j\times\nabla\Omega=0, \tag{5.119}$$

which means that in this case the surfaces j=const, Ω=const and E_I=const coincide.

In a barotropic star let a parameter a be constant on the von Zeipel cylinders and increase monotonically outward being zero on the rotation axis. Define the function

$$F(a)=(1-\Omega l/c^2)\exp\left(\int_0^a (c^2-\Omega l)^{-1}\Omega\frac{dl}{da}\,da\right). \tag{5.120}$$

If the rotation is rigid (Ω=const) or slow ($\Omega l \ll c^2$), $F=1$. The function F permits the first integrals of the relativistic Euler equation to be written in a compact form. The most important ones, and more important von Zeipel theorems, are presented in Table 5.3. $E_I(0)$ denotes the injection energy at the rotation axis.

TABLE 5.3. *Von Zeipel theorems*

Surface	Rigidly rotating star Ω=const	Isentropic star s=const	Barotropic star $p=p(\rho)$
Equipotential surfaces, b=const	n, s $\ln W=\ln ZF$	n, s $\ln W=\ln ZF$	n, s $\ln W=\ln ZF$
Von Zeipel cylinders, a=const	l $j=E_I(0)\,l(c^2-\Omega l)^{-1}$	l, Ω, E_w $j=E_I(0)\,l(c^2-\Omega l)^{-1}F$ $\mathscr{E}=E_I(0)(1-\Omega l/c^2)^{-1}F$	l, Ω

The symbols n, s and the equation $\ln W=\ln ZF$ in the "Isentropic star"/"Equipotential surface" square, for example, mean that (a) The surfaces n=const, s=const (and hence also the surfaces $f=f(n, s)$=const, where f is an arbitrary thermodynamic function of the variables n and s), W=const and ZF=const coincide if the star is isentropic; (b) If the star is isentropic, $W=ZF$. Note that this last equation makes it possible to write the first integral of the relativistic Euler equation in the form $\int_0^p \frac{dp}{\varepsilon+p}=\ln ZF$. If the equation of state $p=p(\varepsilon)$ and the law of differential rotation in the form $\Omega=\Omega(l)$ are known, it can be written explicitly as $p=p(Z, l)$. The table shows how to write the first integrals of the relativistic Euler equation when the rotation law is given in a different form.

The acceleration potential given by relation (5.112) satisfies the equation (cf. (2.105))

$$\ln W = \int_0^p \frac{dp}{\varepsilon + p}. \tag{5.121}$$

Let b=const on the equipotential surfaces. Since the surfaces a=const and b=const have different topologies, we can draw the conclusion that

if a rotating star is barotropic, then any functions $h=h(a)$ and $k=k(b)$ satisfying the equation $h=k$ must be constant. This is known as von Zeipel's lemma. For an isentropic star we take

$$h(a) = \frac{E_I}{F} = \frac{\varepsilon + p}{nZF} = \frac{\varepsilon + p}{nW}, \qquad k(b) = \frac{\varepsilon + p}{nW}, \tag{5.122}$$

and in virtue of von Zeipel's lemma we get

$$\frac{\varepsilon + p}{nW} = \text{const}. \tag{5.123}$$

This result is a generalization of Boyer's theorem (1965): *in an isentropic, rigidly rotating star, the injection energy is constant.*

Von Zeipel's lemma also implies that if a differentially rotating star is in thermodynamic equilibrium in the sense that

$$\left(\frac{\partial S}{\partial \mathscr{E}}\right)_a = 0, \tag{5.124}$$

then the temperature and the chemical potential satisfy the relations

$$T/W = \text{const}, \qquad \mu/W = \text{const}. \tag{5.125}$$

Using the relationship $\mathfrak{N}E_I + \Omega_j = \mathscr{E}$, we can write the equations of motion (5.117) in the form

$$-\frac{T}{Z}\nabla S + \nabla\mathscr{E} - \Omega\nabla j = 0. \tag{5.126}$$

If some region inside the star is in the state of marginal convective stability, every small element of matter (with a fixed particle number $\mathfrak{N}$) can move in this region freely without any change in its energy. This means that in this region $\mathscr{E}$=const and

$$-\frac{T}{Z}\nabla S = \Omega\nabla j. \tag{5.127}$$

According to von Zeipel's lemma, this equality is possible if and only if S=const and j=const. Thus:

If the angular momentum density and the entropy density are constant in a region inside the star, this region is marginally stable against convection. This theorem is a particular case of the convective stability criterion found by Seguin (1975).

A rotating star is unstable to convection when

$$k\cdot\nabla l+(\varepsilon+p)^{-2}\left(\frac{\partial\varepsilon}{\partial S}\right)_p \nabla p\cdot\nabla S\leqslant 0, \tag{5.128}$$

$$\left(\frac{\partial\varepsilon}{\partial S}\right)_p (k\times\nabla p)\cdot(\nabla S\times\nabla l)\leqslant 0, \tag{5.129}$$

where $k=c^2(c^2-\Omega l)^{-2}(\mathcal{R}^{-2}\nabla l-\nabla\Omega)$. The vector k satisfies the inequality $k\cdot\nabla a>0$, i.e., it is directed away from the rotation axis.

We see that, in accordance with our earlier discussion, a local criterion for the occurrence of convective instability has the same form in general relativity (5.128) as in Newton's theory (5.5).

5.7. Slow Rotation. Models of Rotating Neutron Stars

Equations describing slowly rotating relativistic stars were published in 1967 by James B. Hartle.† This work, one of the most important in the theory of relativistic rotating stars, opened a long series of papers by Hartle and his collaborators, and also become a starting point for many other authors. The assumption of slow rotation means that the angular velocity Ω obeys the inequalities

$$\Omega^2\ll\left(\frac{c}{R}\right)^2\frac{GM}{c^2R}, \tag{5.130}$$

$$R\Omega\ll c, \tag{5.131}$$

where c is the speed of light, G the gravitational constant, and M and R are the mass and the radius of the star. As we know, the metric of a spherically symmetric, nonrotating star can be written

$$ds^2=e^{\nu_0}c^2dt^2-e^{\lambda_0}dr^2-r^2(d\theta^2+\sin^2\theta d\varphi^2), \tag{5.132}$$

where the functions ν_0 and λ_0 satisfy the following relations implied by the field equations:

$$e^{\lambda_0}=\left(1-\frac{2Gm}{c^2r}\right)^{-1}, \quad m=m(r)=4\pi\int_0^r \rho_0 r^2 dr, \tag{5.133}$$

$$\frac{d\nu_0}{dr}+\frac{1}{r}-re^{\lambda_0}\left(\frac{1}{r^2}+\frac{8\pi G}{c^4}p_0\right)=0, \tag{5.134}$$

$$2\frac{dp_0}{dr}=-(\varepsilon_0+p_0)\frac{d\nu_0}{dr}. \tag{5.135}$$

Here $\rho_0=\varepsilon_0/c^2$; the index "0" refers to nonrotating configurations. Let us now assume

† A list of errata in this important paper was given by Chandrasekhar and Miller (1974). The Chandrasekhar and Miller paper is not free from printer's errors either. In the last row of their formula (4) the sign before the expression $\sin^2\theta$ should be +.

that—in keeping with inequalities (5.130) and (5.131)—the rotation is so slow that the terms proportional to Ω^3 and higher can be dropped from the expansions in Ω of the functions describing the rotating configuration. There is the question of an appropriate choice of the coordinate system in which to expand these functions.

A rotating star becomes flattened, and the proportional density changes $\Delta\rho/\rho_0$ are infinite in the region shaded in Fig. 5.3. For this reason, the coordinate system will be

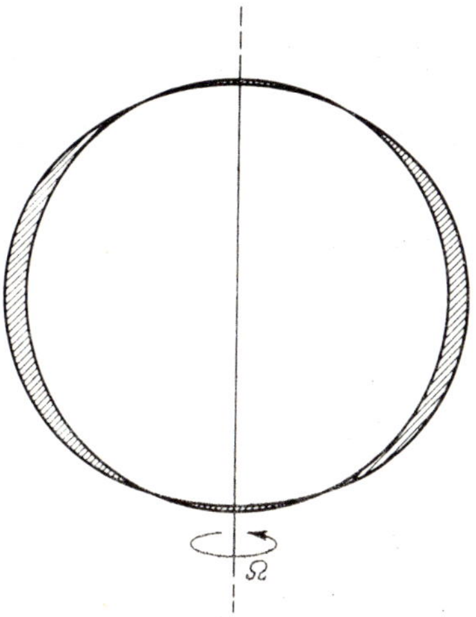

FIG. 5.3. The region near the surface of a rotating star, where density perturbations are infinite compared with the nonrotating configuration.

chosen in the following way. Take a point inside the rotating configuration. It lies on some surface of constant density, $\rho=\rho^*=\text{const}$. Consider the radius of the surface $\rho_0=\rho^*=\text{const}$ in the nonrotating configuration; it will be used as the radial coordinate R of the point we began with. Thus we have

$$\Theta=\theta, \quad \rho[r(\tilde{R},\Theta),\Theta]=\rho(\tilde{R})=\rho_0(\tilde{R}); \tag{5.136}$$

in addition we put

$$r=\tilde{R}+\xi(\tilde{R},\Theta)+o(\Omega^4). \tag{5.137}$$

(Attention: the displacement ξ has nothing to do with the Killing vector ξ^α. In the present considerations the Killing vector ξ^α does not appear at all, and therefore using the same letter for a different quantity should not cause any confusion).

The quantities characterizing a rotating star which under rotation behave as scalars (e.g. some of the metric functions) have expansions

$$F(\tilde{R},\Theta)=F_0(\tilde{R})+F_2(\tilde{R},\Theta)+o(\Omega^4), \tag{5.138}$$

where F_0 relates to the nonrotating star and $F_2\sim\Omega^2$. The function $F_2(R,\Theta)$ can be written

$$F_2(\tilde{R},\Theta)=\sum_l F_2^l(\tilde{R})\,P_l(\cos\Theta); \tag{5.139}$$

where $P_l(\cos\theta)$ is a Legendre polynomial of order l. It can be shown that

$$F_2^l=0 \quad \text{for} \quad l\geqslant 4. \tag{5.140}$$

If the configuration is assumed symmetric about the equatorial plane $\theta=\pi/2$, $F_2^l=0$ for odd l. Therefore, expansion (5.139) will contain only the terms with $l=0$ and $l=2$. The only metric function which under rotation behaves like a vector (the function $g_{t\varphi}$) can be shown to depend on the coordinate $\tilde{R}$ only. Consequently, the metric of the rotating star can be written

$$\begin{aligned} ds^2 &= e^{\nu_0}[1+2h_0(\tilde{R})+2h_2(\tilde{R})P_2(\cos\Theta)]\,c^2dt^2 - e^{\lambda_0}\{1+e^{2\lambda_0}2G/c^2R\,[m_0(\tilde{R}) \\ &+ m_2(\tilde{R})P_2(\cos\Theta)]\}\,dR^2 - \tilde{R}^2[1+2k_2(R)P_2(\cos\Theta)]\,\{d\Theta^2+[d\varphi-\tilde{\omega}(\tilde{R})dt]^2+\sin^2\Theta d\varphi^2\} \\ &+ o(\Omega^3). \end{aligned} \tag{5.141}$$

The functions h_0, h_2, m_0, m_2 and k_2 are of second order in Ω, and the function $\tilde{\omega}$ is of first order in Ω.

Introduce the quantity

$$P=\ln(\varepsilon+p)-\int_0^{\varepsilon}\frac{d\varepsilon}{\varepsilon+p}. \tag{5.142}$$

The expansion of P and that of the velocity v defined by formula (5.58) are

$$P=P_0(\tilde{R})+\delta P_0(\tilde{R})+\delta P_2(\tilde{R})P_2(\cos\Theta), \tag{5.143}$$

$$v=e^{\frac{-\nu_0}{2}}(\Omega-\tilde{\omega})\tilde{R}\sin\Theta.$$

Using these expansions, we can write the first integrals of the equations of motion in the form

$$h_0(\tilde{R})-\tfrac{1}{3}\tilde{R}^2e^{-\nu_0}(\bar{\omega}/c)^2+\delta P_0=A=\text{const}, \tag{5.144}$$

$$h_2(\tilde{R})+\tfrac{1}{3}\tilde{R}^2e^{-\nu_0}(\bar{\omega}/c)^2+\delta P_2=0. \tag{5.145}$$

where

$$\bar{\omega}=\Omega-\tilde{\omega}. \tag{5.146}$$

After considerable calculations one can show that to second order in Ω Einstein's field equations for the metric (5.141) take the form

$$\frac{1}{\tilde{R}^2}\frac{d}{d\tilde{R}}\left(\tilde{R}^4\beta\frac{d\bar{\omega}}{d\tilde{R}}\right)+4\frac{d\beta}{d\tilde{R}}\bar{\omega}=0, \tag{5.147}$$

$$\frac{G}{c^2}\frac{dm_0}{d\tilde{R}}=\frac{4\pi G}{c^4}\tilde{R}^2(\varepsilon+p)\frac{d\varepsilon}{dp}\delta P_0+\frac{1}{12}\tilde{R}^4\beta^2\left(\frac{1}{c}\frac{d\bar{\omega}}{d\tilde{R}}\right)^2-\frac{1}{3}R^3\frac{\bar{\omega}^2}{c^2}\frac{d\beta^2}{d\tilde{R}}, \tag{5.148}$$

$$\begin{aligned} \frac{dh_0}{d\tilde{R}} &= -\frac{d}{d\tilde{R}}\delta P_0+\frac{1}{3}\frac{d}{d\tilde{R}}\left(\tilde{R}^2e^{-\nu_0}\frac{\bar{\omega}^2}{c^2}\right) \\ &= \frac{G}{c^2}m_0e^{2\lambda_0}\left(\frac{1}{\tilde{R}^2}+\frac{8\pi G}{c^4}p_0\right)-\frac{1}{12}e^{\lambda_0}\tilde{R}^3\beta^2\left(\frac{1}{c}\frac{d\bar{\omega}}{d\tilde{R}}\right)^2+\frac{4\pi G}{c^4}\tilde{R}e^{\lambda_0}(\varepsilon+p)\delta P_0, \end{aligned} \tag{5.149}$$

$$\frac{d}{d\tilde{R}}(h_2+k_2)=h_2\left(\frac{1}{\tilde{R}}-\frac{1}{2}\frac{d\nu_0}{d\tilde{R}}\right)+\frac{Gm_2}{c^2\tilde{R}-2Gm}\left(\frac{1}{\tilde{R}}+\frac{1}{2}\frac{d\nu_0}{d\tilde{R}}\right), \tag{5.150}$$

$$h_2+\frac{Gm_2}{c^2\tilde{R}-2Gm}=\frac{1}{6}\tilde{R}^4\beta^2\left(\frac{1}{c}\frac{d\bar{\omega}}{d\tilde{R}}\right)^2-\frac{1}{3}\tilde{R}^3\frac{\bar{\omega}^2}{c^2}\frac{d\beta^2}{d\tilde{R}}, \tag{5.151}$$

$$\frac{1}{\tilde{R}}\left(1-\frac{2Gm}{c^2\tilde{R}}\right)\frac{dh_2}{d\tilde{R}}+\left(1-\frac{2Gm}{c^2\tilde{R}}\right)\left(\frac{1}{2}\frac{d\nu_0}{d\tilde{R}}+\frac{1}{\tilde{R}}\right)\frac{dk_2}{d\tilde{R}}-\frac{Gm_2}{c^2\tilde{R}^2}\left(\frac{d\nu_0}{d\tilde{R}}+\frac{1}{\tilde{R}}\right)$$
$$-\frac{3h_2}{\tilde{R}^2}-\frac{2k_2}{\tilde{R}^2}-\frac{4\pi G}{c^4}(\varepsilon+p)\delta P_2-\frac{1}{12}\tilde{R}^2\beta^2\left(\frac{1}{c}\frac{d\bar{\omega}}{d\tilde{R}}\right)^2=0, \tag{5.152}$$

where for simplicity we have denoted

$$\beta=e^{-\frac{1}{2}(\lambda_0+\nu_0)}. \tag{5.153}$$

In the external region, outside the rotating star, we have

$$\varepsilon=0=p, \quad \beta=1, \quad m(\tilde{R})=M. \tag{5.154}$$

This permits explicit integration of the equations. The results can be written as follows:

$$\bar{\omega}=\Omega-\frac{2GJ}{c^2\tilde{R}^3}, \qquad m_0=\delta M-\frac{GJ^2}{c^4\tilde{R}^3}, \tag{5.155}$$

$$h_0=-\frac{G\delta M}{c^2\tilde{R}-2GM}+\frac{G^2J^2}{c^4\tilde{R}^3(c^2\tilde{R}-2GM)}, \tag{5.156}$$

$$h_2=\frac{G^2J^2}{c^6}\left(\frac{c^2}{GM\tilde{R}^3}+\frac{1}{\tilde{R}^4}\right)+KQ_2^2\left(\frac{\tilde{R}c^2}{GM}-1\right), \tag{5.157}$$

$$h_2+k_2=-\frac{G^2}{c^6}\frac{J^2}{\tilde{R}^4}+\left[\frac{2KGM}{c^2\tilde{R}}\left(1-\frac{2GM}{c^2\tilde{R}}\right)^{-\frac{1}{2}}\right]Q_2^1\left(\frac{\tilde{R}c^2}{GM}-1\right). \tag{5.158}$$

In these formulae δM and K are arbitrary constants, and Q_2^{α} is the associated Legendre function of the second kind.

The boundary conditions for equations (5.147)–(5.152) are: m_2, h_0, h_2 and k_2 vanish and $\bar{\omega}$ is finite at $\tilde{R}=0$. At the surface of the star ($p=0$) the functions just mentioned join up smoothly with the functions given by equations (5.155)–(5.158).

In keeping with the theory presented above, the construction of a numerical model of a slowly rotating neutron star may proceed using the following steps (Hartle and Thorne, 1968):

1. Choose an equation of state.
2. Choose the values of the central density and the angular velocity Ω.
3. Construct a model of a nonrotating star with the central density chosen.
4. Integrate numerically the system of equations (5.147)–(5.152). The results obtained in this way for a particular equation of state (the so-called V_γ equation) are presented in Fig. 5.4 and Fig. 5.5.

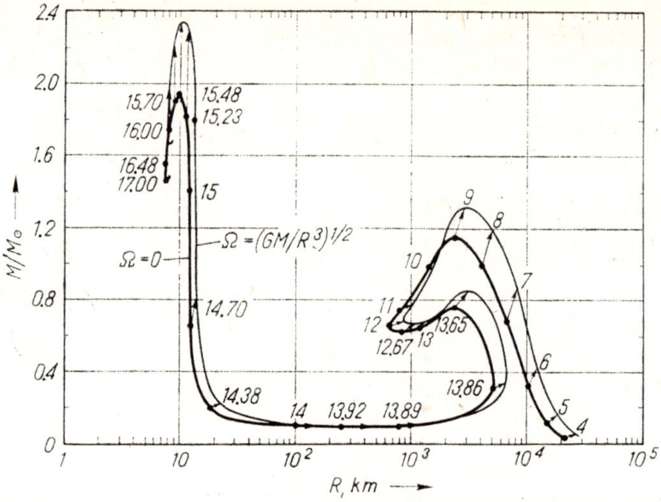

Fig. 5.4. Effect of rotation on mass and radius of an equilibrium configuration satisfying a V_γ equation of state. The heavy line is the mass-radius curve for the non-rotating configuration (Hartle and Thorne, 1968).

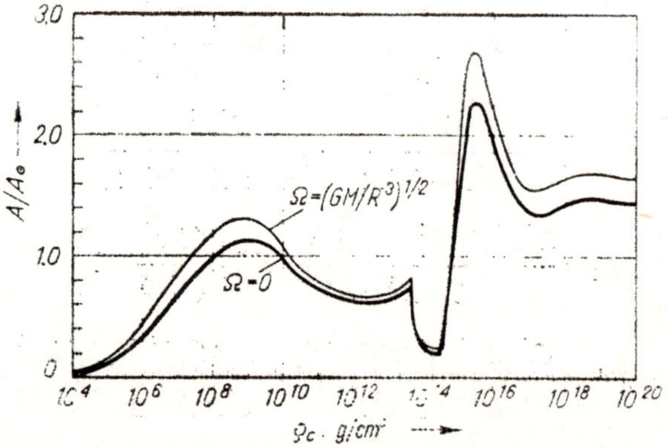

Fig. 5.5. Effect of rotation on the maximum number of baryons that may be contained in an equilibrium configuration. The heavy line is the baryon number-central density relation for nonrotating configurations (Hartle and Thorne, 1968).

Our knowledge of how to construct models of rotating stars without the simplifying assumption that the rotation is slow leaves much to be desired. All the works on models of rapidly rotating stars published so far (Wilson, 1971, 1972, 1973; Bonazzola and Schneider, 1974) use methods which can hardly be regarded as satisfactory.

Literature

Abramowicz, M. and Muchotrzeb, B. (1975) preprint ZA PAN.
Abramowicz, M. and Wagoner, R. (1976) *Astrophys. J.* **204**, 896.
Bardeen, J. M. (1970) *Astrophys. J.* **162**, 71.
Bardeen, J. M., Press, W. and Teulosky, S. (1972) *Astrophys. J.* **178**, 347.

BONAZZOLA, S. and SCHNEIDER, P. (1974) *Astrophys. J.* **191**, 273.
BOYER, R. (1965) *Proc. Camb. Phil. Soc.* **61**, 527.
CARTER, B. (1969) *J. Math. Phys.* **10**, 70.
CARTER, B. and QUINTANA, H. (1973) *Astrophys. J. Lett.* **14**, 105.
CHANDRASEKHAR, S. (1969) *Ellipsoidal Figures of Equilibrium*, Yale University Press, N. H.
CHANDRASEKHAR, S. and LEBOVITZ, N. R. (1962) *Astrophys. J.* **135**, 248; *Astrophys. J.* **136**, 1069.
CHANDRASEKHAR, S. and LEBOVITZ, N. R. (1968), *Astrophys. J.* **152**, 267.
CHANDRASEKHAR, S. and MILLER, J. (1974) *Mon. Not. Roy. Astr. Soc.* **167**, 63.
FRIEDMAN, J. L. (1978) *Comm. Math. Phys.* **62**, 247.
FRIEDMAN, J. and SCHUTZ, B. (1978) *Astrophys. J.* **222**, 281.
HARRISON, B. (1970) *Phys. Rev.* **D1**, 2269.
HARTLE, J. B. (1967) *Astrophys. J.* **150**, 1005.
HARTLE, J. B. (1978) *Physics Reports* **46**, 201.
HARTLE, J. B. and THORNE, K. S. (1968) *Astrophys. J.* **153**, 807.
LEDOUX, P. (1965) *Stellar Stability*, in *Stellar Structure—Stars and Stellar Systems*, VIII, eds.: Aller, L. H., McLaughlin, D. B., Chicago University Press, Chicago.
LEDOUX, P. and WALRAVEN, Z. (1958) *Variable Stars*, in *Handbuch der Physik*, vol. II, ed.: Flügge, S. Springer, Berlin.
LENSE, J. and THIRRING, H. (1918) *Phys. Z.* **19**, 156.
LICHTENSTEIN, L. (1933) *Gleichgewichtsfiguren rotierender Flüsigkeiten*, Springer, Berlin.
LYTTLETON, R. A. (1953) *The Stability of Rotating Liquid Masses*, Cambridge University Press, Cambridge.
OSTRIKER, J. P. (1978) *The Influence of Rotation on Stars and Stellar Systems*, in *Theoretical Principles in Astrophysics and Relativity*, eds.: Lebovitz, N. R., Reid, W. H. and Vandervoort, P. O., The University of Chicago Press.
OSTRIKER, J. P. and BODENHEIMER, P. (1968) *Astrophys. J.* **151**, 1089.
OSTRIKER, J. P. and MARK, J. W. K. (1968) *Astrophys. J.* **151**, 1075.
PAPAPETROU, A. (1948) *Proc. Roy. Irish Acad.* **52**, 11.
RHOADES, C. and RUFFINI, R. (1974) *Phys. Rev. Lett.* **32**, 324.
SCHUTZ, B. F. and SORKIN, R. (1977) *Ann. Phys.* **107**, 1.
SEGUIN, F. (1975) *Astrophys. J.* **197**, 745.
TASSOUL, J. L. (1978) *Theory of Rotating Stars*, Princeton University Press, Princeton, N. J.
THORNE, K. S. (1967) *Relativistic Stellar Structure and Dynamics*, in *High Energy Astrophysics*, Vol. III, eds.: De Witt, C., Schatzman, E., Veron, P., Gordon and Breach, N. Y.
THORNE, K. S. (1971) *Relativistic Stars, Black Holes and Gravitational Waves* (including an in-depth review of the theory of relativistic, rotating stars), in *Gravitation and Cosmology*, ed.: Sachs, R. K., Academic Press, N.Y.
WAGONER, R. and MALONE, R. (1974) *Astrophys. J.* **189**, L 75.
WILSON, J. R. (1971) *Astrophys. J.* **163**, 209.
WILSON, J. R. (1972) *Astrophys. J.* **176**, 195.
WILSON, J. R. (1973) *Phys. Rev. Lett.* **30**, 1082.

CHAPTER 6

NEUTRON STARS AND PULSARS

6.1. Late Stages of Stellar Evolution

From the time it is formed a star goes through a long and complex succession of evolutionary changes. A detailed discussion of the successive stages of this evolution and the physical processes that dominate them can be found in textbooks on astrophysics (Chandrasekhar, Schwarzschild, Frank-Kamenetsky, Cox and Giuli, Tayler, and others). Here we shall

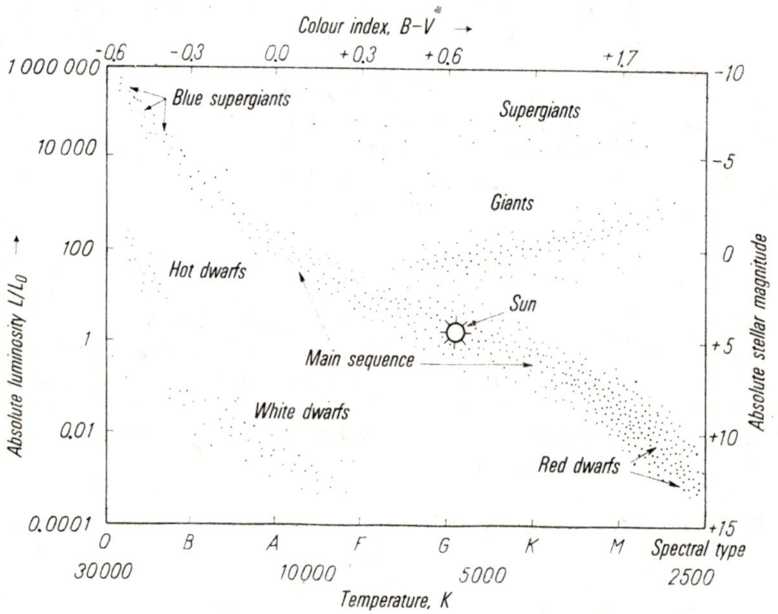

FIG. 6.1. Spectral type and luminosity of stars (Hertzsprung–Russell diagram).

restrict ourselves to recalling the general picture; the last stages of the evolution will be discussed in more detail.

Our knowledge of how a star is formed from hot gases and interstellar matter is not yet very precise. The first stage is the condensation of interstellar matter and formation of a protostar. If the gravitational field of the protostar is strong enough to cause its contraction, its further evolution is uniquely determined. It will continue to contract. Owing to various dissipative processes (shock waves, viscosity, turbulent motions, etc.) part of the kinetic energy of the macroscopic motion is converted into thermal energy. The temperature in the central regions of the protostar steadily rises, and when it reaches 10^7 K, hydrogen is ignited. The star enters the main sequence on the Hertzsprung–Russell diagram (H–R diagram), showing the relationship between temperature and luminosity (Fig. 6.1). Both the rate and the course of the evolution of a star depend on its initial mass

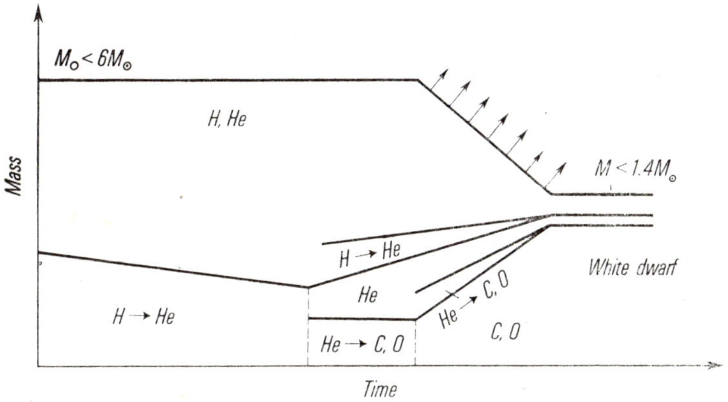

FIG. 6.2. Evolution of a star with an initial mass less than $6M_\odot$. The time-axis is not drawn to scale. Arrows indicate mass loss.

and chemical composition (which in a protostar is homogeneous). The greater the mass of a star, the higher the temperature of its interior and the speed with which hydrogen burns into helium. In stars whose masses are close to the mass of the Sun the burning of hydrogen is slow and lasts for about 10^{10} years. More massive stars evolve much faster. In the following we shall only consider the evolution of single stars of Population I (hydrogen content $X_0 = 0.7$, heavy element content $Z_0 = 0.03$), basing ourselves mainly on the works of Paczyński, Iben and Stothers.

As the burning of hydrogen progresses, its amount in the central parts of the star slowly decreases and the chemical composition of the star becomes more and more heterogeneous. With all hydrogen exhausted in the centre, the burning of hydrogen is confined to a thin layer surrounding a helium core. The temperature in the core is still too low for the helium to ignite. If the mass of the star is less than $6M_\odot$, the helium core is practically isothermal, and its temperature rises slowly as the mass of the core increases. However, once its mass has increased to about 0.1 of the total mass of the star, the isothermal core cannot support the external layers. It then contracts rapidly, and when its temperature reaches about 10^8 K, the helium begins to burn while the combustion of hydrogen still continues in the surrounding layer. The central core now consists mainly of carbon and oxygen. At this stage the star suffers an extensive loss of mass. The core shrinks while the external parts expand. The outer envelope of the star may be blown off and form a planetary nebula. The process of mass loss is so effective that at the final stages of the evolution the mass of the star falls below $1.4M_\odot$ and the star becomes a white dwarf. Nuclear reactions come to a halt and the star slowly radiates away its store of thermal energy. The cooling takes a long time, typically about 10^9 years, after which the white dwarf emits almost no light. The evolution of such a star is shown schematically in Fig. 6.2.

The evolution of a star with a mass in the interval $6M_{\odot} \leqslant M_0 \leqslant 9M_{\odot}$ follows a somewhat different course. The helium core which forms after all the hydrogen in the central part of the star is burned out cannot be isothermal, and in order to produce a temperature gradient ensuring dynamic stability of the star the core shrinks. Its temperature rises and finally the helium is ignited. The process continues as in the case of less massive stars until a carbon-oxygen core is formed. In spite of the mass loss, which is most effective during the burning of helium in the layer surrounding the carbon–oxygen core, the mass of the core eventually reaches $1.4M_{\odot}$. Its own energy sources exhausted, the core loses its stability and implodes. The temperature and density at the centre grow to such an extent that violent ignition of carbon takes place. The energy released in this process is so enormous that part of the mass of the star is ejected. The star becomes a supernova. The energy released in such supernova explosion is estimated to be about 10^{51} erg. The remaining core collapses, to form a neutron star or a white dwarf (Fig. 6.3). It is still unknown what exactly the minimum initial mass of the star should be for the supernova explosion to end with the formation

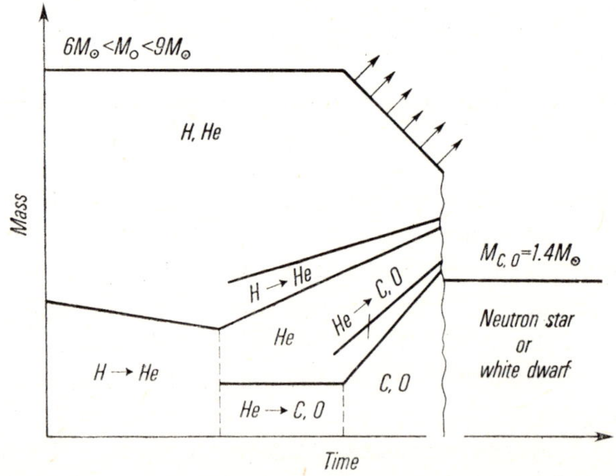

FIG. 6.3. Evolution of a star with an initial mass of 6 to $9M_{\odot}$. The time-axis is not drawn to scale Arrows indicate mass loss. The wavy line indicates the moment of violent ignition of carbon. which sets off a supernova explosion. The central parts of the star collapse, to form a white dwarf or a neutron star. The energy released in the explosion is of order 10^{51} erg.

of a neutron star. The process of explosion itself is also far from being fully understood. To all interested in these problems we recommend an extensive review article by Sugimoto and Nomoto.

The initial stages of the evolution of stars with masses in the interval $9M_{\odot} \leqslant M_0 \leqslant 40M_{\odot}$ are similar to those of lighter stars, but the rate of the evolution is much faster. This case is illustrated in Fig. 6.4. Such a massive star may lose mass as early as the hydrogen burning stage. The ignition of carbon in the core has a quiet start compared with that in lighter stars, and further nuclear processes eventually lead to the formation of a core composed mostly of iron. As the mass of the core grows, the central density and temperature also rise. The core becomes unstable when its mass is still below the maximum possible mass of a neutron star. The instability sets in when the central density reaches a value of about 10^9 g/cm^3 and the process of neutronization begins, i.e. the capture of electrons by iron nuclei with the simultaneous conversion of protons into neutrons and emission of neutrinos, or when the temperature becomes high enough to cause photodisintegration of iron nuclei. Both processes reduce the central pressure. The core collapses. The neutrinos escaping from it are absorbed by external layers. This additional transmission of energy results in the expulsion of the envelope, or in other words, triggers a supernova explosion. If the stability loss is caused by photodisintegration of iron nuclei, the released energy may be so high that the star is completely fragmented. It is not certain yet whether stars with masses in the range considered always survive supernova explosion phase, with the core of the star imploding to form a neutron star, or whether the explosion ends with a complete fragmentation of the star. It is estimated that the energy released in the explosion is of order of 10^{53} erg.

Stars with initial masses greater than $40M_\odot$ evolve faster still and lose mass throughout the process. Their evolution is schematically shown in Fig. 6.5. At the final stages, the mass of the iron core exceeds the maximum mass of neutron stars. The core becomes unstable and collapses, giving rise to a black hole. There follows a supernova explosion accompanied by an enormous release of energy, reaching 10^{54}–10^{55} erg.

The most important information about the final stages of stellar evolution is collected in Table 6.1.

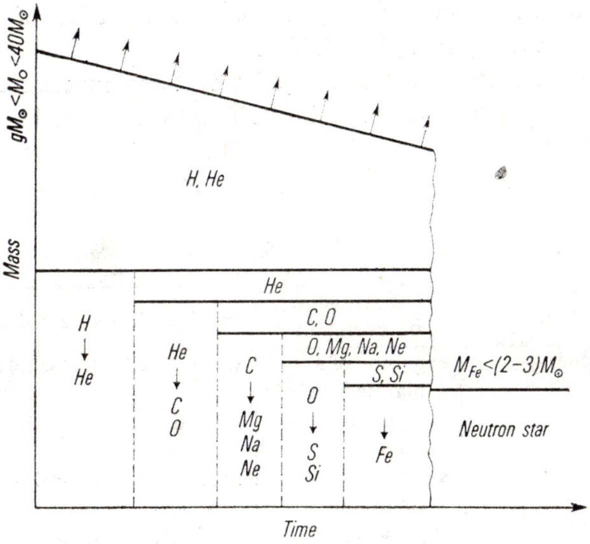

FIG. 6.4. Evolution of a star with an initial mass of 9 to $40M_\odot$. The time-axis is not drawn to scale. Arrows indicate mass loss. The wavy line marks the moment when the star loses stability whereupon its central parts collapse to form a neutron star and the envelope is expelled. The energy released in this catastrophic process is of order 10^{53} erg.

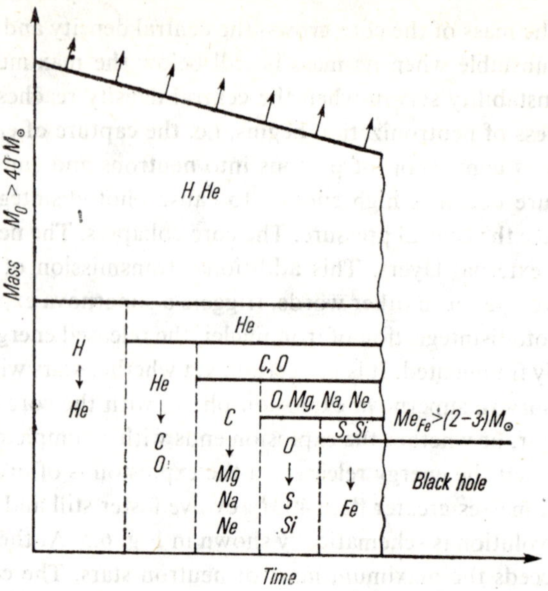

FIG. 6.5. Evolution of a star with an initial mass greater than $40M_\odot$. The time-axis is not drawn to scale. Arrows indicate the mass loss. The wavy line marks the moment when the star loses stability. The internal layers collapse, giving rise to a black hole, and the external layers can be expelled. The energy released in this catastrophic process is estimated to be of order 10^{54-55} erg

Evolution of stars in binary systems is somewhat different. In general, they evolve much faster than single stars. Extensive information about the evolution of binary stars can be found in the survey article by Paczyński (1980).

We now turn to cold stars at the end point of thermonuclear evolution, hence to cold white dwarfs, neutron stars and black holes. Thanks to the works of Harrison, Wakano, Wheeler and Thorne we know the relationship between the mass and the central density of a star at temperatures close to absolute zero or at zero entropy. It may be expected

TABLE 6.1

Initial mass	The last stage of evolution	Final product	Final mass
$<6M_\odot$	expulsion of a planetary nebula	white dwarf	$<1.4M_\odot$
$(6-9)M_\odot$	supernova explosion	white dwarf or neutron star	$<1.4M_\odot$
$(9-40)M_\odot$	supernova explosion	neutron star	$<(2-3)M_\odot$
$>40M_\odot$	supernova explosion	black hole	$>(2-3)M_\odot$

that a similar curve will describe systems whose total entropy is small. Examples of such curves are shown in Fig. 6.6. We shall assume that essentially no loss of mass occurs at the final stages of cooling. The history of the configuration is then a horizontal curve. We assume furthermore that the critical Chandrasekhar mass M_{Ch} is less than the maximum possible mass of a neutron star M_L (the Landau mass). Three different situations are possible. Light stars, with masses below the critical mass of a white dwarf, will cool down as white dwarfs, gradually becoming "red" dwarfs and "black" dwarfs as their surface temperature falls. The process of cooling is a long-lasting one, and takes about

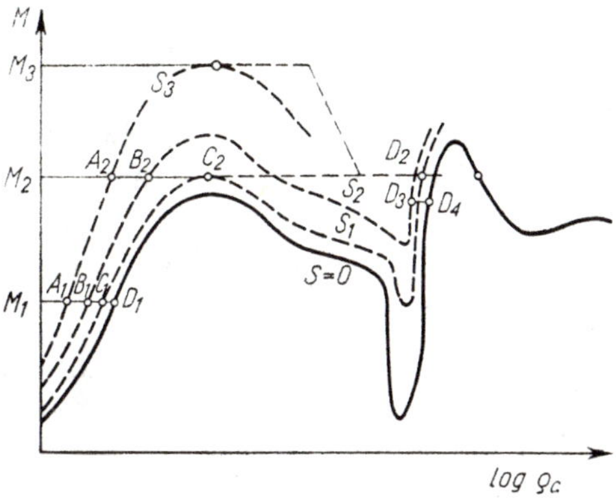

FIG. 6.6. Stellar mass vs. central density for different values of the total entropy (adapted from Zeldovich and Novikov, 1971).

10^9 years. If the mass of the star is between M_{Ch} and M_L, the star ends its evolution as a neutron star. Stars whose masses at the late stages of evolution exceed the Landau mass, will go on contracting forever—undergoing gravitational collapse. The asymptotic, final result of this contraction is called a black hole. An analysis of the Schwarzschild metric implies that the radius of a spherically symmetric black hole is equal to its gravitational radius. The properties of black holes will be taken up in more detail in Chapters 7 and 8. Observational data relating to the late stages of the evolution of stars are scarce. There are only few theoretical models. As follows from the mass versus central density curve at fixed entropy, a star whose evolution progresses quietly can for long periods of time remain in metastable states corresponding to local maxima of the curve $M=M(\rho_c)$. We know practically next to nothing about the structure of such stars and the rate of their evolution near the critical points. Studies of stellar evolution are however one of the rapidly developing fields of astrophysics, and it may be hoped that our knowledge of what happens to ageing stars will soon be broadened considerably.

6.2. Neutron Stars

We have concluded that a stable configuration of cold catalyzed matter can take one of only two forms. It may be either a white dwarf—a star with a mass less than $1.4M_\odot$ and central density up to 10^9 g/cm^3, or a neutron star of central density reaching 10^{16} g/cm^3. The structure of white dwarfs is well known. The first theoretical model of such stars was worked out by Chandrasekhar, on the assumption that the matter throughout the star can be described by a polytropic equation of state. More realistic equations of state change predictions of the Chandrasekhar model only quantitatively. The basic features, namely the existence of a maximum mass and the fact that the pressure in the core of such a star is mostly produced by a degenerate electron gas, remain unchanged. Gravitational fields generated by white dwarfs are weak, and Newton's theory of gravitation is quite sufficient to describe them. To see this is so, note that a degenerate electron gas is relativistic when the electron momentum becomes comparable to $m_e c$, where m_e is the electron rest-mass. This happens at a density

$$\rho_{cr} = \frac{m_p \mu m_e^3 c^3}{3\pi^2 \hbar^3} = 0{\cdot}97 \cdot 10^6 \mu \text{ g/cm}^3; \tag{6.1}$$

here μ is the number of baryons per electron. In white dwarfs $\mu \approx 2$. For densities above ρ_{cr}, the equation of state of the electron gas can be written

$$p = \frac{\hbar c}{4}(3\pi^2)^{\frac{1}{3}}(m_p \mu)^{-\frac{4}{3}}\rho^{\frac{4}{3}}. \tag{6.2}$$

Now using (4.11) and (4.21) and the fact that $\xi^2 \left.\frac{d\theta_3}{d\xi}\right|_{\xi=\xi_1} = -2.018$, we find $M = 5.87\,\mu^{-2} M_\odot$. The radius of such a star is equal to

$$R = \frac{1}{2}(3\pi)^{\frac{1}{2}}\xi_1 \frac{\hbar^{\frac{3}{2}}}{c^{\frac{1}{2}} G^{\frac{1}{2}} m_e m_p \mu}\left(\frac{\rho_{cr}}{\rho_c}\right)^{\frac{1}{3}}, \tag{6.3}$$

where ξ_1 is the first zero of the function θ_3. From the tables of Emden functions we find $\xi_1 = 6.89685$. Inserting all the numerical data gives $R = 5.3 \cdot 10^9\, \mu^{-1}\left(\frac{\rho_{cr}}{\rho_c}\right)^{\frac{1}{3}}$ cm. The system is now characterized by μ and the central density ρ_c. Physical considerations permit us to set limits on these parameters. Namely, it turns out that at a density $\rho \approx 10^9$ g/cm³ it becomes energetically favourable for electrons to be captured by nuclei, with one proton turned into a neutron and a simultaneous emission of a neutrino in every such capture. As a result, the number of baryons per electron increases, which reduces pressure and the maximum possible mass of the system. The mass is maximum when $\rho_c \approx 10^9$ g/cm³; then $\mu = 56/26$ and $M = 1.46 M_\odot$, with $R = 3.1 \cdot 10^8$ cm. At the surface of the star the ratio of the gravitational potential to the speed of light is then $\varepsilon = \frac{GM}{c^2 R} \approx 10^{-4}$, and this value is typical throughout the star. Thus, general relativity is of little relevance to the structure of white dwarfs.

In stars of higher central densities the situation becomes more complicated. Matter can no longer be regarded as a perfect gas because interactions between particles can no longer be neglected. Even if we restrict ourselves to configurations at absolute zero, the only way to derive the equation of state is to minimize the system's internal energy at a fixed number of baryons, on the assumption that the whole system is electrically neutral. Consider, qualitatively, the composition of matter in relation to density. Suppose the initial system, which we shall gradually compress so as to make the density rise, consists of ^{56}Fe nuclei intermixed with a degenerate electron gas. As we mentioned before, at a density of about 10^9 g/cm³ the Fermi momentum of the electrons will have increased enough to make their capture by iron nuclei possible. As the density continues to rise, more and more neutron-rich nuclei are produced. Under normal conditions they would be unstable to beta decay. Under present conditions, however, they are stable because all the states that a beta-electron could normally assume are already occupied in the phase space. Fig. 6.7 shows the atomic masses and charges of the most stable nuclei as function of the density. In this density range, matter still consists of nuclei and relativistic electrons.

When the density reaches about $3 \cdot 10^{11}$ g/cm^3, the binding energy of neutrons falls to zero. The neutron-rich nuclei begin to drip neutrons. Free neutrons now appear as a new

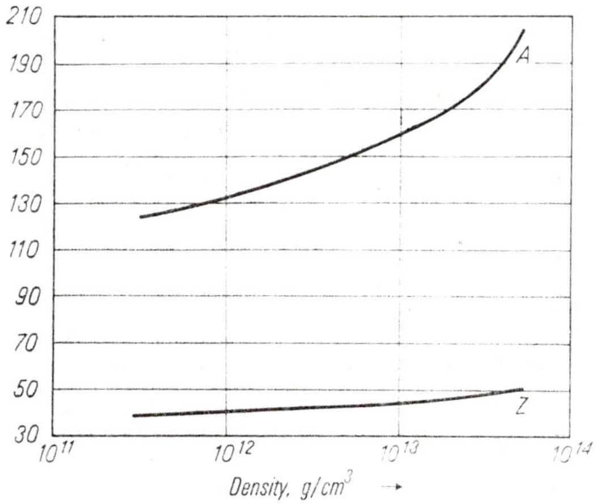

FIG. 6.7. Charge and mass of the most stable nuclei in relation to density (adapted from Cameron, 1970).

component of the gas. Their number increases, and as the density continues to rise they form a degenerate neutron gas. The composition of matter becomes very complex. It now consists of a relativistic degenerate electron gas, whose share in the total pressure is still dominant, of a non-relativistic degenerate neutron gas, and of the nuclei that have remained.

At densities of order $4 \cdot 10^{13}$ g/cm^3 the neutron Fermi momentum becomes comparable to the electron Fermi momentum. At about the same range of densities the binding energy of protons in the nuclei becomes comparable to the energy of interaction between free protons and the neutron gas. When the density reaches $6 \cdot 10^{13}$ g/cm^3, the nuclei practically disappear. The material of the star is then formed of electron, proton and neutron gases, the dominant contribution to pressure coming from the nonrelativistic degenerate neutron gas. It is easy to estimate the electron–proton and the proton–neutron ratios in the equilibrium state. The system is electrically neutral, $n_e = n_p$, and from the condition of thermodynamic equilibrium we find that the neutron Fermi momentum is equal to the sum of the proton and the electron Fermi momenta; hence $n_n^{1/3} = n_p^{1/3} + n_e^{1/3}$, and thus $n_e : n_p : n_n = 1 : 1 : 8$. The above argument is of course only approximate because we have neglected the energy of mutual interactions. More detailed calculations of the electron and the proton number densities at densities $\approx 10^{14}$ g/cm^3, which are typical of nuclear matter, show that $n_e = n_p = 0.03 n_n$.

At still higher densities, when the electron Fermi momentum divided by c exceeds the muon rest-mass, it becomes energetically favourable for relativistic electrons to be replaced by muons of zero kinetic energy. From that moment, the dense matter will also contain negative muons. One can now see what is to be expected at even higher densities, when the neutron Fermi momentum divided by c becomes equal to the mass difference between the neutrons and some of the hyperons: heavier and heavier hyperons will appear in the

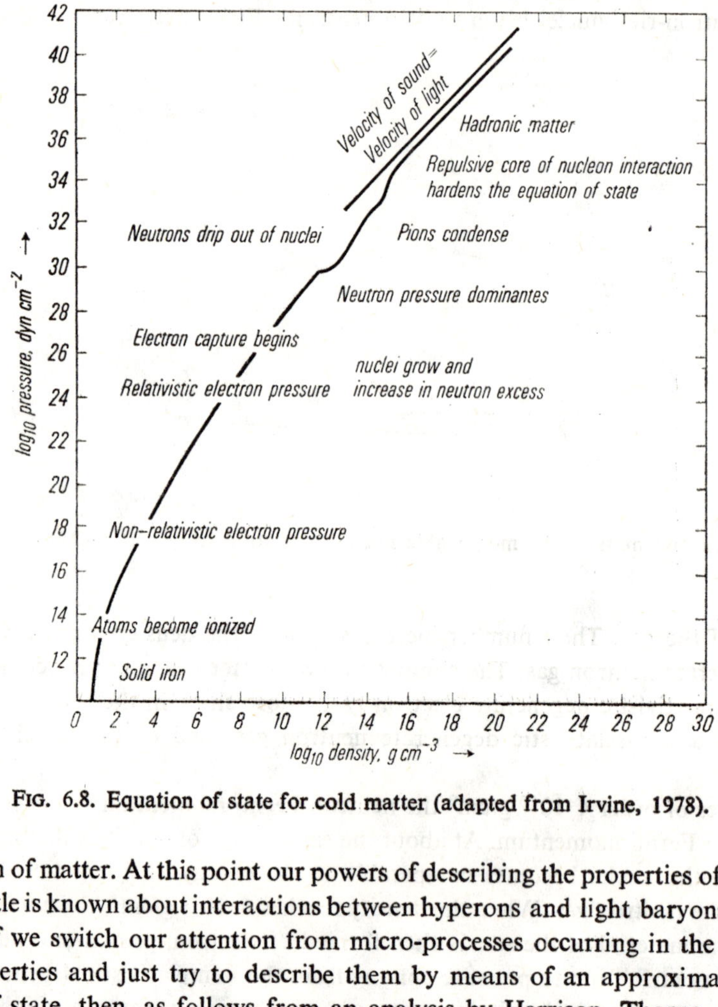

FIG. 6.8. Equation of state for cold matter (adapted from Irvine, 1978).

composition of matter. At this point our powers of describing the properties of dense matter fail: very little is known about interactions between hyperons and light baryons and leptons. However, if we switch our attention from micro-processes occurring in the system to its global properties and just try to describe them by means of an approximate relativistic equation of state, then, as follows from an analysis by Harrison, Thorne, Wakano and Wheeler, we shall find that matter of density higher than about $5.6 \cdot 10^{15}$ g/cm^3 admits no stable spherical configurations. At such densities the special- and general-relativistic effects come into prominence and become responsible for instability. The relationship between density and pressure for cold matter is illustrated in Fig. 6.8.

The discovery of pulsars and their identification with rotating neutron stars caused a surge of interest in the structure of neutron stars. Explanation and analysis of the internal structure of neutron stars have become one of the pressing tasks of the theory of nuclear matter. Densities that occur in the interiors of neutron stars may exceed typical densities of atomic nuclei, which are of order $2.8 \cdot 10^{14}$ g/cm^3. At such high densities nuclear matter exhibits new properties, for example superfluidity, condensation of pions, phase transition to quark matter, or transition to an "abnormal" state when the clouds of virtual mesons surrounding the nucleons affect the internucleonic interactions so strongly that the nucleons can be considered practically massless. The relationship between pressure and density in neutron stars is shown in Fig. 6.9.

In the last few years many new equations of state for very dense nuclear matter were proposed. They differ in the details of the description of interactions between nucleons at high densities. It is generally accepted that the proposed equations are sufficiently accurate for densities $\rho \leqslant 10^{15}$ g/cm^3, whereas the properties of nuclear matter of higher density are known only approximately (Börner (1973), Baym and Pethick (1976), Canuto (1974), Irvine (1978), Källman (1979)). The pressure versus density relations for some of the proposed equations of state are shown in Fig. 6.10.

Given the equation of state, one can proceed to integrate the equations describing a spherically symmetric equilibrium configuration. The methods of integration were already discussed in Chapter 4 and need not be repeated here. Let us only recall that the solution is constructed numerically, starting from the centre, with the central density prescribed and

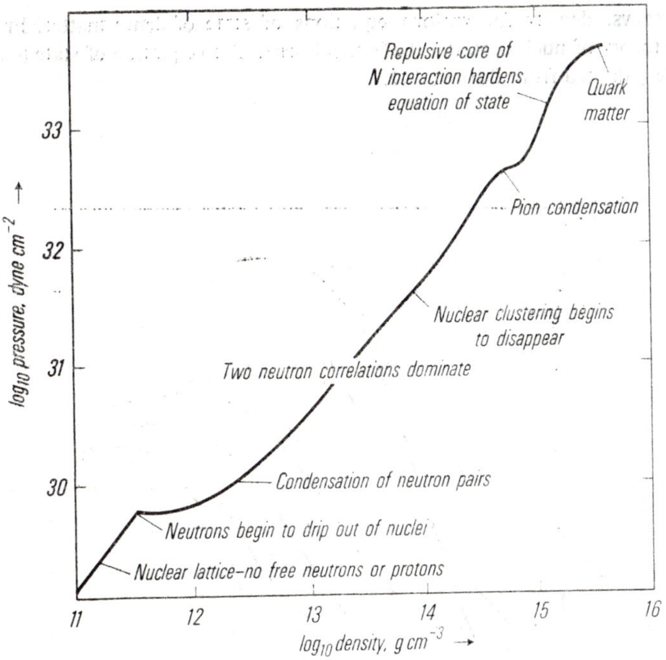

FIG. 6.9. Equation of state for cold matter at high densities (adapted from Irvine, 1978).

$m(0)$ assumed to be zero. The results obtained for a few different equations of state are presented in Figs. 6.11 and 6.12. The figures show the dependence of the gravitational mass of a neutron star on the central density and on the radius of the star for the same equations of state. Notice the large differences in the maximum mass of a neutron star for different equations of state. The maximum mass increases as we pass to more rigid equations of state,

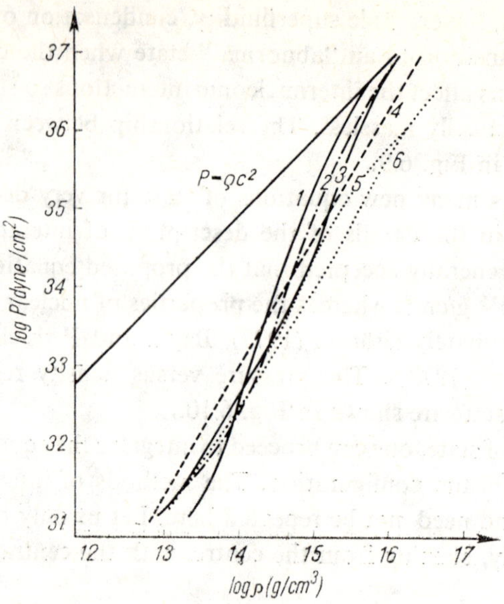

Fig. 6.10. Pressure vs. density for various equations of state of dense matter. From the viewpoint of the theory of nuclear matter the Bethe–Börner–Sato equation of state (curve 2) is the most reliable (adapted from Börner, 1973).

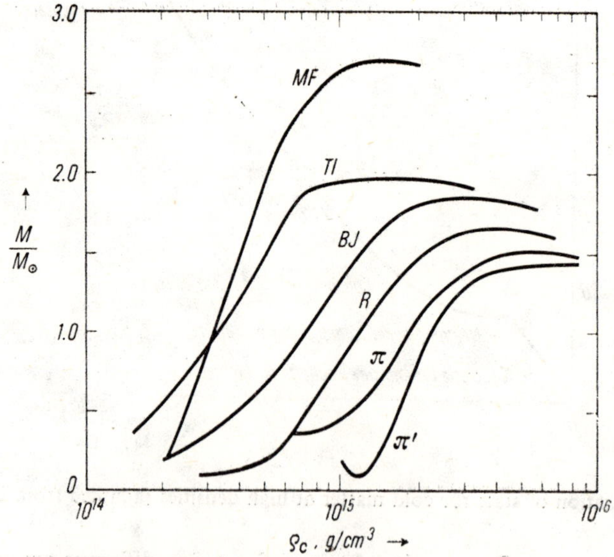

Fig. 6.11. Mass vs. central density for different equations of state. ***MF*** is the Pandharipande–Smith mean field theory model, ***TI*** is the Pandharipande–Smith tensor-interaction model, ***BJ*** is the Bethe–Johnson equation of state, ***R*** is the pure neutron equation of state with the Reid potential, π is the Reid equation of state modified by charged pion condensation; in π' the pion condensation is somewhat stronger (adapted from Baym and Pethick, 1979).

but it always remains less than $3M_\odot$, and for most equations it does not exceed $2M_\odot$. It seems therefore that neutron stars with masses greater than $(2–3)M_\odot$ cannot exist. The radius also varies considerably according to the adopted equations of state, taking greater values for more rigid equations. Detailed parameters of the neutron star models calculated on the basis of the equation of state proposed by Bethe, Börner and Sato are presented in Table 6.2. As the central density rises, the mass of the equilibrium configuration initially

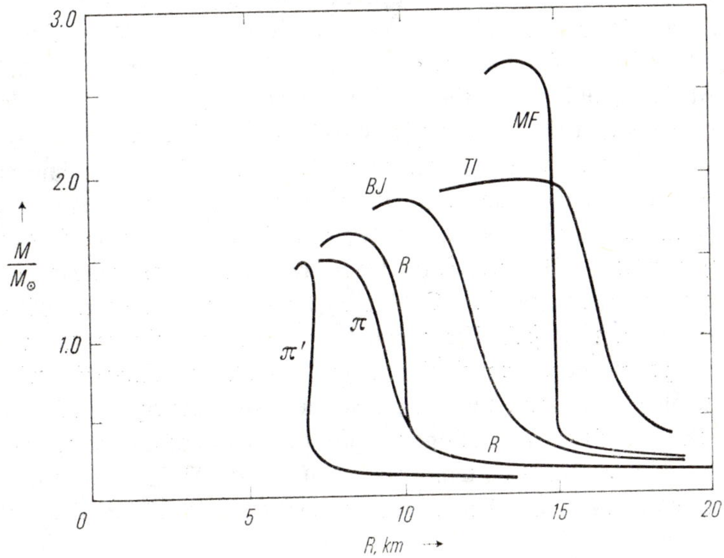

FIG. 6.12. Mass–radius relation for neutron stars with different equations of state (adapted from Baym and Pethick, 1979).

TABLE 6.2. *Parameters of neutron star models*

$\log \rho_c$ g/cm³	Mass/10^{33} g	Specific mass/10^{33} g	Radius km	Binding energy/10^{33} g
15.6	3.45	4.69	8.74	1.24
15.5	3.44	4.56	9.13	1.12
15.4	3.35	4.31	9.54	0.96
15.3	3.17	3.95	9.92	0.78
15.2	2.88	3.46	10.2	0.58
15.1	2.48	2.87	10.5	0.40
15.0	2.01	2.25	10.6	0.25
14.9	1.53	1.68	10.7	0.14
14.8	1.09	1.16	10.7	0.07
14.7	0.73	0.77	11.0	0.03
14.6	0.52	0.54	11.6	0.02
14.5	0.37	0.38	12.5	0.01
14.4	0.25	0.25	15.1	0.004
14.3	0.14	0.14	47.9	0.001
14.2	0.37	0.45	341	0.03
14.1	0.95	1.03	401	0.03
14.0	1.10	1.17	349	0.02

decreases and then rises monotonically to the maximum value of $1.73M_{\odot}$. The radius of a neutron star of maximum mass, a mere 8.74 km, is very small for an astronomical object. Such a star is very strongly bound gravitationally, the binding energy amounting to 26% of the total rest-mass energy of the constituents. Very dense matter is packed almost uniformly throughout the volume of the star and only near the surface does the density fall sharply. The density distribution in relation to radius is illustrated in Fig. 6.13.

Four regions differing in physical properties can be distinguished in a neutron star (Fig. 6.14). The star is surrounded by a very thin atmosphere (just a few metres high) composed mainly of iron. As the surface is approached, the density of the atmosphere rises rapidly. At the altitude where the density is 10^4 g/cm^3, atoms are completely ionized. Physically this means that the density of electrons is so high that the energy of electrons whose momentum is equal to the Fermi momentum exceeds the binding energy of the lowest-placed electron in the iron atom. Internuclear distances are still larger than the Debye radius so the Coulomb interaction between nuclei can be neglected. As we go deeper and the density increases to 10^7 g/cm^3, the Debye radius becomes comparable to the internuclear distances and the energy of the Coulomb interactions becomes much higher than the energy of thermal motion kT. The result is a phase transition: the nuclei form a lattice. This perhaps unexpected conclusion follows from the melting-point criterion, known from solid-state physics. A solid body begins to melt when the square of the amplitude of thermal oscillations is of the order of one tenth of the square of the mean distance between the nuclei. At realistic neutron-star temperatures of about 10^8 K, the iron nuclei will form a crystalline structure when the density reaches $7 \cdot 10^6$ g/cm^3. We shall call the sphere at

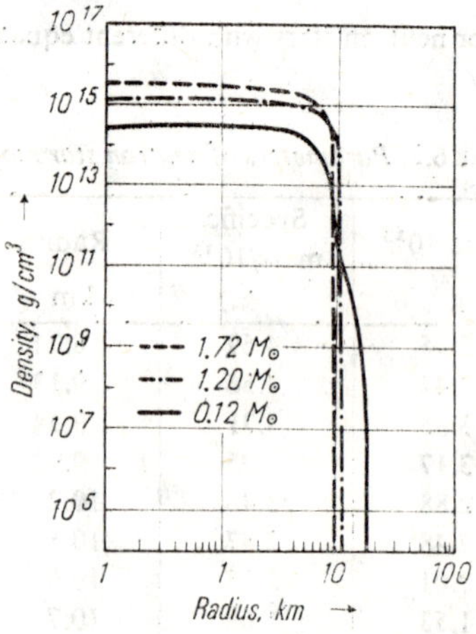

Fig. 6.13. Density of a nutron star as a function of radius for different models based on the Bethe–Börner–Sato equation of state (adapted from Börner, 1973).

which the density is $7 \cdot 10^6$ g/cm^3 the surface of the neutron star. The crystal crust can be divided into two regions: an outer and an inner one. As we move to higher and higher densities within the outer crust, nuclei of increasing atomic mass and neutron content become stable against beta decay. At the border between the two regions, at a density of $3 \cdot 10^{11}$ g/cm^3, free neutrons begin to appear. Their number increases very quickly, and accordingly the mean distance between the neutrons decreases, so the energy of their mutual interactions must be taken into account. The effective potential of neutron interaction is strongly repulsive at small distances but attractive at large ones. When the mean distance between neutrons is still much larger than the range of the repulsive forces, the attractive forces may produce a superfluid state.

The crystal crust encloses the core filled with free neutrons, protons and electrons. The distances between the particles in the core are so small that matter can be regarded as a mix-

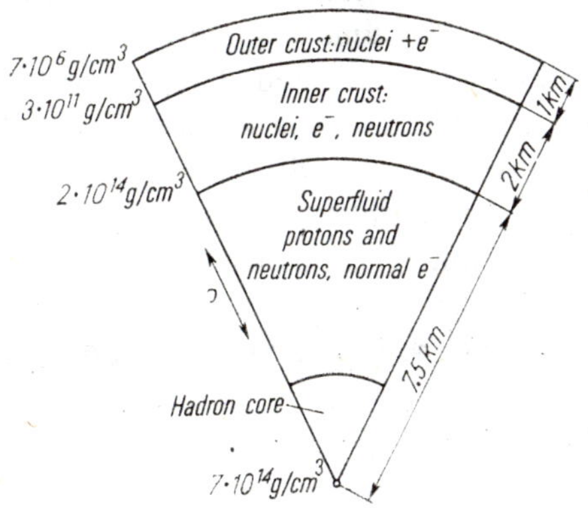

FIG. 6.14. Cross-section of a neutron star of mass less than the Solar mass (adapted from Pines, 1970).

ture of three degenerate quantum fluids. Neutrons remain in the superfluid state up to a density much higher than the nuclear density. The interactions between protons at this density range are qualitatively very similar to those between neutrons. Like neutrons, protons are in a superfluid state. Except for the extreme cases of very heavy and very light neutron stars, this region, filled with quantum fluids of neutrons, protons and electrons, is the main part of the star, and almost the entire mass of the star is concentrated in it. Very light neutron stars may have no such liquid part but only the crystal phase. Very massive stars, on the other hand, contain in addition an innermost core where besides neutrons, protons, electrons and heavier hadrons are also present.

Neutron stars are formed as a result of gravitational collapse of the central parts of stars undergoing supernova explosions. The temperature of such an object is very high; at the centre it may be as high as 10^{11}K. In the initial period the cooling of the remnant is mainly due to the emission of neutrino-antineutrino pairs. The ν, $\tilde{\nu}$ pairs can be produced in three processes: the decay of plasmons, the URCA process, in which a neutron breaks up into a proton, an electron and an antineutrino whereupon the electron and proton recombine to form a neutron and a neutrino, and bremsstrahlung. The neutrino processes dominate the energy transfer at high temperatures and cause a rapid cooling of the star. Condensation of pions, if it occurs, also accelerates the cooling. Further cooling is due to the emission of photons from the surface of the star. About 10^8 years after the formation of the star, its temperature will have fallen very low. This account of the thermal history of a neutron star is correct only if the star is left alone and if the accretion of matter onto the star can be neglected. The cooling will follow a different course if the star is surrounded by a cloud of interstellar dust or belongs to a binary system. We shall discuss this in more detail in the chapter dealing with the astrophysical significance of neutron stars and black holes.

One can hardly expect a neutron star emerging from a dramatic supernova explosion to be spherically symmetric and stationary immediately after its formation. More likely than not the explosion will excite all possible vibrational modes of the system. Nonradial oscillations were analysed by Thorne. He investigated the first few possible quadrupole excitations and found that they die out very rapidly owing to the emission of gravitational waves. The oscillation period of the lowest quadrupole excitations is a fraction of a microsecond, and their characteristic decay time is about one second. Periods of radial oscillations are in the microsecond range as well. Radial oscillations are also damped by dissipative processes. One of the possible damping mechanisms is the emission of neutrinos. The Fermi energy of neutrons, protons and electrons in such an oscillating star is also subject to oscillations about the equilibrium level. Owing to this, neutrons can break up into protons, electrons and antineutrinos while the star expands, and electrons and protons can recombine producing neutrons and antineutrinos while it contracts. The period of decay of radial oscillations in this process is estimated to be about 100 years.

TABLE 6.3. *Parameters of rotating neutron star models*

$\log \rho_c$	Mass/10^{33}	Crust mass/10^{31}	Moment of inertia/10^{45}	Dragging of inertial frames at the surface
g/cm³	g	g	g·cm²	Ω_s/ω
15.6	3.45	0.790	1.19	0.26
15.5	3.44	0.995	1.26	0.25
15.4	3.35	1.30	1.30	0.22
15.3	3.17	1.72	1.27	0.19
15.2	2.88	2.31	1.16	0.16
15.1	2.48	3.11	0.98	0.13
15.0	2.01	4.12	0.75	0.093
14.9	1.54	5.31	0.53	0.064
14.8	1.09	6.70	0.33	0.040
14.7	0.73	8.35	0.20	0.022
14.6	0.52	9.56	0.13	0.012
14.5	0.37	10.4	0.08	0.0065
14.4	0.25	10.8	0.05	0.0021
14.3	0.14	11.5	0.04	0.0054
14.2	0.37	36.5	76.2	0.0003
14.1	0.95	94.7	378	0.0009
14.0	1.10	110	396	0.001

Oscillations of other kinds may also occur. The crystal crust, for example, may be subject to torsional vibrations with a period of about 0.1 s.

To close this section, let us consider two more attributes of neutron stars, very important from the astrophysical viewpoint, namely their magnetic fields and rotation. When a star with a typical internal magnetic field (about 100 Gs) collapses gravitationally turning to a neutron star, the law of conservation of magnetic flux implies that the magnetic field to be expected inside the neutron star will be of order 10^{12} Gs. The material of the neutron star is a very good conductor and therefore the dissipation of the magnetic field is practically negligible. The characteristic period of decay of the magnetic field is much longer than the age of the universe.

Similarly, by applying the law of conservation of angular momentum, one can show that neutron stars should rotate. Some data concerning the moment of inertia of neutron stars and the rate of dragging of inertial frames at the surface, based on the Bethe–Börner–Sato equation of state, are collected in Table 6.3. For stars with very high central densities the relativistic effect of the inertial frame dragging may contribute one fourth of the angular velocity of the star. Just as the gravitational mass and the radius, the moment of inertia of a neutron star also depends in an essential way on the equation of state: the more rigid the equation of state, the greater the moment of inertia. The fact that the basic parameters of neutron stars depend on the equation of state leads us to expect that in a not too distant future, when more observational data become available, it will be possible to verify some of the assumptions adopted in the theory of nuclear matter concerning the interactions between nucleons at very high densities (Carter, Quintana, 1972). The maximum possible mass of a rotating neutron star is not known but, presumably, it should not differ from the maximum mass of a nonrotating neutron star by more than about 10% (Hartle, 1978).

Owing to their rotation and strong magnetic fields, neutron stars can interact very effectively with the surrounding matter.

6.3. Pulsars

General Properties. Observational Data

Like many recent astronomical discoveries, pulsars were first observed accidentally. A Cambridge group of radioastronomers investigated scintillations of radio waves emitted by distant radiosources. The scintillations were observed only when the angular dimensions of the source were small. The method permitted detection of quasars, whose dimensions come within that range. Scintillations can only be observed by means of an antenna suitable for reception of weak, rapidly-varying signals. A radiotelescope of this kind was constructed and put into operation in Cambridge by Hewish and his collaborators. In November 1967 Jocelyn Bell observed very peculiar signals. They were composed of short, periodic pulses. In February 1968, after a close examination of two more sources of a similar type, Hewish and his collaborators reported the discovery of a new kind of astronomical objects – pulsars.

So far pulsars have been the subject of several hundred papers, a few dozen review articles and a number of books. The state of our knowledge about pulsars as for 1977 was summed up in a book by Manchester and Taylor, and another one by Smith. The reminiscences of Jocelyn Bell and Anthony Hewish about the discovery of the first pulsar also make very interesting reading. More than 340 radio pulsars have been found since, and their basic properties can now be discussed in more detail. The main parameters of some 70 pulsars are set together in Table 6.4.

TABLE 6.4. *Characteristics of some of the known pulsars**)

PSR	*Right Ascension* (1950)	*Declination* (1950)	*Period* (s)	*Period Derivative* (10^{-15})
0011+47	00^h11^m	+47°29	1.2406987577	0.55
0100+65	01 00	+65 12	1.67916129	
0105+65	01 05	+65 52	1.2386529262	13.16
0136+57	01 36	+57 59	0.27244563408	10.6867
0138+59	01 38	+59 54	1.22294826723	0.3904
0148−06	01 48	−06 49	1.4646643399	0.445
0149−16	01 49	−16 52	0.83274112609	1.300
0154+61	01 54	+61 57	2.351653037	188.99
0320+39	03 20	+39 34	3.0320716577	0.71
0340+53	03 40	+53 11	1.934606	
0402+61	04 02	+61 30	0.59457139808	5.575
0447−12	04 47	−12 53	0.4380141010	0.103
0450+55	04 50	+55 38	0.34072820672	2.3626
0458+46	04 58	+46 49	0.6385489565	5.583
0523+11	05 23	+11 12	0.35443756681	0.071
0559−05	05 59	−05 27	0.39596855140	1.309
0621−04	06 21	−04 23	1.0390760459	0.846
0626+24	06 26	+24 17	0.47662189633	1.990
0647+80	06 47	+80 59	1.21443894	
0727−18	07 27	−18 30	0.51015078278	18.948
0751−32	07 51	+32 39	1.4423490281	1.074
0756−15	07 56	−15 19	0.68226433641	1.617
0906−17	09 06	−17 27	0.40162534295	0.6709
0919+06	09 19	+06 51	0.43061431165	13.7248
0942−13	09 42	−13 40	0.57026410323	0.0462
1112+50	11 12	+50 46	1.65643808033	2.4929
1319+83	13 19	+83 40	0.6700367	
1530+27	15 30	+27 55	1.1248353580	0.82
1540−06	15 40	−06 11	0.70906364986	0.8830
1612+07	16 12	+07 45	1.2068002120	2.357
1648−17	16 48	−17 04	0.97339208101	3.042
1718−02	17 18	−02 09	0.47771530393	0.087
1737+13	17 37	+13 13	0.80304971623	1.454
1738−08	17 38	−08 39	2.0430815106	2.27
1745−12	17 45	−12 59	0.39413270428	1.212

*) Based on Backus, Taylor, and Damashek (1982).

Radio pulsars are designated with the abbreviation PSR followed by a four-figure number denoting their right ascension and a two-figure number denoting their declination, for example PSR 1919+21 (the first pulsar discovered).

One of the most important features of the radiation emitted by pulsars is its periodicity. Other pulse parameters are highly variable. The amplitude and the shape change from

TABLE 6.4 (continued)

PSR	*Right Ascension* (1950)	*Declination* (1950)	*Period* (s)	*Period Derivative* (10^{-15})
1804−08	18 04	−08 48	0.16372736083	0.02868
1811+40	18 11	+40 12	0.9310877849	2.553
1821+05	18 21	+05 48	0.75290644846	0.225
1831−04	18 31	−04 29	0.29010815629	0.197
1834−10	19 34	−10 10	0.5627056550	11.775
1839+09	18 39	+09 09	0.38131888123	1.0916
1842+14	18 42	+14 51	0.37546250760	1.866
1851−14	18 51	−14 25	1.1465926813	4.171
1905+39	19 05	+39 57	1.2357572279	0.53
1907+03	19 07	+03 53	2.330260471	4.53
1907−03	19 07	−03 14	0.50460347381	2.189
1923+04	19 23	+04 25	1.0740771471	2.465
1924+16	19 24	+16 42	0.5798116728	18.003
1940−12	19 40	−12 44	0.97242816020	1.659
1941−17	19 41	−17 57	0.8411572623	0.980
1953+50	19 53	+50 51	0.51893741261	1.366
2003−08	20 03	−08 15	0.5808713142	0.040
2028+22	20 28	+22 18	0.6305121827	0.886
2043−04	20 43	−04 32	1.54693750023	1.476
2044+15	20 44	+15 29	1.1382856067	0.185
2053+36	20 53	+36 18	0.22150803351	0.3648
2106+44	21 06	+44 29	0.41487049185	0.0863
2113+14	21 13	+14 01	0.44015295481	0.290
2148+63	21 48	+63 15	0.38014034472	0.1681
2224+65	22 24	+65 20	0.68253370807	9.671
2228+61	22 28	+61 54	0.443289059	
2255+58	22 55	+58 53	0.36824365392	5.7501
2306+55	23 06	+55 31	0.47506759150	0.202
2310+42	23 10	+42 36	0.34943363975	0.1155
2315+21	23 15	+21 33	1.4446526747	1.05
2323+63	23 23	+63 00	1.4363084791	2.89
2351+61	23 51	+61 39	0.9447768664	16.226
1913+16	19 13	+16 01	0.05902999	0.00864
1937+21	19 37	+21 28	0.00155780	0.000126

The first column contains the name of the pulsar, the second and the third its position referred to 1950, the fourth gives the pulse period in seconds and the fifth the time-derivative of the period.

one pulse to another as well as over longer periods of time, reaching hours and days. However, when we examine the pulse shape averaged over a few minutes, it turns out that this average pulse is stable and characteristic for a given pulsar. A few examples are shown in Fig. 6.15. It has been noticed that the average pulse shape of some pulsars (e.g. PSR 0329+54 and PSR 1237+25) changes abruptly from time to time, with the new shape maintained for several tens or hundreds of pulses and than returning to the original form in an equally abrupt manner. The duration of the average pulse is approximately proportional to the pulsar period. With respect to the average pulse shape, pulsars can be divided into two classes.

Pulsars of type *S* have simple average pulse shapes, always with a single, clearly marked maximum. Their periods are usually below one second. Pulses are strongly linearly polarized with the polarization angle changing continuously along the pulse profile. The average pulse shape for pulsars of type *C* is more complex, with two or more comparable maxima. Periods of pulsars of type *C* are usually shorter than those of pulsars of type *S*. Their pulses are weakly polarized and the polarization angle may vary at random along the pulse profile.

A high time-resolution analysis of pulsar radiation has revealed that every pulse contains one or more subpulses which occur at different moments of the pulse duration. For most pulsars the moments at which the subpulses occur in successive pulses are uncorrelated. In a few cases, however, the subpulses were observed to shift slowly relative to the average pulse, appearing earlier in every next pulse and, on reaching the left end of the pulse profile, reappearing at the right end, whereupon the whole process is repeated. Pulsars whose pulses exhibit such regular shifts of subpulses are referred to as type *D*, and the periodicity of these changes is called periodicity of the second kind. The second-kind

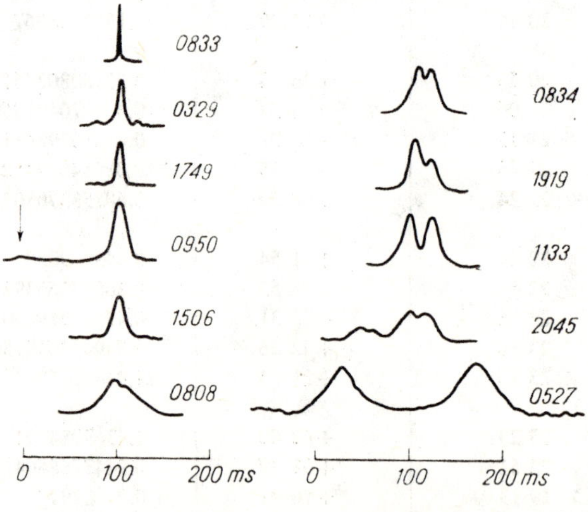

FIG. 6.15. Characteristic shapes of averaged pulsar pulses (adapted from Hewish, 1970).

period will be denoted by P_2. Data for the pulsar PSR 0809+74 are presented in Fig. 6.16.

The periods of known pulsars range from 0.0015 s to about 4.3 s, most of them falling between 0.5 and 1 s (Fig. 6.17). As we have mentioned, the period of a pulsar is extremely stable. More precise measurements of the fundamental period *P* have shown, however, that it increases, very slowly but systematically. For all pulsars for which it was measured, the first time-derivative of the period is positive and usually of order $10^{-15}\,\mathrm{s\cdot s^{-1}}$, though in a few cases it is much lower, for example for PSR 1913+16 $P=8.6\cdot10^{-18}\,\mathrm{s\cdot s^{-1}}$ and for PSR 1952+29 $\dot{P}=1.9\cdot10^{-18}\,\mathrm{s\cdot s^{-1}}$. In models of pulsar radiation mechanisms, the

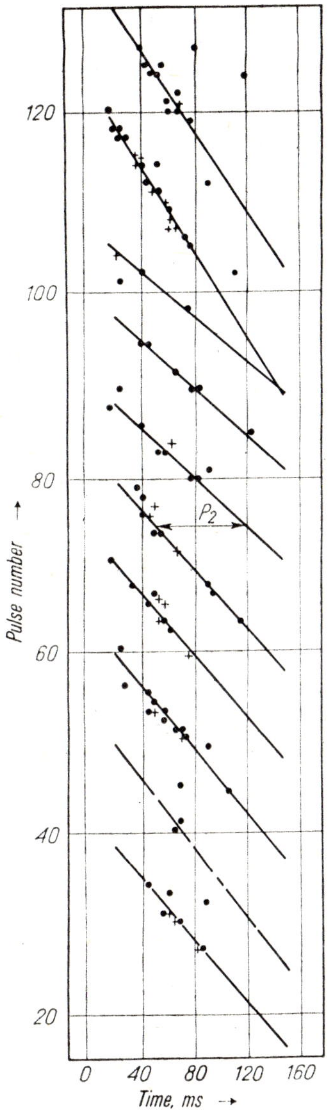

FIG. 6.16. The second-kind period P_2 for the pulsar PSR 0809+74 (adapted from Cole, 1970).

slow increase of the period is often described by the formula

$$\dot{P}=(2\pi)^{n-1}KP^{2-n}, \tag{6.4}$$

where K is a positive constant and

$$n=1-P\ddot{P}/\dot{P}^2 \tag{6.5}$$

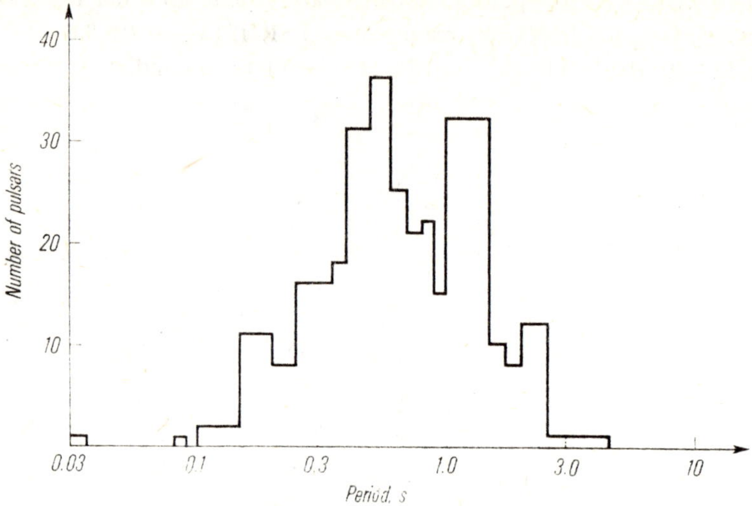

FIG. 6.17. Period distribution of pulsars.

is the so-called breaking index. In most models $n \approx 3$. In the last few years the second derivative of the period was determined for a few dozen pulsars. The results show that it can be positive as well as negative, and that in some cases the absolute value of the breaking index is much larger than 3.

Analysis of the frequency-dependence of pulsars established the following facts. Firstly, the pulses are dispersed, and the lower the frequency the later they reach us. Secondly,

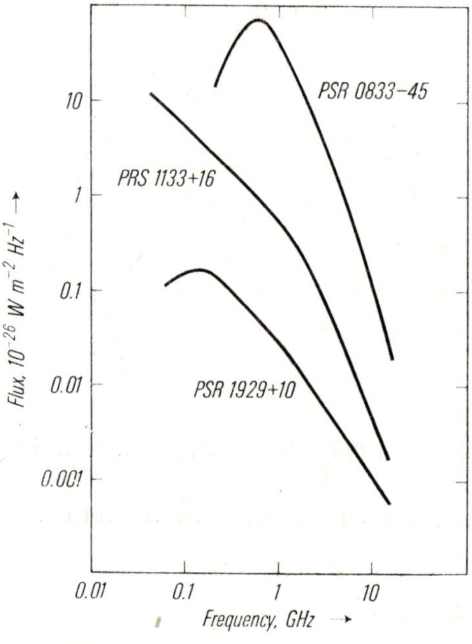

FIG. 6.18. Typical energy spectra of pulsars (adapted from Smith, 1977).

as the frequency falls, the pulses become increasingly blurred, to the extent that at frequencies below 38MHz separate pulses can no longer be distinguished. Finally, the average pulse shape changes with frequency.

For a few pulsars it was found possible to determine the energy spectrum. It varies considerably from pulsar to pulsar (see Fig. 6.18), showing however one common feature. At frequencies higher than 1GHz there always occurs a sharp drop in energy.

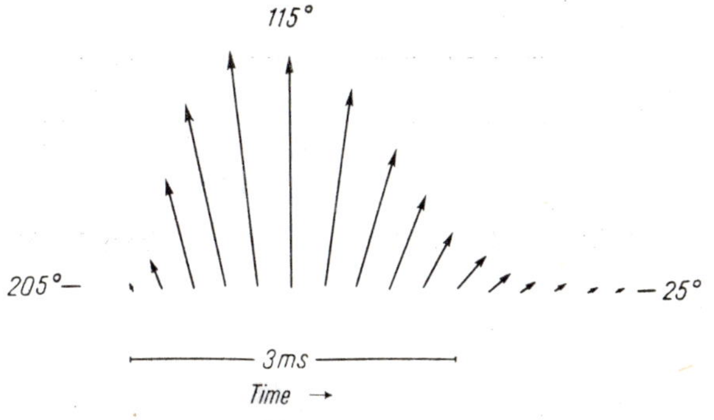

FIG. 6.19. Changes of the polarization plane in the PSR 0833-45 radiation (adapted from Radakrishnan *et al.*, 1969).

A further insight into the mechanism of pulsar radiation can be gained by studying polarization of pulses. The pulsar pulses are strongly polarized, individual pulses can be hundred per cent elliptically polarized. Linear polarization dominates, but for some pulsars the degree of circular polarization reaches 20%. The plane of linear polarization changes along the pulse profile. Changes of the polarization plane for the pulsar PSR 0833-45 are shown in Fig. 6.19. In the radiation of pulsars of type *S* the polarization angle can undergo abrupt changes. Subpulses are also polarized very strongly, the character and degree of their polarization varying from one subpulse to another as well as during the single subpulse. Polarization of pulses changes with time, but in contrast with what is observed for ordinary radiosources its degree usually decreases with rising frequency.

The best-studied pulsar is PSR 0532+21 in the Crab Nebula. It was observed for the first time in 1968 by Staelin and Reifenstein, and a good deal of information about it has accumulated since. The period of the Crab pulsar is the second shortest observed.† Both its first and its second derivative have been measured. If midnight 15th November 1968 is taken as the starting point for the time count then

$$P=[33091121.05+(36.47\pm0.32)\tau-(0.60\pm0.63)\cdot10^{-4}\tau^2]\pm2.09 \text{ ns}; \qquad (6.6)$$

if time t is measured in seconds, $\tau=t/86400$. Stability of this period is comparable to that of atomic clocks. If we define the characteristic time as $\tau_0=P\left(\frac{dP}{dt}\right)^{-1}$, for PSR 0532+21 we get $\tau_0=2500$ years. Three sudden changes in the period of PSR 0532+21 occurred since 1968, and each time $\Delta P/P\sim3\cdot10^{-9}$. Unfortunately the pulsar was not observed

† In 1982 a fast pulsar PSR 1937+21 was discovered with a period $P=0.001558$ s.

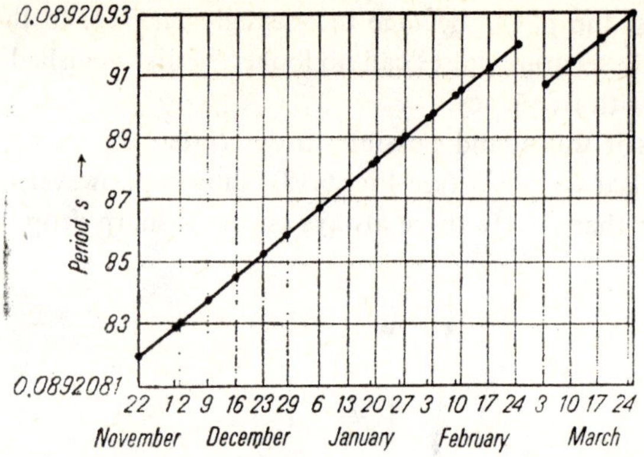

FIG. 6.20. Sudden change in the period of PSR 0833-45 (adapted from Reichley and Downs, 1969)

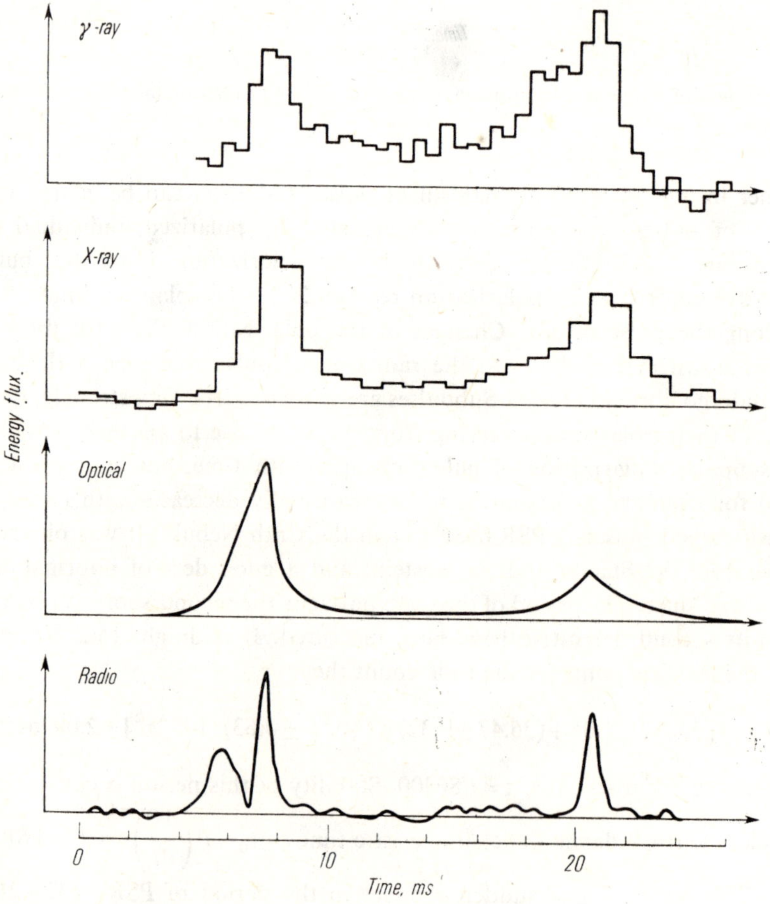

FIG. 6.21. Integrated pulse profile of the Crab pulsar at radio, optical, X-ray and γ-ray frequencies (adapted from Manchester and Taylor, 1977).

during those changes, and no details of the changes themselves are known. It is thought they were of a few hour duration. Similar changes were also reported three times for the pulsar PSR 0833-45 (Fig. 6.20).

So far PSR 0532+21 is the only pulsar whose pulses were observed in the radio, optical, infrared, X-ray and gamma regions of the spectrum. The average pulse shape in these regions is shown in Fig. 6.21 and the radiation spectrum in Fig. 6.22. In the optical spec-

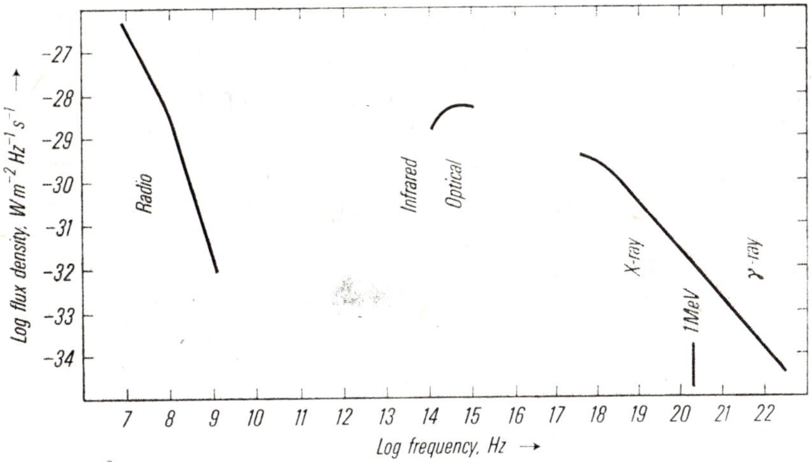

FIG. 6.22. Energy spectrum of the Crab pulsar (adapted from Smith, 1977).

trum the radiation of PSR 0532+21 exhibits the main pulse and an interpulse which lags 13.37 ms behind the main pulse. In the radio spectrum there is an additional pulse leading the main pulse by 1.64 ms. In both regions the main pulse is very sharp. In the radio range its half-length is less than 300 μs; in the optical range the main pulse reveals no structure even if observed with a time resolution of 32 μs. It follows that the active radiation region has very small dimensions, less than 10 km. Now if we take into account the fact that the luminosity of the pulsar in the radio is at least 10^{25} $erg \cdot s^{-1}$, we must conclude that for such an energy flux to be emitted by so small a region the effective temperature of the latter must be of order 10^{23} K. This means that the radio pulses of the pulsar in question can only be produced by coherent radiation of a multi-particle system. This information is of great importance for the construction of models of the pulsar radiation mechanism.

Another special feature of the Crab Nebula pulsar is that once in every 10^3 or so pulses a single pulse appears whose energy is about 10 times higher than the average pulse energy. There also occur, though much more rarely, pulses of energy 1000 times higher than the average.

At the end of 1974 Hulse and Taylor discovered an extraordinary pulsar (PSR 1913 +16). Its period is approximately 59 ms, it means the third shortest known. The period of PSR 1913+16 undergoes pericdic changes. The observed rate of those changes permits the conclusion that the pulsar belongs to a binary system with an eccentric orbit and an orbital period of 27906.9 s (~7h45′). It is therefore a close binary. Fig. 6.23 shows changes in the radial velocity of PSR 1913+16. It is the first radio pulsar observed to occur in a binary system. The system probably consists of two compact stars, which may well be neutron stars, close to each other, and therefore is an excellent object with which to verify various effects predicted by the general theory of relativity. The subject was treated by many authors (Blandford and Teukolsky, Will, Wagoner, Eardley, Brumberg *et al.*, Smarr and Blandford, Epstein, and others), who showed that precise observation of PSR 1913+16 combined with general-relativistic analysis would make it possible to determine many important parameters of this system.

Indeed, very precise observations concluded shortly afterwards by Taylor and his collaborators permitted measurement of some relativistic effects beyond the post-Newtonian approximation. So far the following relativistic parameters of the system in question are known: perihelion motion (periastron motion), transverse Doppler effect, the inclination of the orbit to the plane of the sky and the rate of change of the orbital period. These and other parameters of the system which can be determined from observation are listed in Table 6.5. If the general theory of relativity is a correct theory of gravitation and if

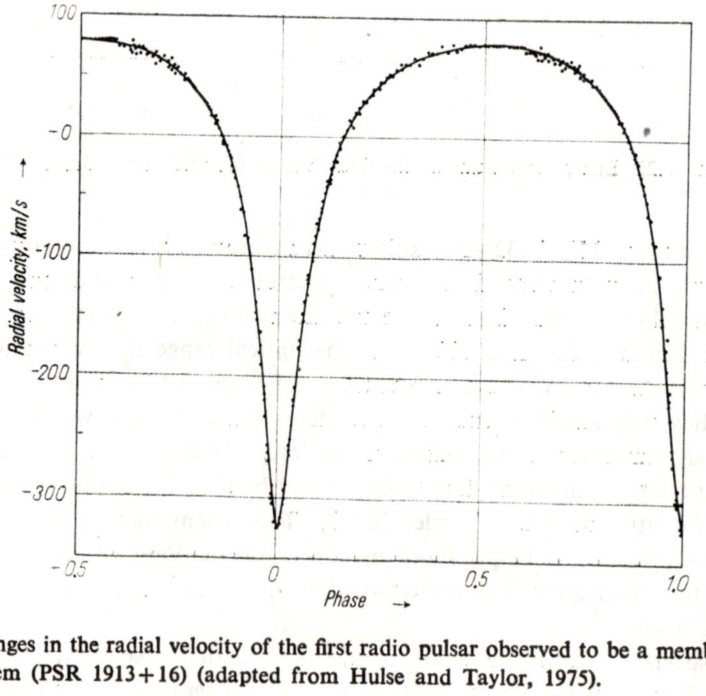

FIG. 6.23. Changes in the radial velocity of the first radio pulsar observed to be a member of a binary system (PSR 1913+16) (adapted from Hulse and Taylor, 1975).

the periastron motion is caused by relativistic effects only, one can calculate the masses of the two components. It turns out that the mass of the pulsar is $M_p = 1.39 \pm 0.15\ M_\odot$ and that of its companion is $M_c = 1.44 \pm 0.15\ M_\odot$. Finally the most important information: knowing the masses of both components, we can test the general theory of relativity by using it to calculate the changes in the orbital period due to the emission of gravitational

TABLE 6.5. *Parameters of the binary pulsar* PSR 1913+16 *derived from timing data*

Right ascension (1950.0)	$\alpha = 19h\ 13\ min\ 12.474\ s \pm 0.004\ s$
Declination (1950.0)	$\delta = 16°\ 01'08''.02 \pm 0''.06$
Period	$P = 0.059029995269 \pm 2\ s$
Derivative of period	$\dot{P} = (8.64 \pm 0.02)\ 10^{-18}\ s \cdot s^{-1}$
Projected semimajor axis	$a_1 \sin i = 2.3424 \pm 0.0007$ light s
Orbital eccentricity	$e = 0.617155 \pm 0.000007$
Binary orbit period	$P_b = 27906.98172 \pm 0.00005\ s$
Longitude of periastron	$\omega_0 = 178.864 \pm 0.002°$
Time of periastron passage	$T_0 =$ JD $2442321.433206 \pm 0.000001$
Rate of advance of periastron	$\dot{\omega} = 4.226 \pm 0.002\ deg\ yr^{-1}$
Transverse Doppler and gravitational redshift	$\gamma = 0.0047 \pm 0.0007\ s$
Sine of inclination angle	$\sin i = 0.81 \pm 0.16$
Derivative of orbital period	$\dot{P}_b = (-3.2 \pm 0.6)\ 10^{-12}\ s \cdot s^{-1}$

waves. The observed value agrees with the predictions of general relativity to a factor of 1.3 ± 0.3. Thus far, this is the only, indirect, evidence for the existence of gravitational waves. The implications of this result and the controversies it raises will be discussed in more detail in Chapter 9.

In 1979 Crane, Nelson and Tyson reported a possible optical identification of the companion of the pulsar PSR 1913+16. They discovered a very weak star ($m_R = 20.9$),

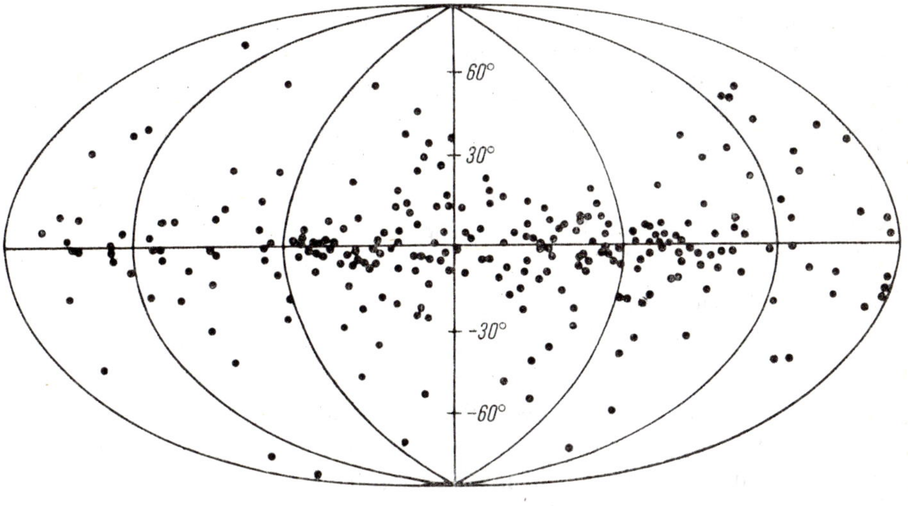

FIG. 6.24. Spatial distribution of pulsars in the galactic coordinates.

most likely a helium star, whose position exactly coincides with that of the pulsar. The probability that a star unconnected with the pulsar happens to lie exactly on the pulsar's line of sight is small, but such a case is not impossible. The problem can only be solved by further, more precise radio and optical observations.

As mentioned before, more than 340 pulsars are known at the present time. We are thus in a position to draw some general conclusions concerning their distribution in space (Fig. 6.24). The basic parameters are the position on the celestial sphere and the distance.

Positions are determined by radioastronomical methods. Distances can be estimated by measuring signal delays in relation to frequency. Passing through clouds of interstellar plasma, radio signals undergo dispersion. If a source emits a short pulse, which is a wave packet, then the rate of frequency change in passing through a cloud whose plasma frequency is much lower than the frequency of the radio signal is given by

$$\dot{\nu}=\frac{1{\cdot}23\cdot 10^{-4}\nu^3}{\int n_e dl}\ \mathrm{MHz\,s^{-1}}. \tag{6.7}$$

The number of electrons that the signal meets on its way is equal to $\int n_e dl$ and is called the dispersion measure. The concentration of electrons in the galaxy in a given direction can be calculated from the measurement of the intensity of the 21 cm lines. Thus, knowing the dispersion measure, we can estimate the distance of the source. It turns out that the known pulsars are concentrated near the galactic plane and are markedly more numerous in the neighbourhood of the Sun. This may be a selective effect connected with the greater difficulty in observing more distant pulsars, which appear as weaker sources. Based on the number of pulsars within 1 kpc of the Sun, the spatial density of pulsars is calculated to be about 50 pulsars per 1 kpc^3. It follows that our galaxy should contain from 10^4 to 10^5 pulsars.

Pulsar Clock Mechanism

The first data to be used in interpreting the clock mechanism of pulsars is provided by signal analysis. The pulses which a pulsar signal is composed of last from a few dozen to several hundred milliseconds. The source of such pulses can only be an object whose dimensions are small by astronomical standards. The extraordinary stability of the periods of pulsars suggests that the *periodicity of pulses should be attributed to the mechanical properties of the source.* Binaries can at once be excluded because they do not ensure the observed degree of stability of the period. The only possibilities that remain are rotation or oscillation of a small object—a white dwarf or a neutron star. It is known that the maximum angular velocity of a star with an average density ρ is of order $\sqrt{G\rho}$. For a white dwarf with an average density $\rho\approx 10^6\ \mathrm{gcm^{-3}}$, $\omega_{cr}=\sqrt{G\rho}\approx 0.25\ \mathrm{sec^{-1}}$, while for the Crab Nebula pulsar, for example, $\omega=190\ \mathrm{s^{-1}}$. Thorne's analysis of oscillations of white dwarfs and neutron stars showed that the period of fundamental oscillations of white dwarfs cannot be shorter than 2 s, even allowing for relativistic corrections, whereas for neutron stars it is of the order of milliseconds. Thus neither rotation nor oscillation of a white dwarf can account for the observed periodocity of signals emitted by pulsars. And so, quite unexpectedly, the only candidate left is the neutron star. The average density of a neutron star is comparable to nuclear density; putting $\rho=10^{14}\ \mathrm{g/cm^3}$, we obtain the critical angular velocity $\omega_{cr}=2500\ \mathrm{s^{-1}}$. The period of the fundamental radial oscillations of a neutron star is very short. Nonradial oscillations are damped very quickly. We thus come to the conclusion that *only a rotating neutron star can be a pulsar.* In order to explain the observed radio pulses a plausible assumption was that the star is endowed with a strong magnetic field. The intensity of the magnetic field to be expected at the surface of a neutron star can be estimated from conservation of magnetic flux, $HR^2=\mathrm{const}$.

Neutron stars arise from gravitational collapse of stars whose initial masses are greater than about 6 $M_\odot$ and whose radii shrink in the process by a factor of 10^5. If we assume that the intensity of the magnetic field at the surface of such a star is 1 gauss, the magnetic field at the surface of the resulting neutron star will be very strong, with intensity of about 10^{10} gauss.

Further arguments indicating that pulsars are rotating neutron stars are provided by energy considerations. If I is the moment of inertia of a pulsar, the energy of its rotational motion is $E_{rot}=\frac{2\pi^2}{P^2}I$. The period of the pulsar increases gradually, so the rotational energy decreases. The rate of change of the rotational energy is

$$\frac{dE_{rot}}{dt}=-\frac{4\pi^2}{P^3}I\frac{dP}{dt}. \tag{6.8}$$

The moment of inertia of a typical neutron star is of order 10^{44} g·cm². For the Crab pulsar we find $dE_{rot}/dt\approx 5\cdot 10^{37}$ erg·s^{-1}. The observed energy flux emitted by the Crab Nebula is $4\cdot 10^{37}$ erg·s^{-1} in the X-ray range, and $3\cdot 10^{36}$ erg·s^{-1} in the optical range. The rotational energy of the pulsar is probably transmitted to the nebula through the pulsar's magnetic field. This, incidentally, would solve a problem that has long disturbed radioastronomers: where does the Crab Nebula find the enormous amounts of energy it radiates continually?

The rotating neutron star model of pulsars makes it also possible to explain the sudden changes observed in the periods of pulsars PSR 0532+21 and PSR 0833-45. Dyson and Ruderman have advanced the hypothesis that the abrupt changes in period are caused by the cracking and displacement of the crust of the star.

The general picture of this process would be as follows. As the angular velocity decreases, the free surface of the superfluid core changes. The weight of the crust, which so far has been partly balanced by the fluid pressure, is now supported by internal stresses. The stresses grow, and when they exceed a critical value, the crust cracks and assumes a new shape. In the case of the pulsar PSR 0833-45 the crust would be displaced about 1 cm, and in the case of the Crab Nebula pulsar merely 10^{-2} cm. Baym and Pines have calculated the time interval between two successive "pulsar-quakes". One of the parameters involved is the pulsar mass; the larger it is, the longer is the time interval between the successive quakes. The method can be used to estimate the mass of the pulsar; the value obtained for PSR 0532+21 is 0.3 $M_\odot$.

The discovery of pulsars became an indirect proof of the existence of neutron stars. Baade and Zwicky suspected as early as 1932 that the central star in the Crab Nebula is a neutron star. Extensive observations of the nebula were subsequently conducted by Baade and Minkowski. It had long been known that the Crab Nebula is an expanding gas cloud formed as a result of a supernova explosion in 1054. The central star is a peculiar one. Its spectrum contains no lines, either emission lines or absorption lines. Back in the 1930's it was difficult to forsee that significant new information would be provided by an analysis of the rapid changes in the intensity of the optical radiation of the star. But the fact is hat pulsars could have been discovered by optical methods as far as forty years ago.

Models of the Pulsar Radiation Mechanism

Before we go on to discuss models for the radiation mechanism of radio pulsars, let us recall the most important facts to be taken into account when attempting to construct such models. A pulsar is a rotating neutron star endowed with a strong magnetic field. If we assume that this magnetic field is a dipole field, with the dipole moment inclined to the axis of rotation at an angle β, the system will emit dipole magnetic radiation. It is known from observation that pulsars slow down their rotation ($\dot{P}>0$). If the energy loss due to the emission of electromagnetic waves is taken as responsible for this slowing down then

$$\frac{dE_{rot}}{dt}=-\frac{2}{3}\frac{1}{c^3}\left(\frac{2\pi}{P}\right)^4 d_{\perp}^2, \tag{6.9}$$

where $d_{\perp}=\frac{B_0 R^3}{2}\sin\beta$, B_0 is the surface intensity of the magnetic field, and R is the star's radius. Now using (6.8) we find B_0 to be given by

$$B_0=\left(\frac{Ic^3}{4\pi^2\alpha R^6}\right)^{\frac{1}{2}}(P\dot{P})^{\frac{1}{2}}, \tag{6.10}$$

where $\alpha=\frac{1}{6}\sin^2\beta$. P and $\dot{P}$ can be measured directly, and the other quantities occurring in (6.10) can be taken from neutron star models. The accepted standard values for the radius and the moment of inertia of a neutron star with a mass close to the solar mass are $R\approx 10^6$ cm and $I\approx 10^{45}$ gcm^2. Assuming furthermore that β is small, we obtain the following limits for B_0: $2\cdot 10^{10}$ G$\leqslant B_0\leqslant 2\cdot 10^{13}$ G. Recently, radioastronomers have succeeded in carrying out a direct measurement of the magnetic field at the surface of the neutron star in Her X-1 (X-ray pulsar) and it has been found that $B_0=6\cdot 10^{12}$ G. Thus, it may be safely assumed that the magnetic fields at the surface of pulsars are very strong.

Next, there is the question of whether a neutron star can have an atmosphere. With the aid of the barometric formula

$$\rho(z)=\rho_0\, e^{-\frac{mgz}{kT}}, \tag{6.11}$$

where g is the gravitational acceleration at the surface of the star, z is the distance from the surface, and m the mass of the particles forming the atmosphere, we can characterize the thickness of the atmosphere by determining the altitude H at which the density is e times lower than at the surface. We have

$$H=\frac{kT}{mg}. \tag{6.12}$$

Substituting the values typical for neutron stars, $T\approx 10^6$K, $g\approx 10^{14}$ cm s^{-2}, and taking as m the mass of a hydrogen molecule, $m=3\cdot 10^{-24}$ g, we obtain $H\approx 1$ cm. Now if we suppose that the density of the atmosphere at the surface is $7\cdot 10^6$ g cm^{-3}, then at the altitude of 50 cm it will be only 10^{-15} g cm^{-3}. The processes occurring in the atmosphere can therefore hardly be considered responsible for pulsar radiation. It should be borne in mind, however, that neither rotation nor the presence of a strong magnetic field have been taken into account in the above reasoning.

Goldreich and Julian have shown that a rotating neutron star endowed with a strong magnetic field must have a magnetosphere. If we accept that the star is a very good electric conductor, then every free charge e in the star is subject to the force $e\left(\frac{\boldsymbol{\Omega}\times\boldsymbol{r}}{c}\right)\times\boldsymbol{B}$, where $\boldsymbol{\Omega}$ is the angular velocity vector of the star. Under this force, the charges are displaced, which gives rise to an electric field. An equilibrium will be established when

$$\boldsymbol{E}=-\left(\frac{\boldsymbol{\Omega}\times\boldsymbol{r}}{c}\right)\times\boldsymbol{B}, \tag{6.13}$$

i.e. when the total force acting on the charges vanishes. If the star is surrounded by vacuum, we can use the Maxwell equations to calculate the electric and the magnetic fields outside the star. We shall assume that the external magnetic field is a dipole field and that the potential of the external electric field satisfies Laplace's equation. At the surface of the star the tangential component of the electric field and the normal component of the magnetic field should be continuous. As is well known, a surface discontinuity of the normal component of the electric field is connected with the surface charge layer, and that of the tangential component of the magnetic field with the surface current. We shall assume that the magnetic field is continuous at the surface of the star, so that there is no surface current. The potential of the electric field outside the star written in the spherical coordinates takes the form

$$\varphi=-\frac{B_0\Omega R^5}{3cr^3}P_2(\cos\theta), \tag{6.14}$$

where $P_2(\cos\theta)$ is the Legendre polynomial of second degree. From the jump of the normal component of the electric field at the surface we can find the surface charge density

$$\sigma=-\frac{B_0\Omega R}{4\pi c}\cos^2\theta. \tag{6.15}$$

It can be seen that charges will be concentrated about the poles.

Inside the star the electric and the magnetic fields are such that $\boldsymbol{E}\cdot\boldsymbol{B}=0$. Outside the star, if the star is in vacuum and has a dipole magnetic field, we get

$$\boldsymbol{E}\cdot\boldsymbol{B}=-\left(\frac{\omega R}{c}\right)\left(\frac{R}{r}\right)^7 B_0^2\cos^3\theta\neq 0. \tag{6.16}$$

Across a thin layer at the surface of the star $\boldsymbol{E}\cdot\boldsymbol{B}$ must therefore change continuously from zero to the value it takes outside. At the outer boundary of this layer the electric force in the direction of the magnetic-field lines will exceed the gravitational force in this direction by $5\cdot 10^8 B_{12}R_6^3\frac{\cos^2\theta}{PM}$ for protons and by $8\cdot 10^{11}B_{12}R_6^3\frac{\cos^3\theta}{PM}$ for electrons, where $R_6=R\cdot 10^{-6}$, M is the mass of the star in the Solar mass units, P the period, and $B_{12}=B\cdot 10^{-12}$. It follows that the surface charge layer cannot be in a state of dynamic equilibrium.

Charges will be pulled out from the surface of the star. *A rotating neutron star endowed with a strong dipole magnetic field cannot be surrounded by vacuum.* The plasma surrounding the pulsar will be carried by the magnetic field, forming a rotating magnetosphere. The distribution of charges in a stationary magnetosphere must be such that $\boldsymbol{E}\cdot\boldsymbol{B}=0$.

The structure of magnetospheres of pulsars is not yet exactly known, and the existing models are not fully satisfactory (Sturrock, Julian, Scharleman and Wagoner, Ruderman and Sutherland, Michel and others). Certain general properties of the magnetosphere follow from Ferraro's theorem, which states that in the magnetosphere of a rotating star with an axisymmetric magnetic field the regions where $\boldsymbol{E}\cdot\boldsymbol{B}=0$ rotate with the star at an angular velocity Ω^* which is constant along the magnetic-field lines. In other words, in the cylindrical coordinates ρ, z, φ, the angular velocity Ω^*, which may depend on ρ and z, satisfies the equation

$$(\boldsymbol{B}\cdot\nabla)\Omega^*=0=B_\rho\frac{\partial\Omega^*}{\partial\rho}+B_z\frac{\partial\Omega^*}{\partial z}. \tag{6.17}$$

Thus in the regions where $\boldsymbol{E}\cdot\boldsymbol{B}=0$ we have

$$\boldsymbol{E}=-\left(\frac{\boldsymbol{\Omega}^*\times\boldsymbol{r}}{c}\right)\times\boldsymbol{B}. \tag{6.18}$$

If $\boldsymbol{E}\cdot\boldsymbol{B}=0$ along the magnetic-field lines connecting the magnetosphere with the star, then Ω^* is equal to the angular velocity of the star. Relation (6.18) allows us to determine the charge density ρ_e in the magnetosphere, namely

$$\rho_e=\frac{\nabla\cdot\boldsymbol{E}}{4\pi}=-\frac{\boldsymbol{\Omega}^*}{2\pi c}(\boldsymbol{B}-\tfrac{1}{2}\boldsymbol{r}\times(\nabla\times\boldsymbol{B})); \tag{6.19}$$

near the star surface, where $\boldsymbol{\Omega}^*$ can be assumed equal to $\boldsymbol{\Omega}$,

$$\rho_e=-\frac{\boldsymbol{\Omega}\cdot\boldsymbol{B}}{2\pi c}. \tag{6.20}$$

This important result shows that the magnetosphere divides into three zones. In the simplest case where Ω is parallel to $\boldsymbol{B}$ (Fig. 6.25) the two polar zones are filled with negative charges and the equatorial zone with the positive ones. Plasma in the magnetosphere is thus charge-separated, a phenomenon which may occur only because the star rotates and has a huge magnetic field. The density of plasma in the magnetosphere is comparatively high, the density of electrons near the poles is $n\approx 10^{11}\frac{B_{12}}{P}$ cm^{-3} $(B=10^{12}B_{12})$. Also, it is a relativistic plasma, because the energy density of the magnetic field is much higher than the rest-mass energy density of the plasma. In contrast to situations typical for a normal nonrelativistic plasma, the current flowing in the magnetosphere of a pulsar is of a convective nature, and electric forces are comparable to magnetic forces. Near the surface of the star, in spite of the presence of the magnetosphere, the magnetic field has practically the same properties as in a vacuum. As we go away from the surface, the effect of the magnetosphere on the shape of the magnetic field becomes more and more marked. At a certain distance from the rotation axis, in order to corotate with the star, the plasma has to move with the speed of light. The cylinder on which $R_c\Omega = c$ is called the *light cylinder*. Outside the light cylinder the plasma can no longer rotate rigidly with the star. Within the light cylinder the magnetosphere can be divided into two regions: a quiet one, where the lines of magnetic force are closed, and an active one, where the lines of forces do not close. The

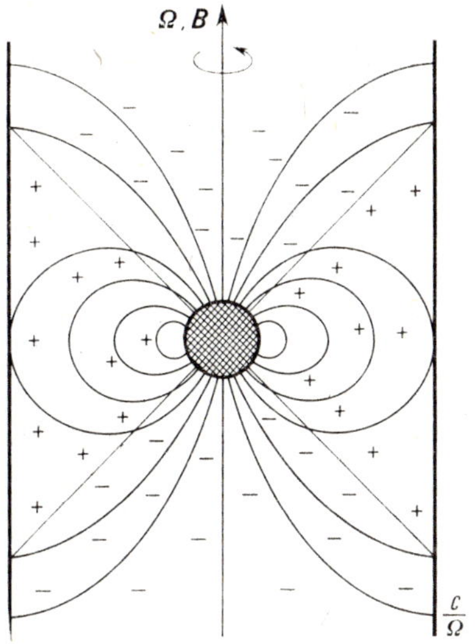

FIG. 6.25. Charge-separated magnetosphere of a pulsar.

quiet magnetosphere corotates rigidly with the star nearly all the way up to the light cylinder. Near the cylinder the magnetic field differs considerably from a vacuum field. It is much weaker here than close to the surface, and inertial forces become dominant, driving charged particles across the lines of force of the magnetic field and out of the light cylinder.

The active magnetosphere sustains a continuous outflow of plasma along the magnetic field lines crossing the light cylinder. This means that the regions near the poles, the so-called polar caps, whose radii are estimated as $R_p \approx R\left(\frac{\Omega R}{c}\right)^{\frac{1}{2}}$, are sources of electric currents.

Recall that in the situation considered above the angular velocity vector was assumed to be parallel to the magnetic dipole moment of the star. It turns out, however, that the structure of the magnetosphere does not change in any essential way when the magnetic dipole moment is inclined to the axis of rotation (Mestel, 1971).

Models of the pulsar radiation mechanism can be divided into two groups: models in which radiation is supposed to originate near the light cylinder (light cylinder models) and models which explain pulsar radiation in terms of phenomena which occur near the poles (polar cap models).

The first pulsar model was proposed in 1968 by Gold, but a year earlier, before pulsars were discovered, Pacini had considered a rotating neutron star with a dipole magnetic field inclined to the axis of rotation and predicted that such a system should be a strong source of electromagnetic waves. In Gold's model, known as the *search light model*, a group of charged particles near the light cylinder is supposed to move with relativistic velocities. As is well known, relativistic particles emit radiation mainly in the direction of their instantaneous velocity. Such a beam of radiation will sweep the celestial sphere, and an observer who happens to come within its reach will receive periodic pulses (Fig. 6.26). Gold's idea was subsequently analysed and developed by many authors (Smith; Ferguson; Ginsburg, Zheleznyakov and Zaitsev; Zheleznyakov; McCrea; and others). In all these models an almost isotropic source of radiation moves near the light cylinder at a relativistic velocity, consequently, a distant observer sees the radiation as collimated in a narrow beam. The radiation mechanism of the source is different in different models, to mention for example cyclotron radiation and the maser effect. The cylinder models encounter considerable difficulties. The group of particles emitting the radiation is unstable, if only because of radiation reaction. It has to be explained, therefore, how high-energy particles are continually supplied to the neighbourhood of the light cylinder. The light cylinder models are currently out of favour; it does not mean, however, that they are clearly in conflict with observation.

In author's opinion the most ingenious pulsar model was proposed by Ruderman and Sutherland (RS model); it has recently been improved by Cheng and Ruderman (also see

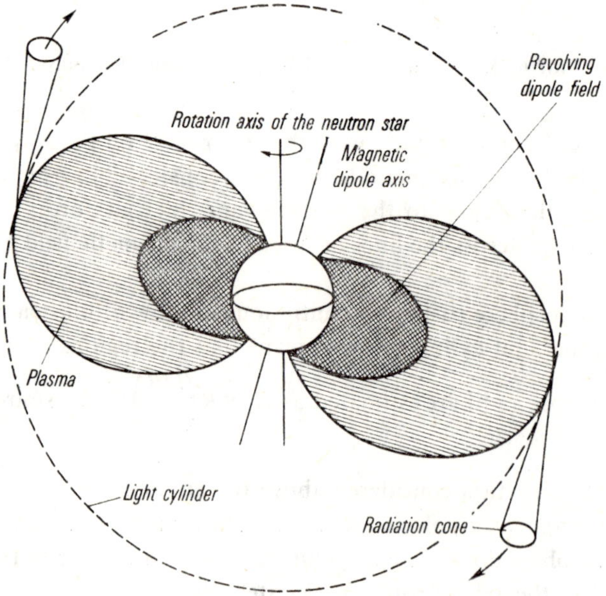

FIG. 6.26. The search-light model of pulsar radiation mechanism proposed by Gold.

Sutherland, 1979). In point of fact, it is an elaboration of Sturrock's idea that the radiation of a pulsar originates close to the polar caps.

The huge magnetic field at the surface of a neutron stars has a drastic effect on the structure of matter. Under such conditions, electron shells form long thin cylinders oriented along the lines of force of the magnetic field (Ruderman). The atoms which compose the surface, probably mainly iron atoms, form long threads of a one-dimensional ion lattice surrounded by a thin cylindrical electron sheath. In this way a new, strongly bound structure of high-density matter is formed. The binding energy of the ions may reach several

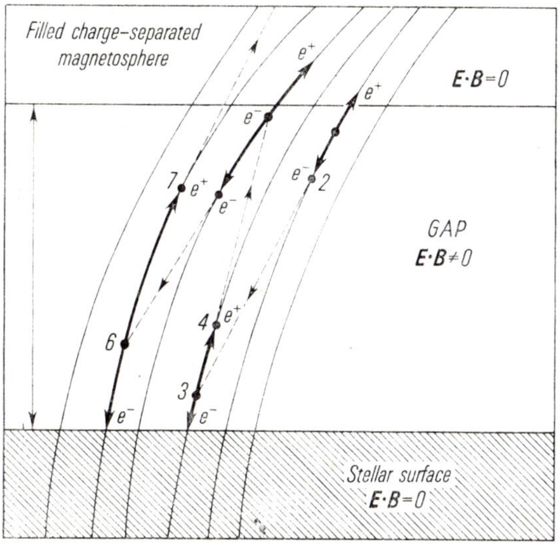

FIG. 6.27. Discharge at the polar gap (adapted from Ruderman and Sutherland, 1975).

keV. In the RS model it is assumed to be high enough to practically prevent the ions from being torn out from the surface.

Ruderman and Sutherland consider the case where the magnetic dipole moment of the star is antiparallel to the angular velocity vector. In this situation the positive charges on the surface of the star drift towards the polar cap. From there, the charges can flow away along the open lines of force of the magnetic field. If the binding energy of the ions is high so that they cannot be torn out from the surface, the charge outflow will produce a charge-free polar gap (Fig. 6.27), which will in turn give rise to a potential difference

$$\Delta\varphi = 2\pi\rho_e h^2 = \Omega B_0 \frac{h^2}{c}; \tag{6.21}$$

here ρ_e is the positive charge density and h the polar gap thickness. This formula is approximate and valid only for small h. It is possible, however, to derive an exact formula valid for $R_p \ll h \ll R$ (R_p and R are the radii of the polar cap and the neutron star, respectively), from which one can find that the potential difference is largest along the central line of force and that its maximum value is

$$\Delta\varphi_{\max} = \frac{\Omega B_0 R_p^2}{2c}. \tag{6.22}$$

The gap thickness h grows very rapidly but the potential difference, which is proportional to h^2, increases still faster. Finally, a discharge takes place, sparking off an avalanche of electron-position pairs. The strong component of the electric field directed along the magnetic field accelerates the charged particles of both signs up to relativistic velocities. The charged particles travelling along the lines of force of the magnetic field emit curvature radiation. The electrons and positrons accelerated to energies exceeding 10^{11} eV can emit photons capable of producing pairs. The result is a cascade growth of the discharge. The electrons are accelerated towards the surface of the star whereas the positrons move outwards and eventually hit the region where $\boldsymbol{E}\cdot\boldsymbol{B}=0$. High-energy photons also enter this region, where they can create pairs. The pairs created in the region where $\boldsymbol{E}\cdot\boldsymbol{B}=0$ move inertially (are not accelerated). There are thus two plasma components in this region: neutral electron-positron plasma and positrons which originated near the polar cap. Interaction between these two components results in a two-stream instability, charged particles collect in groups which emit coherent curvature radiation, strongly collimated in the direction of the magnetic dipole of the star. To obtain the effect of pulsation, it is enough to assume that the magnetic dipole is inclined at a small angle to the rotation axis of the star.

The RS model and its modified version have some weak points, and thus the problem of how pulsars operate is still open.

Literature

BAADE, W. and ZWICKY, F. (1934) *Phys. Rev.* **45**, 138.
BACKUS, P. R., TAYLOR, J. H. and DAMASHEK, M. (1982) *Astrophys. J.* **255**, L 63.
BAYM, G. and PETHICK, Ch. (1979) *Ann. Rev. Astron. Astrophys.* **17**, 415.
BAYM, G. and PINES, D. (1971) *Ann. Phys.* **66**, 816.
BLANDFORD, R. and TEUKOLSKY, S. A. (1976) *Astrophys. J.* **205**, 580.
BÖRNER, G. (1973) *On the Properties of Matter in Neutron Stars*, Springer Tracts in Physics, Springer, Berlin.
BRUMBERG, V. A., ZELDOVICH, YA. B., NOVIKOV, I. D. and SHAKURA, N. I. (1975) *Astr. Letters* **1**, 5.
CAMERON, A. (1970) *Ann. Rev. Astron. Astrophys.* **8**, 179.
CANUTO, V. *Ann. Rev. Astron. Astrophys.* **12**, **167** (1974), **13**, 335 (1975).
CANUTO, V. and CHIU, H. (1968) *Phys. Rev.* **173**, 1229.
CARTER, B. and QUINTANA, H. (1972) *Proc. Roy. Soc. London* **A 331**, 57.
CHANDRASEKHAR, S. (1957) *An Introduction to Study of Stellar Structure*, Dover, N. Y.
CHENG, A. and RUDERMAN, M. *Astrophys. J.* **212**, 800 (1977); **214**, 588 (1977); **216**, 865 (1977); **229**, 348 (1979).
COLE, T. W. (1969) *Nature*, **221**, 29.
COX, J. P. and GIULI, R. T. (1968) *Principles of Stellar Structure*, Gordon and Breach, N. Y.
CRANE, P., NELSON, J. E. and TYSON, J. A. (1979) *Nature*, **280**, 367.
DYSON, F. (1971) *Neutron Stars and Pulsars*, Fermi Lectures 1970, Acad. Nazionale Lincei, Roma.
EARDLEY, D. M. (1975) *Astrophys. J. Lett.* **196**, L59.
EPSTEIN, R. (1977) *Astrophys. J.* **216**, 92.
FERGUSON, D. C. (1971) *Nature Phys. Sci.* **234**, 86.
FERGUSON, D. C. (1973) *Astrophys. J.*, **183**, 977.
FRANK-KAMENETSKY, (1962) *Physical Processes in Stellar Interiors*, Israel Program for Scientific Translations, Jerusalem.
GINSBURG, V. L., ZHELEZNYAKOV, V. V and ZAITSEV, V. V. (1969) *Astrophys. Space Sci.* **4**, 464.
GOLD, T. (1969) *Nature* **221**, 25.
GOLDREICH, P. and JULIAN, W. (1968) *Astrophys. J.* **175**, 869.
HARRISON, B. K., THORNE, K. S., WAKANO, M. and WHEELER, J. A. (1965) *Gravitation Theory and Gravitational Collapse*, The University of Chicago Press, Chicago.

HARTLE, J. B. (1970) *Astrophys. J.* **161**, 111.
HARTLE, J. B. (197) *Physics Reports* **46**, 201.
HEWISH, A. (1970) *Ann. Re . Astron. Astrophys.* **8**, 265.
HEWISH, A. (1975) *Rev. Mod. Phys.* **47**, 5 7.
HULSE, R. and TAYLOR, J. (1975) *Astrophys. J.* **195**, L51.
IBEN, I. (1969) *Stellar Evolution from Main Sequence to Red Giants*, in *Stellar Astronomy*, Vol. II, eds.: Chiu, H. Y., Warasila, R. T., Remo, J. L., Gordon and Breach, N.Y.
IRVINE, J. M. (1978) *Neutron Stars*, Clarendon Press, Oxford.
JULIAN, W. I. (1973) *Astrophys. J.*, **183**, 967.
KÄLLMAN, C. G. (1979). *Fu damentals of Cosmic Physics*, **4**, 167.
MANCHESTER, R. N. and TAYLOR, J. H. (1977) *Pulsars*, Freeman and Comp., San Francisco.
MESTEL, L. (1971) *Nature*, *Phys. Sci.* **233**, 149.
McCREA, W. H. (1972) *Mon. Not. Roy. Astron. Soc.*, **157**, 359.
MICHEL, F. C. (1982) *Rev. Mod. Phys.* **54**, 1.
MISNER, C. W., THORN , K. S. and WHEELER, J. A. (1973) *Gravitation*, Freeman and Comp., San Francisco.
PACINI, F. (1967) *Nature* **216**, 567.
PACZYŃSKI, B. *Acta Astronomica* **20**, 47 (197); **21**, 271 (1971); **21**, 417 (1971).
PACZYŃSKI, B. (1974) *Evolution of Stars with* $M \leqslant 8M_{\odot}$, in *Last Stages of Stellar Evolution*, ed. Taylor, R. J., D. Reidel Publ. Co., Do drecht.
PACZYŃSKI, B. (1980) *Stellar Evolution and Close Binaries*, in *Highlights of Astronomy*, Vol. 5, D. Reidel Publ. Co., Dordr cht.
PINES, D. (1970) Proc. 12th Inter. Conf. Low Temp. Physics, Kyoto.
RADHAKRISHNAN, V. an COOKE, D. (1969) *Astrophys. Lett.* **3**, 225.
REICHLEY P. E. and DOWNS G. S. (1969) *Nature*, **222**, 229; (1971) *Nature Phys. Sci.* **234**, 48.
RUDERMAN, M. (1972) *A n. Rev. Astron. Astrop ys.* **10**, 427.
RUDERMAN, M. (1974) *Matter in Superstrong Magnetic Fields*, in *Physics of Dense Matter*, ed. Hansen, C. T., Reidel Publ. Co., Dordrecht.
RUDERMAN, M. and SUTHERLAND, P. G. (1975) *Astrophys. J.* **196**, 51.
SCHARLEMAN, E. T. and WAGONER, R. V. (1973) *Astrophys. J.* **182**, 9 1.
SCHWARZSCHILD, M. (1965) *Structure and Evolution of the Stars*, Dover, N. Y.
SMITH, F. G. (197) *Mon. Not Roy. Aastr. Soc.* **149**, 1.
SMITH, F. G. (1977) *Pulsars*, Cambridge University Press, Cambridge.
SMITH, F. G., and HEWISH A. (1968) *Pulsating Stars*, Nature reprints, Macmillan, London.
SMARR, L. L. and BLANDFORD, R. (1976) *Astrophys. J.*, **207**, 574.
STAELIN, D. H. and REIF NSTEIN, E. C. (1968) *Science*, **162**, 1481.
STOTHERS, R. (1969) *Massive Stars*, in *Stellar Astronomy*, Vol. II, eds.: Chiu, H. Y., Warasila, R. T., Remo, J. L., Gordon and Breach, N. Y.
STURROCK, P. A. (1971) *Astrophys. J.* **164**, 529.
SUTHERLAND, P. G. (1979) *Fundamentals of Cosmic Physics*, **4**, 95.
SUGIMOTO, D. and NOMOTO, K. (1980) *Space Sci. Rev.* **25**, 155.
TAYLER, R. J. (1970) *The Stars: Their Structure and Evolution*, Wykeham Publications, London.
TAYLOR, J. H., FOWLER, L. A. and MCCULLOCH, P. M. (1979) *Nature* **277**, 437.
TAYLOR, J. H. and MANCHESTER, E. (1975) *Astronomical J.* **80**, 794.
TER HAAR, D. (1972) *Physics Reports* **36**, March.
TER HAAR, D. (1975) *Contemp. Phys.* **16**, 243.
THORNE, K. S. (1969) *A trophys. J.* **158**, 1.
WAGONER, R. V. (1975) *Astrophys. J. Lett.* **196**, L63.
WILL, C. M. (1975) *Astrophys. J. Lett.* **196**, L3.
ZELDOVICH, YA. B. and NOVIKOV, I. D. (1971) *Relativistic Astrophysics*, Vol. I: *Stars and Relativity*, The University of Chicago Press, Chicago.
ZHELEZNYAKOV, V. V. (1971) *Astrophys. Space Sci.*, **13**, 87.

CHAPTER 7

GRAVITATIONAL COLLAPSE AND BLACK HOLES

7.1. Spherically Symmetric Collapse of a Dust Cloud

Consider the evolution of a spherically symmetric cloud of particles whose mutual interactions are purely gravitational. The pressure inside such a system is zero, and therefore the system cannot be in equilibrium. Depending on the initial conditions, the cloud will either contract or expand. The first possibility was already treated to some extent in Chapter 3, in which the general Oppenheimer–Snyder solution describing the evolution of a spherical dust cloud was presented. Let us recall the most important results. If R_0 is the initial radius of the cloud then it follows from (3.17) that a distant observer will see the cloud radius R change with time according to

$$R=\frac{R_0}{2}(1+\cos\eta),$$

$$ct=r_g\ln\left[\frac{(R_0/r_g-1)^{\frac{1}{2}}+\tan\eta/2}{(R_0/r_g-1)^{\frac{1}{2}}-\tan\eta/2}\right]+r_g(R_0/r_g-1)^{\frac{1}{2}}\left[\eta+\frac{R_0}{2r_g}(\eta+\sin\eta)\right]. \tag{7.1}$$

As $\tan\eta/2\to(R_0/r_g-1)^{\frac{1}{2}}$, $ct\to\infty$, while $R\to r_g$, where r_g denotes the gravitational radius of the cloud. From our distant observer's point of view, it will take an infinite time for the cloud surface to contract to its gravitational radius (Fig. 7.1). An observer comoving with the surface will see this process differently. According to his clock, the time after which the cloud surface will have reached the gravitational radius is finite and equal to

$$\tau=\frac{1}{c}\int\limits_0^\infty\left[c^2\left(1-\frac{r_g}{R}\right)-\left(1-\frac{r_g}{R}\right)^{-1}\left(\frac{dR}{dt}\right)^2\right]^{\frac{1}{2}}dt$$

$$=\frac{R_0}{2c}\left(\frac{R_0}{r_g}\right)^{\frac{1}{2}}\arccos\left(\frac{2r_g}{R_0}-1\right)+\frac{R_0}{c}\left(1-\frac{r_g}{R_0}\right)^{\frac{1}{2}}. \tag{7.2}$$

From his point of view, the process of contraction will not stop when the gravitational radius is reached but will continue until the cloud radius reduces to zero and the dust density becomes infinite. In this last phase, not only the density but also the curvature of the space grows to infinity. The region of the spacetime where the curvature is infinite is called the singularity. For the distant observer, the process of contraction, caused by the gravitational forces, will asymptotically lead to the formation, after an infinite time, of a spherically symmetric object whose radius is equal to the gravitational radius. We call such an object a *black hole.* From the point of view of the distant observer the question what happens to this object later would not make much sense as the process of its formation alone

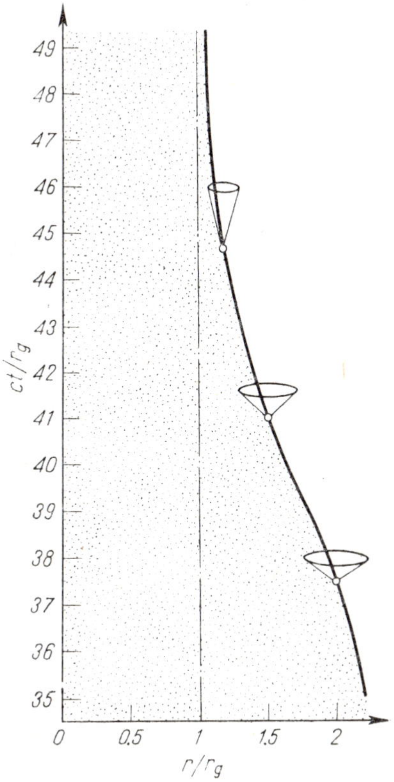

FIG. 7.1. Gravitational collapse of a star as seen by a distant observer. The stellar surface asymptotically contracts to the gravitational radius (adapted from Misner, Thorne and Wheeeler, 1973).

is infinitely long. In order to find out what the further evolution of the cloud is like we have to move together with the cloud. Suppose we undertake such an expedition. We can easily predict the fate of the expedition from the properties of the Schwarzschild space-time. As was shown in Chapter 3, the gravitational field inside our dust cloud is described by the Friedman metric and that outside the cloud by the Schwarzschild metric. It will be convenient, therefore, to represent the evolution of the cloud on a Kruskal diagram (Fig. 7.2). Using the relationship between the Schwarzschild and the Kruskal coordinates, we can trace the evolution of the cloud in the extended Schwarzschild space-time. The surface $r=r_g$ divides the space into two parts with entirely different properties. In region I ($r>r_g$), information can be sent to a distant observer, while no information can cross from region II ($r\leqslant r_g$) to region I. The boundary $r=r_g$ is a null surface, which means that at every point it is tangent to the light cone. A surface having such properties is called a *horizon.* Region I, where $r > r_g$, is a static region, and a timelike Killing vector exists there. No such vector exists in region II, which is nonstationary. Light signals sent radially away from the centre are bent in the strong gravitational field and turned back towards the centre. All timelike geodesics and timelike lines are bent still more, and even those directed outwards hit the central singularity after a finite proper time (Fig. 7.3). The observer travelling in our rocket will never be able to escape from region II once he enters it, and, what is worse, after a finite

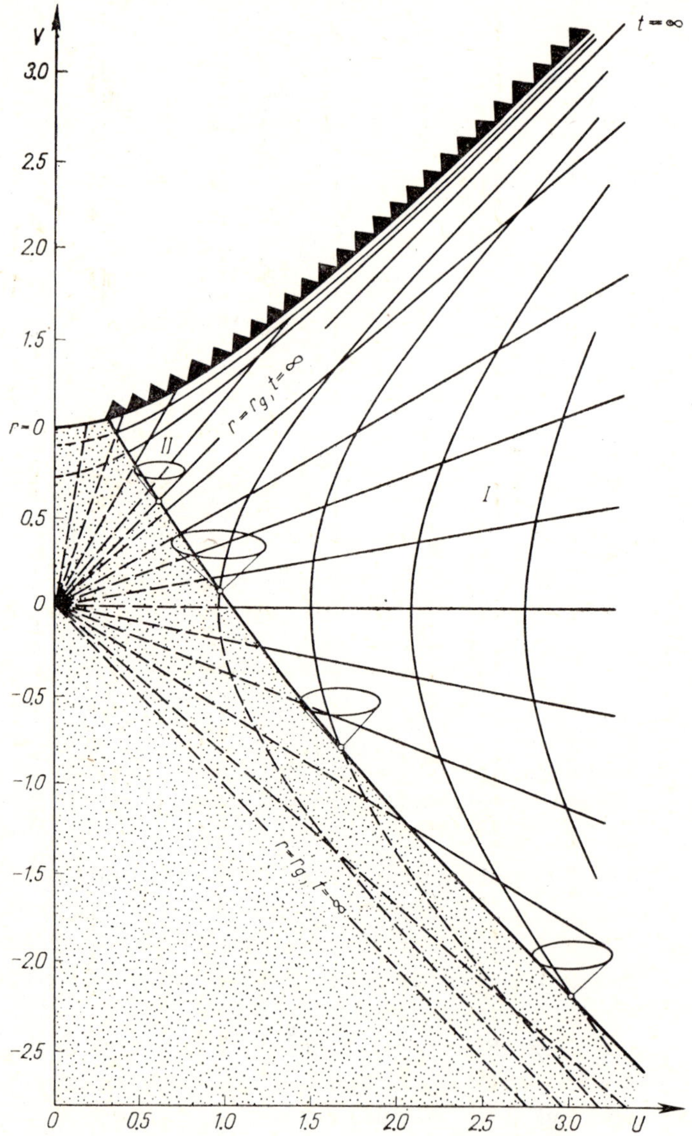

FIG. 7.2. Gravitational collapse of a star in the Kruskal coordinates. As seen by an observer at the surface of the star, after a finite proper time this surface reaches the gravitational radius and then contracts further until the formation of a central singularity, which is invisible to observers beyond the horizon (adapted from Misner, Thorne and Wheeler, 1973).

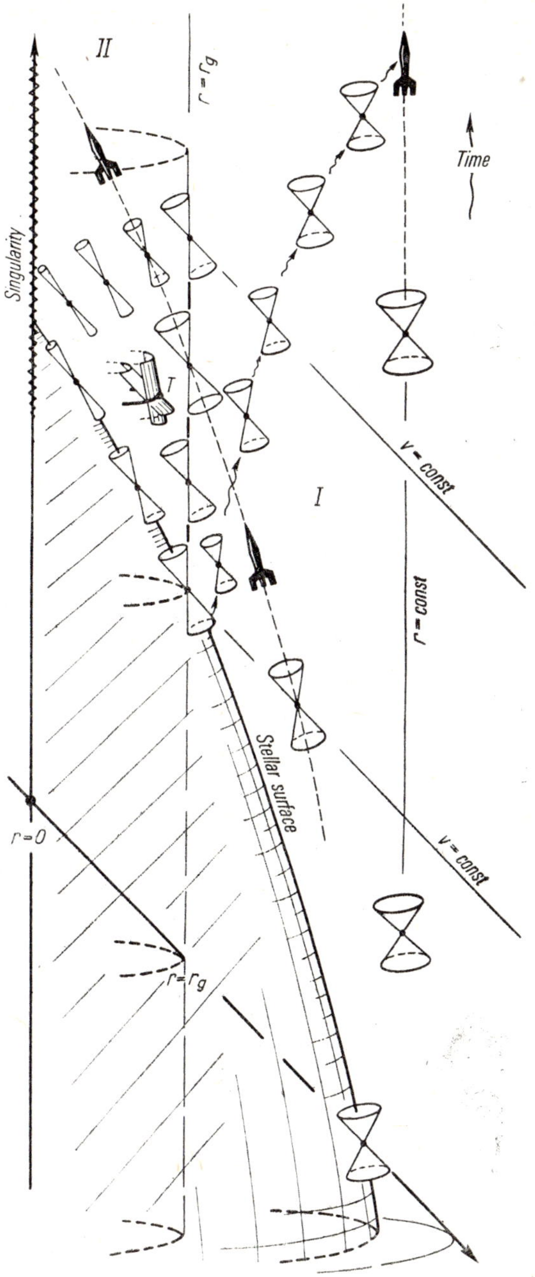

FIG. 7.3. An observer on a rocket wants to watch a collapsing star. If he is so reckless as to cross the horizon, his fate is determined: the strong gravitational field will draw the rocket into the singularity. From the moment the observer crosses the horizon he is not able to send any information to region I. An observer that remains in region I can escape to infinity (adapted from Penrose, 1969).

time, reckoned by his clock, he will be drawn into the central singularity, where the gravitational field will pull the rocket apart, a tragedy that an external observer will never learn about.

From the point of view of a comoving observer, two stages can be distinguished during the gravitational collapse of a spherical dust cloud. In the first, the cloud contracts until a horizon is formed, the moment of this formation not being distinguished by anything in particular. The second stage is a further contraction, which eventually results in the formation of a real physical singularity at the centre of the star, where the density and the curvature of the space approach infinity. As we shall see later, these two stages are distinguishable in any process of catastrophic collapse.

Finally, let us consider how the gravitational collapse of a star would be seen by a distant observer. The model we are considering is of course only a rough approximation of the actual situation; we have neglected pressure and radiation emitted by the star. However, when the star radius is close to the gravitational radius, these effects can be regarded as small corrections. At the final stages of the collapse the luminosity of the star falls exponentially with time. A detailed analysis carried out by Ames and Thorne has given the formula

$$L=L_0\, e^{-\frac{2}{3\sqrt{3}}\frac{ct}{r_g}}. \tag{7.3}$$

The characteristic decay time of the luminosity, $\tau=\frac{3\sqrt{3}}{2}r_g/c$, is very short. For stars of mass comparable to that of the Sun it equals 10^{-5}s. Besides the changes in the luminosity, there also occurs a red shift of photons emitted radially from the surface of the star. The relative change of the wavelength increases exponentially according to the formula

$$z=z_0\, e^{\frac{ct}{2r_g}}. \tag{7.4}$$

When the star radius becomes less than $\frac{3}{2}r_g$, i.e. less than the radius of the last circular photon orbit, the stream of photons that reach the distant observer will be dominated by photons which once formed a photon cloud of radius $\frac{3}{2}r_g$. The last circular photon orbit is not stable and these photons are either caught by the collapsing star or escape to infinity; for the latter z is close to 2. The spectral lines become thicker as z approaches 2 and are brightest at $z=2$.

The above picture is simplified as it takes no account of the envelope of the star. Usually only the central part of the star will collapse. As a result of contraction, enormous amounts of energy can be released in the core of the star. As in a supernova explosion, part of this energy is transferred to the envelope, which instead of contracting may then be blown off. If this happens, the distant observer will hardly be able to tell a gravitational collapse from a catastrophic explosion. A real collapsing star will not simply disappear from his sight.

7.2. Adiabatic Spherical Collapse

The next step towards a more realistic description of the process of gravitational collapse is to take pressure into consideration. We assume, as before, that the configuration in question is spherically symmetric. To simplify the problem we shall assume in addition that the contraction proceeds adiabatically. Equations which describe the evolution of such a system cannot be solved analytically; they provide, however, valuable qualitative information.

The field equations for a spherically symmetric system were already considered in Chapter 3. Our interest was then focused on equilibrium configurations. Now we shall deal with an entirely different situation. The gravitational field inside the matter will depend on time. To describe the evolution of the system we shall use comoving coordinates. The metric of the space-time takes the form

$$ds^2 = e^{2\phi} c^2 dt^2 - e^{\lambda} dr^2 - R^2(d\theta^2 + \sin^2\theta \, d\varphi^2), \tag{7.5}$$

where ϕ, λ and R are arbitrary functions of t and r. We shall find them from the field equations. Assume that the energy-momentum tensor is of the form

$$T_{\mu\nu} = (\varepsilon + p) u_\mu u_\nu - p g_{\mu\nu}, \tag{7.6}$$

i.e. regard matter as a perfect fluid. In comoving coordinates the velocity four-vector has only one non-zero component:

$$u^\mu = (e^{-\phi}, 0, 0, 0). \tag{7.7}$$

The thermodynamic properties of matter are described by the equation of state

$$\varepsilon = \varepsilon(n, s), \tag{7.8}$$

where n is the particle number density and s the entropy per particle. Pressure can be found from the thermodynamic relation

$$p = n\left(\frac{\partial \varepsilon}{\partial n}\right)_s - \varepsilon, \tag{7.9}$$

and the enthalpy is given by

$$h = \left(\frac{\partial \varepsilon}{\partial n}\right)_s. \tag{7.10}$$

Combining (7.10) and (7.9), we find that

$$h = \frac{\varepsilon + p}{n}. \tag{7.11}$$

From the first law of thermodynamics (2.10), which can be written in the equivalent form

$$d\varepsilon = nT ds + (\varepsilon + p)\frac{dn}{n}, \tag{7.12}$$

we obtain the relationship

$$\frac{dh}{h} = \frac{dp}{\varepsilon + p} + \frac{T ds}{h}. \tag{7.13}$$

Assume for simplicity that the energy density ε depends only on the particle number density, $\varepsilon = \varepsilon(n)$. This automatically ensures adiabaticity of the contraction process. From the assumed equation of state and the first law of thermodynamics it follows that the specific entropy (the entropy per particle) is constant along the world-line of the element of matter.

Using (7.13), we can now write the hydrodynamic equations of motion

$$u^{\alpha}{}_{;\beta}u^{\beta}=(g^{\alpha\beta}-u^{\alpha}u^{\beta})\frac{p_{,\beta}}{\varepsilon+p} \tag{7.14}$$

in the form

$$u^{\alpha}_{;\beta}u^{\beta}=(g^{\alpha\beta}-u^{\alpha}u^{\beta})(\ln h)_{,\beta}. \tag{7.15}$$

Inserting the velocity four-vector (7.7) in (7.15), we get

$$\frac{\partial\phi}{\partial r}=-\frac{1}{\varepsilon+p}\frac{\partial p}{\partial r}. \tag{7.16}$$

The field equations and the equations of motion have to be supplemented with additional conditions which will guarantee the regularity of the solution in the whole space. The centre of the star $r=0$ will be regular if

$$R(0,t)=0 \tag{7.17}$$

and if for small r, $2\pi R$ is the circumference of the infinitesimal circle with the centre at the centre of the star and with the radius $e^{\lambda/2}dr$, that is if

$$e^{\lambda}=\left(\frac{\partial R}{\partial r}\right)^2_{r=0}. \tag{7.18}$$

At the surface of the star the internal metric should regularly match the Schwarzschild metric. Let the stellar surface be given by the equation

$$R=R_s(t). \tag{7.19}$$

The line element will be continuous at this surface if

$$ds^2=\left(1-\frac{2GM}{c^2R_s}\right)c^2dt^2-\frac{\dot{R}_s^2dt^2}{1-\dfrac{2GM}{c^2R_s}}-R_s^2d\Omega^2=(e^{2\phi})_s c^2dt^2-R^2(r_s,t)\,d\Omega^2, \tag{7.20}$$

whence we get the relations

$$R=R_s(t)=R(r_s,t) \tag{7.21}$$

and

$$(e^{\phi})_{r=r_s}=\left(1-\frac{2GM}{c^2R_s}\right)\left(1+\frac{u_s^2}{c^2}-\frac{2GM}{c^2R_s}\right)^{-\frac{1}{2}}, \tag{7.22}$$

where

$$u_s=(e^{-\phi})_s\dot{R}_s=\left(e^{-\phi}\frac{\partial R}{\partial t}\right)_s; \tag{7.23}$$

u_s is the rate of change of the star radius R_s, measured relative to the proper time of a comoving observer.

The field equations divide naturally into two groups. The first group comprises the equations which contain only the first time-derivatives of the functions R and λ. They are called the *constraint equations* because they restrict the initial conditions. The second group are the equations involving the second time-derivatives of the functions R and λ. These equations determine the dynamics of the system. It can be shown that if the constraint equations are satisfied at a certain moment of time, the dynamic equations propagate them so that they are satisfied at all times. Denoting

$$U = D_t R = e^{-\phi}\left(\frac{\partial R}{\partial t}\right)_r = e^{-\phi}\dot{R}, \tag{7.24}$$

we can write the equation $\frac{8\pi G}{c^4} T_1^0 = G_1^0$ as

$$e^{-\phi}\dot{\lambda} = 2\frac{U'}{R'}, \tag{7.25}$$

where a prime denotes differentiation with respect to r. The equation $\frac{8\pi G}{c^4} T_0^0 = G_0^0$ does not contain any second time-derivatives either. Using (7.25), we can eliminate $\dot{\lambda}$; the result is

$$\frac{8\pi G\varepsilon R^2}{c^4} = 1 + \frac{U^2}{c^2} + \frac{R\partial U^2}{c^2\partial R} - e^{-\lambda}(2RR'' + R'^2) - (e^{-\lambda})' RR'; \tag{7.26}$$

we have used the equality

$$\frac{\partial}{\partial R} = \frac{1}{R'}\left(\frac{\partial}{\partial r}\right)_t. \tag{7.27}$$

Equation (7.26) is a linear first-order equation for $e^{-\lambda}$. For convenience of interpretation we denote

$$m(r, t) = \int_0^R 4\pi R^2(\varepsilon/c^2)\, dR. \tag{7.28}$$

The solution of equation (7.26) satisfying the boundary conditions (7.18) is given by

$$e^{\lambda(r,t)} = \left[1 + \frac{U^2}{c^2} - \frac{2Gm(r, t)}{c^2 R}\right]^{-1} \left(\frac{\partial R}{\partial r}\right)^2. \tag{7.29}$$

Note that U is the radial velocity of the collapsing element of matter and $m(r, t)c^2$ is the total energy contained in the sphere of radius $R(r, t)$. Using the fact that an element of the proper volume on the surface $t = \text{const}$ can be represented in the form $dV = 4\pi R^2 e^{\lambda/2}\, dr$, we can write $m(r, t)$ as

$$m(r, t) = \int_0^r \frac{\varepsilon}{c^2}\left(1 + \frac{U^2}{c^2} - \frac{2Gm(r, t)}{c^2 R}\right)^{\frac{1}{2}} dV. \tag{7.30}$$

It is seen that not only the energy density ε but also the kinetic energy and the gravitational potential energy contribute to the total energy.

The field equation containing second time-derivatives can be written

$$D_t U = -c^2 \left[\frac{1+\dfrac{U^2}{c^2}-\dfrac{2Gm}{c^2 R}}{\varepsilon+p} \right] \left(\frac{\partial p}{\partial R} \right)_t - \frac{G(m+4\pi R^3 p/c^2)}{R^2} . \tag{7.31}$$

For a complete description we only need to know the function ϕ. We can find it from equation (7.16), which after integration with the boundary conditions (7.22) leads to

$$e^{\phi} = \frac{1}{h(r,t)} \frac{1-\dfrac{2GM}{c^2 R_s(t)}}{\left(1+\dfrac{U_s^2(t)}{c^2}-\dfrac{2GM}{c^2 R_s(t)}\right)^{\frac{1}{2}}} ; \tag{7.32}$$

the enthalpy is normalized by the condition $h=1$ for $r=r_s$. Equation (7.25), which describes the change of the function λ with time, after substitution of (7.29) and use of (7.32), reduces to

$$D_t m = 4\pi R^2 \frac{p}{c^2} U . \tag{7.33}$$

We can now list all the equations describing the dynamics of the system. We have: three equations containing the first time-derivatives

$$D_t R = U , \tag{7.34a}$$

$$D_t m = -4\pi R^2 \frac{p}{c^2} U , \tag{7.34b}$$

$$D_t U = -c^2 \left[\frac{1+\dfrac{U^2}{c^2}-\dfrac{2Gm}{c^2 R}}{\varepsilon+p} \right] \left(\frac{\partial p}{\partial R} \right)_t - \frac{G(m+4\pi R^3 p/c^2)}{R^2} , \tag{7.34c}$$

two equations without any time-derivatives, (7.32) and (7.28), and the continuity equation $(nu^{\alpha})_{;\alpha}=0$, which can be written

$$\frac{4\pi R^2 n}{\left(1+\dfrac{U^2}{c^2}-\dfrac{2Gm}{c^2 R}\right)^{\frac{1}{2}}} \frac{\partial R}{\partial r} = \left(\frac{\mathrm{d}A}{\mathrm{d}r} \right)_{t=0} = \text{const.} \tag{7.35}$$

For a unique description of the evolution of the system, we have to set initial conditions by giving $R(r, 0)$, $m(r, 0)$ and $U(r, 0)$. Differentiating relation (7.28) with respect to R at a fixed time, we obtain

$$\left(\frac{\partial m}{\partial R} \right)_t = 4\pi R^2 \frac{\varepsilon}{c^2} ; \tag{7.36}$$

hence we can find $\varepsilon(r, 0)$. From the equation of state and the thermodynamic relations we can then find p, n and h. This information is sufficient to determine the time-derivatives of the functions R, m and U. A complete solution can be obtained by iterations. Note that the evolution of the system is thus determined without resort to the continuity equation (7.35). This is possible because the continuity equation is a first integral of the dynamic equations. It determines the initial particle density distribution, and the dynamic equations propagate this distribution in such a way that the continuity equation is satisfied at every subsequent moment of time.

In addition to the initial conditions, which can be given arbitrarily, the boundary conditions ensuring that the internal solution is smoothly matched with the Schwarzschild metric should also be satisfied. We saw in Chapter 3 that for this matching to be smooth the pressure at the boundary must be assumed zero. Equation (7.34b) then implies that $m(r_s, t)=M=\text{const}$, where M is the mass occurring in the Schwarzschild solution.

The complete system of equations given above was first found by Misner and Sharp. In general the system cannot be solved analytically. A number of important conclusions can be drawn by analysing the general form of the equations. Let us find out whether pressure alters the two basic effects which we observed while discussing collapse of a spherical cloud of dust. There, a particle which crossed the horizon was bound to hit the central singularity after a finite proper time, and light rays directed radially away from the centre were bent inwards and focused in the centre. We shall now prove that if the radius of a given spherical layer becomes smaller than the gravitational radius, $R(r, t)<\dfrac{2Gm(r, t)}{c^2}$, and if the pressure is positive, the layer will contract to zero after a finite proper time, i.e. its radius will fall to zero and the density at the centre will increase to infinity.

From equations (7.34a) and (7.34b) it follows that D_tR and D_tm have opposite signs. During collapse $D_tR<0$, so $D_tm>0$. Assume that the layer is already inside its gravitational radius and $1-\dfrac{2Gm(r, t)}{c^2R}=-e^2$, where e is some small number. As follows from (7.29), $1+\dfrac{U^2}{c^2}-\dfrac{2Gm(r, t)}{c^2R}\geqslant 0$. For this expression to be strictly positive, $U<-|e|c$. The radius of the layer will thus continue to fall and after a finite proper time it will drop to zero, with the density at the centre increasing to infinity.

Examining the behaviour of light rays is more difficult. We first need to assign an invariant meaning to the statement "a light ray directed radially outward moves away from the centre" (Hernandez and Misner). The form of the metric implies that, owing to the spherical symmetry, the function $R(r, t)$ has an invariant sense as a position coordinate. Indeed, given any space-time event, the set of all events equivalent to it with respect to the group of rotations forms a 2-sphere. Its surface area, $4\pi R^2$, defines the value of R to be assigned to the given event. The change of $R(r, t)$ along a light ray measures the ray's progress outward. If

$$D_kR=e^{-\phi}\left(\frac{\partial R}{c\partial t}\right)_r+e^{-\frac{\lambda}{2}}\left(\frac{\partial R}{\partial r}\right)_t>0, \tag{7.37}$$

the ray moves away from the centre. The equality in (7.37) can be rewritten as

$$D_k R = \frac{1-\dfrac{2Gm}{c^2 R}}{\left(1+\dfrac{U^2}{c^2}-\dfrac{2Gm}{c^2 R}\right)^{\frac{1}{2}}-\dfrac{U}{c}}. \tag{7.38}$$

In the region where $R(r, t) < \dfrac{2Gm}{c^2}$, inside the collapsing star ($U<0$), even ou'going light rays are falling in, because then $D_k R < 0$. Moreover, it can be shown that once $D_k R$ becomes negative along the light ray, it remains negative thereafter. This follows from the equation

$$D_k^2 R = -4\pi \frac{RG}{c^4}(\varepsilon + p) < 0 \tag{7.39}$$

and from the fact that $D_k R$ always decreases along the ray. Thus, the presence of pressure does not change the general picture of gravitational collapse. As before, a horizon is formed, enclosing part of the space from which no signal can be sent to an observer at infinity. The same conclusion was obtained by May and White, who integrated numerically the system of equations describing an adiabatic spherically symmetric collapse. Depending on the initial conditions, the first layer of matter for which $R(r, t) < 2Gm(r, t)/c^2$ lies somewhere between the centre and the surface of the configuration. The region where this condition is satisfied expands on both sides. Very quickly a singular region is formed, into which the remaining matter is drawn. Pressure is important only at the initial stages of the collapse; after the first layer disappears below the horizon, it practically no longer plays any dynamic role.

7.3. Nonspherical Collapse. Trapped Surfaces and Apparent Horizons

A real collapsing star is never exactly spherical. Every star rotates a little and has a magnetic field. Einstein's equations for such a star form a very complex nonlinear system, which makes it very difficult to derive even general qualitative information. There are two ways out: we can either consider nearly spherically symmetric configurations and apply approximate perturbation methods or introduce several restricting assumptions and then try to answer very general questions. Both approaches have their merits and shortcomings. The first permits an intuitive explanation of the processes occurring during a gravitational collapse, but the range of its applicability is limited. The second enables us to draw some general conclusions but leaves a number of physically important questions unanswered. However, a combination of the two methods makes it possible to predict with reasonable accuracy the course of the gravitational collapse of a real star.

In the spherically symmetric case, a star of mass greater than the critical mass, at the late stages of its evolution, loses stability against collapse and undergoes a catastrophic contraction. The collapse is accompaniend by the formation of a horizon and a central singularity. The question is in what way realistic conditions alter this picture. Rotation of the star and its magnetic field will certainly have an effect on the value of the maximum mass stable against collapse. Rapid rotation and a strong magnetic field will also affect the very process of contraction. A rapidly rotating star tends to form a thin disc, which, depending on the initial conditions, undergoes fragmentation or a further gravitational collapse. Slow rotation and a weak magnetic field, which can be treated as small perturbations, will not change the process of collapse qualitatively. The influence of small per-

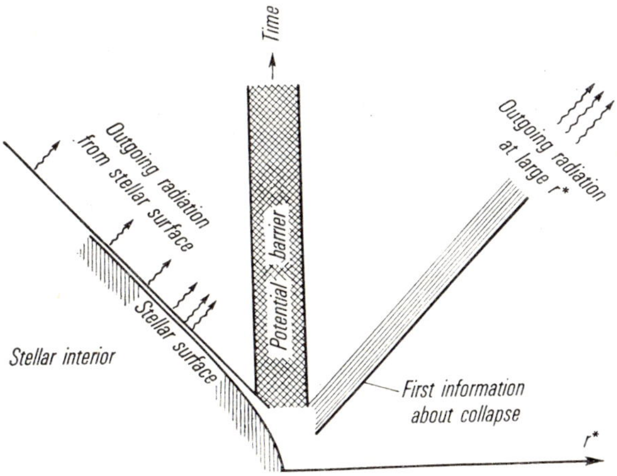

FIG. 7.4. Perturbations of a spherically symmetric gravitational collapse of a star, produced by a scalar field whose source is the collapsing star. When the surface of the star approaches the horizon, a potential barrier is formed, penetrable only to the shortwave component of radiation. The longwave component, carrying information about the stationary component of the field, is screened. As seen by a distant observer, the scalar field dies away (adapted from Price, 1972).

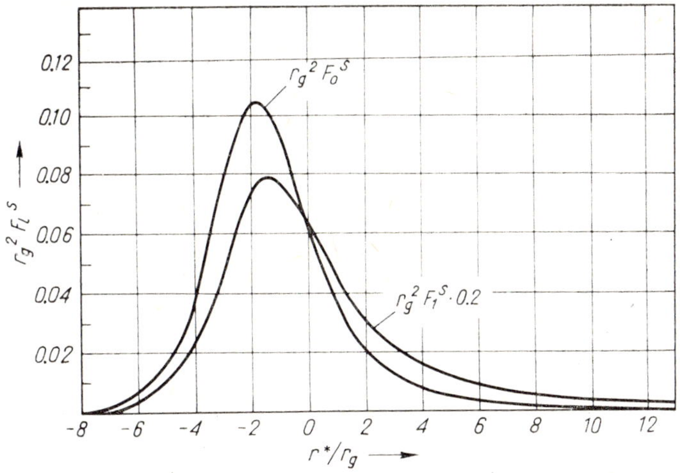

FIG. 7.5. Effective potential as a function of r^* for $l=0$ and $l=1$. As l increases, the maximum of the effective potential shifts towards $(3/2)r_g$ (adapted from Price, 1972).

turbations on gravitational collapse in the first linear approximation has been investigated thoroughly. Zeldovich, Doroshkevich and Novikov showed that in a comoving coordinate system small perturbations of a spherically symmetric collapsing star remain small when the surface of the star crosses the horizon. A full dynamical analysis of this process was given by Price. Let us outline its general idea and main results. There are several kinds of perturbations one can consider. It may be a scalar, electromagnetic or gravitational field. The situation we consider is that of a star charged with an integer-spin massless scalar field. Owing to the spherical symmetry of the original space, the perturbations can be resolved into scalar harmonics. Such a multipole decomposition permits us to observe the evolution of those multipole moments which may be emitted. They can be characterized by means of a scalar potential satisfying the wave equation

$$-\frac{1}{c^2}\psi_{,tt}+\psi_{,r^*r^*}-F_l(r^*)\psi=0, \tag{7.40}$$

where $r^*=r+r_g\ln\left(\frac{r}{r_g}-1\right)$ and $F_l(r^*)$ is the effective potential. The effective potential $F_l(r^*)$ decays very quickly as the horizon is aproached, i.e. as $r\to r_g$ or $r^*\to-\infty$, and dies away at infinity, when $r\to\infty$ or $r^*\to\infty$. The effect of the potential barrier on the scalar waves ψ depends on their wavelength. Short waves with $\lambda\ll r_g$, starting from the region $r<\frac{3}{2}r_g$, penetrate the barrier almost unaffected, waves with $\lambda\approx r_g$ are partly let through and partly reflected, and those with $\lambda\gg r_g$ are almost completely reflected.

We choose the initial conditions so that the field is static before the process of collapse begins. The instant the collapse sets off the scalar field becomes time-dependent, for we assume that the star is the source of this field. Using the internal solution, we determine the value of the scalar field at the surface of the star. The scalar field in the whole space is found by solving equation (7.40) subject to the boundary conditions at the surface of the star, where the field changes as $\psi=Q_0+Q_1e^{-u/2r_g}$ (u is the retarded time), and at the null surface $u=u_0=\text{const}$, along which the information about the onset of collapse propagates (Fig. 7.4). The surface of the star begins to contract, and when its radius is down to just a few times the gravitational radius, there appears a *potential barrier*. The curve $F_l(r^*)$ is shown in Fig. 7.5. The potential barrier reaches a maximum near $r=(3/2)\,r_g$. Until that moment, the waves starting from the surface of the star could propagate freely, practically unchanged by the effective potential. When the star radius falls below $(3/2)\,r_g$, the potential barrier acts as a filter, letting through only the shortwave component of the spectrum. According to the conditions given at the surface of the star, the amplitude of these waves decreases as $e^{-u/2r_g}$. The static component of the field corresponds to waves of infinite length. They are completely reflected by the potential barrier. A distant observer receives only the shortwave part of the spectrum with the amplitude falling off exponentially. Besides waves that pass directly through the barrier, a distant observer also receives waves scattered by the tail of the potential, at finite values of r^*. This radiation falls off much more slowly, namely

$$\psi\sim t^{-(2l+2)}, \tag{7.41}$$

when the l-th multipole moment is non-zero at the initial moment, and

$$\psi\sim t^{-(2l+3)}, \tag{7.42}$$

when it is zero. In this way Price has shown that a *distant observer will see small perturba-*

tions of a spherically symmetric collapsing star die away with time at least as fast as $t^{-(2l+2)}$, so that asymptotically the field is spherically symmetric again. All information about the asymmetry of the field is screened off by a potential barrier near $r=(3/2)\,r_g$. Thus, small perturbations do not destroy the horizon.

The question of whether a horizon is always formed in the gravitational collapse of a star remains open. A few examples have been given where this is not so, but all of them are pathological in the sense that they involve some very special choice of the initial conditions, which lead to the appearance of a singularity on the horizon. This property is not stable against small changes in the initial conditions. That the horizon of a spherically symmetric collapsing star cannot be annihilated by small departures from spherical symmetry follows quite clearly from simple intuitive considerations. Imagine that when the radius of the star comes close to the gravitational radius, we introduce small perturbations. This may be realized by changing a little the initial conditions on the surface $t=\text{const}$ when r_s gets close to r_g. Suppose the perturbation field is a massless scalar field. The field propagates along light cones. The outgoing null geodesics diverge (Fig. 7.6). The scalar field is thus washed away and its energy density near the horizon decreases, and so the field is unable to bring about any drastic changes in the properties of the space-time in that region. Penrose advanced the so-called *cosmic censorship hypothesis*, according to which every realistic process of gravitational collapse leads to the formation of a horizon. No one has yet managed to prove it, but nobody has been able to give a counter-example either.

From the point of view of a comoving observer, a spherically symmetric collapse leads inevitably to the formation of a central singularity. The first answer to the question

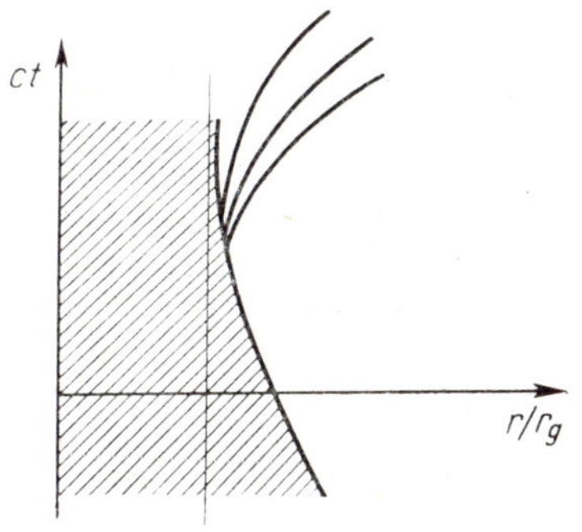

FIG. 7.6. Light rays sent outwards from the surface of a collapsing star diverge.

of whether a singularity is always formed in gravitational collapse was given by Penrose. Studying the global properties of the space-time region $r<r_g$ in the Kruskal diagram, Penrose introduced the notion of a *trapped surface*. Suppose that in the extended Schwarzschild space-time we choose a sphere of radius $r<r_g$ and at a given moment of time we send radial light signals towards and away from the centre from every point of this sphere. The situation is illustrated in Fig. 7.7. The surface area of the ingoing wavefront will decrease and the light wave will reach the centre within a finite affine distance measured along the rays. The area of the "outgoing" wavefront will also decrease and will also

reduce to zero. A trapped surface is a closed, compact two-surface such that the light beams orthogonal to it converge locally in future directions. Using this notion, Penrose showed that if the evolution of the system is completely determined by the initial conditions at a Cauchy hypersurface and if the total energy density is locally positive, then

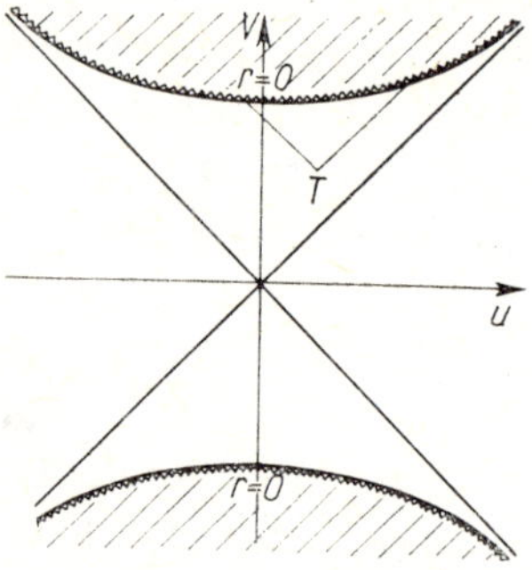

FIG. 7.7. In the region $r<r_0$ every two-sphere is a trapped surface. Both families of null geodesics orthogonal to it converge to the central singularity.

either some of the future-directed timelike geodesics (viz. null geodesics) are incomplete, i.e. cannot be extended to arbitrary values of the affine parameter, or fundamental causality relations cease to apply, for example there may appear closed timelike curves.

A few years later Hawking and Penrose proved a much more general theorem which implies that *gravitational collapse always leads to the formation of a singularity.*† Since the proof of this theorem requires several definitions that would lead us far beyond our present considerations, we shall restrict ourselves to formulating the theorem itself. The Hawking–Penrose theorem reads:

A space-time M cannot be causally geodesically complete if in addition to Einstein's equation the following assumptions are satisfied:

(a) There are no closed timelike curves in M,

(b) For every unit timelike vector t^α

$$T_{\mu\nu}t^\mu t^\nu \geqslant \tfrac{1}{2}T^\mu{}_\mu, \tag{7.43}$$

(c) Every causal geodesic γ contains at least one point at which

$$k_{[\mu}R_{\nu]\alpha\beta[\rho}k_{\sigma]}k^\alpha k^\beta \neq 0, \tag{7.44}$$

where k_μ is the tangent vector to γ,

(d) M contains at least one of the following:

(1) a trapped surface,

(2) a point p such that every past light cone from p reconverges,

(3) a compact spacelike hypersurface.

By *causal geodesics* we mean timelike or null geodesics. Let us explain the physical sense

† More precisely, we should speak here of a singular space-time, meaning a space-time which either admits incomplete timelike or null geodesics, or in which the fundamental causality conditions are violated.

of the assumptions (b) and (c). Since the energy-momentum tensor is a symmetric tensor, it can be reduced, at every point, to the diagonal form $T_{\mu\nu}=\text{Diag}\,\|\,\varepsilon, p_1, p_2, p_3\|$, where ε is the local density of energy and p_1, p_2 and p_3 are the principal values of anisotropic pressure. The assumption (b) is equivalent with the requirement that

$$\varepsilon+\sum_{i=1}^{3} p_i \geqslant 0 \tag{7.45}$$

and

$$\varepsilon+p_i \geqslant 0 \tag{7.46}$$

for $i=1, 2, 3$. The assumption c) is more abstract it means that every causal geodesic either passes through a non-empty region of the space, or hits a region where the tangent vector departs from the principal direction determined by the curvature tensor.† This assumption is not too restrictive and will be satisfied in all but pathological cases. The Hawking–Penrose theorem is an example of the application of global methods. If we restrict ourselves to the first possibility in d), we see that in the case of a collapsing body all the assumptions of the theorem will be satisfied. Thus whenever there exists a trapped surface, then according to the Hawking–Penrose theorem some of the causal geodesics are incomplete or there appear closed timelike curves. In the spherically symmetric case a catastrophic gravitational collapse leads to the formation of a region where the density of matter and the curvature are infinite. In the general case we only know that a singularity will occur. The Hawking–Penrose theorem, which implies the existence of a singularity, does not specify the character of the singularity. This problem still remains open.

In describing a nonspherical collapse it is convenient to introduce the notion of an apparent horizon. Let $\Sigma(t_0)$ be a surface of constant time. The boundary of the region of $\Sigma(t_0)$ containing trapped surfaces lying in $\Sigma(t_0)$ is called an apparent horizon. The position of the apparent horizon can easily be determined on any spacelike hypersurface. On the other hand, to find the position of the event horizon one has to know the asymptotic form of the solution. As shown in Fig. 7.8, an apparent horizon does not, in general, coincide with the event horizon. Consider the following simple example: let a spherically symmetric, collapsing dust cloud be surrounded by a thin, spherical and also collapsing dust shell. When the cloud shrinks to its gravitational radius, the surface $r=r_g$ is an apparent horizon. Outgoing light rays emitted from the point $r_g+\Delta r$ will move away. For sufficiently small Δr, as soon as the light rays cross the collapsing shell, they will become slightly bent and will move along the horizon $r=\frac{2G}{c^2}(M+\delta M)=r_H$, where M is the mass of the cloud and δM the mass of the shell. From this moment the apparent horizon will coincide with the event horizon. The apparent horizon underwent a sudden jump from $r=r_g$ to $r=r_H$ at the moment when the whole system – the cloud and the shell – got surrounded by a common horizon.

We can now sum up our considerations. If the cosmic censorship hypothesis is true, the gravitational collapse of a nonspherical star can be divided into three stages: a loss

† In general, the curvature tensor determines, at every point, four principal null directions.

of stability, a process of contraction until the formation of a horizon and a further contraction up to the formation of a singularity. To avoid misconception, let us recall that from the point of view of a distant observer it takes an infinite time for the collapsing star to approach the horizon, and only comoving observers have a chance to watch

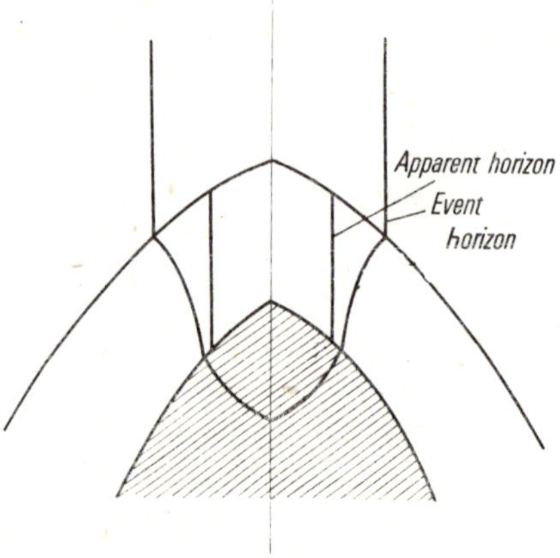

FIG. 7.8. Example of a situtation where an apparent horizon does not coincide with the event horizon. A collapsing dust cloud is surrounded by a thin, collapsing spherical dust layer. Part of light rays sent from the surface of the cloud, which would normally esc pe to infinity, are bent by the gravitational field of the layer and form a global horizon. The horizon formed in the collapse of the cloud is no longer an absolute horizon; here it plays the role of an apparent horizon (adapted from Hawking, 1973).

the process of collapse up to the moment when the singularity is formed. Comoving observer can notice that he entered a space-time region with different properties when he passes through apparent horizon. The apparent horizon lies within the global horizon or coincides with it. Our observer is thus doomed to hitting the singularity. It is also to be stressed that the external solution describing the properties of the space-time outside the collapsing matter is not a Schwarzschild solution. In general, such a space-time is extremely complicated. As follows from Price's analysis, a nonspherically collapsing body will radiate away all multipoles. If the collapsing body is not electrically neutral, it will become a strong source of electromagnetic radiation, and in any case it will be a strong source of gravitational waves.

7.4. General Properties of Black Holes

The cosmic censor, if one exists, makes every process of catastrophic collapse lead to the formation of a horizon. Discussion of the spherically symmetric collapse has equipped us with certain intuitions about the properties of a horizon. We should now formulate them without any reference to spherical symmetry. For example, we might try to define a horizon as the boundary of the region from every point of which information can be sent to an observer at infinity. This, however, would not be sufficiently precise. First, we would have to define the causal conditions to be satisfied in the space-time region of interest; secondly we would have to explain what is meant by "sending information to an observer at infinity". A moment's reflection is enough to show us that the proposed definition would be good for just one black hole. To give a satisfactory definition of a black hole is not a simple task; we have to resort to the global properties of the space-time.

As in the spherically symmetric case, we restrict ourselves to an asymptotically flat space-time. In such a space-time the notions of null, timelike and spacelike infinities are well defined. To eliminate pathological causal properties, we shall assume that every point has a nonvoid neighbourhood which is not intersected by any causal geodesic more than once. This assumption precludes the existence of closed null or timelike geodesics. In an asymptotically flat space-time satisfying this causality condition we define a black hole to be a simply connected space-time region, in which no causal geodesic has any common points with future null infinity $\mathscr{I}^+$. The boundary of this region is called the event horizon.

Using these definitions and assumptions, Hawking has proved a number of theorems which describe the general properties of horizons and black holes. To discuss these properties we first need to introduce the notion of a black hole on a simply connected spacelike hypersurface Σ. The usual choice of hypersurfaces of constant time is not convenient in investigating black holes. In asymptotically flat space-times all such surfaces converge to spacelike infinity I^0. If we want to watch the evolution of a black hole, it will be more convenient to use spacelike hypersurfaces Σ which become null at infinity, so that they interesect $\mathscr{I}^+$ – future null infinity. Suppose we are given a family $\Sigma(\tau)$ of such surfaces which are disjoint and such that $\Sigma(\tau_2)$ lies to the future of $\Sigma(\tau_1)$ for $\tau_2 > \tau_1$. A black hole $B(\tau)$ on a hypersurface $\Sigma(\tau)$ is defined to be a simply connected region of $\Sigma(\tau)$ from which no future-directed causal geodesic can reach future null infinity $\mathscr{I}^+$.

We shall now prove that a black hole, once formed, can never disappear, nor can it bifurcate. Let $B(\tau)$ be a black hole on a hypersurface $\Sigma(\tau)$. Consider a later hypersurface $\Sigma(\tau_1)$, $\tau_1 > \tau$. Every future-directed causal geodesic from an arbitrary point $p \in \Sigma(\tau)$ will either reach $\mathscr{I}^+$ † or intersect $\Sigma(\tau_1)$. According to the definition of the black hole, a future-directed causal geodesic from a point $p \in B(\tau)$ cannot intersect $\mathscr{I}^+$, and therefore it must intersect $\Sigma(\tau_1)$ at some point q. From q, however, $\mathscr{I}^+$ cannot be reached; hence $q \in B(\tau_1)$. The fact that black holes cannot bifurcate is implied by the following reasoning. Consider future-directed causal geodesics from a point $p \in B(\tau)$. Each of them intersects $\Sigma(\tau_1)$. Choose any two geodesics from this family. Either of them can be continuously deformed into the other. This implies that the intersection of the future cone of the point p with $\Sigma(\tau_1)$ is simply connected. Similarly the common part of the future region of the black hole $B(\tau)$ and the surface $\Sigma(\tau_1)$ forms a simply connected region, which therefore must be contained in a single black hole $B(\tau_1)$. Black holes can merge together, and new black holes may be formed; if there are n black holes on a hypersurface $\Sigma(\tau)$, a later hypersurface may also have n black holes, or less or more than n, but always at least one.

We now turn to the general properties of the horizon. As mentioned before, the horizon is the boundary of a region whose every point can be connected with infinity by a future-

† More precisely: I^+ or $\mathscr{I}^+$.

directed causal geodesic. We recall from the special theory of relativity that the future of a given space-time point p is the set of all points q which can be reached from p by future-directed null or timelike geodesics. The boundary of this region is the future-directed part of the light cone of p. Similarly, we define the past of the point p. We denote the future of p by $J^+(p)$, and its past by $J^-(p)$. If we are given a space-time region Σ, for example a hypersurface, we define the future $J^+(\Sigma)$ (the past $J^-(\Sigma)$) as the set of points that can be joined with Σ by future (past)-directed causal geodesics. Thus the event horizon can be defined as the boundary of the past of future null infinity, i.e. $\mathscr{H}=\dot{J}^-(\mathscr{I}^+)$. It follows at once that the event horizon is a null surface generated by null geodesics which have no future end-points. That the event horizon is a null surface is one of its fundamental properties.

The null geodesics that generate the event horizon can be described by means of their expansion θ and shear σ. The definitions of these kinematic parameters are analogous to those for a family of timelike geodesics (see Chapter 2). Consider a spacelike 2-surface S lying in the horizon. The null generators of the horizon are orthogonal to S. Suppose there exists a point $p \in S$ such that the null generators from p have a negative expansion θ. The surface S can be deformed a little to become partly contained in $J^-(\mathscr{I}^+)$,† the expansion of the null geodesics orthogonal to it still remaining negative. Those null geodesics which lie in $J^-(\mathscr{I}^+)$ and are orthogonal to S will begin to intersect each other for finite values of the affine parameter. To show this, consider a small surface element ΔS of area A. If we move the element ΔS an affine distance $\delta\lambda$ up the orthogonal null geodesics, the area A will change by

$$\frac{\delta A}{\delta\lambda}=2A\theta; \tag{7.47}$$

the second derivative is equal to

$$\frac{\delta^2 A}{\delta\lambda^2}=-2A(\sigma\bar{\sigma}+\tfrac{1}{2}R_{\mu\nu}l^\mu l^\nu), \tag{7.48}$$

where l^μ is the tangent vector to the null geodesics. If $\theta<0$ at the point p and the condition $R_{\mu\nu}l^\mu l^\nu \geqslant 0$ is satisfied, as assumed, then within a finite affine distance A will fall to zero, which means that neighbouring null geodesics will intersect. Such geodesics cannot be generators of the surface $\dot{J}^+(S)$ all the way out to $\mathscr{I}^+$. This establishes a contradiction which shows that on the horizon $\theta \geqslant 0$. The horizon is formed from null geodesics which do not converge. Therefore, the surface area of a spacelike cross-section of the horizon cannot decrease. As we shall see later this last conclusion is very important. For example, it can be used to estimate the maximum amount of energy released in the collision of two black holes.

Our knowledge of the topological properties of the event horizon is much poorer. What is known is that the event horizon is simply connected and compact, but, in general,

† $J^-(\mathscr{I}^+)$ is the set of all points from which $\mathscr{I}^+$ can be reached by future-directed timelike geodesics.

its spacelike cross-sections are not topologically equivalent to 2-spheres. The apparent horizon, which is not a null surface in general, is topologically equivalent to a 2-sphere (Hawking). It follows from the definition that the apparent horizon is contained within or coincides with the event horizon. No theorems are known, however, which would imply that the area of a spacelike cross-section of an apparent horizon should increase monotonically.

To close this section, let us summarize the most important general properties of black holes and event horizons. Black holes cannot be annihilated, nor can they bifurcate. New black holes can arise from the collapse of a star or a cluster of stars, and they can merge together. An event horizon is always a null surface. It is formed from null geodesics which do not converge. The area of a spacelike cross-section of the event horizon cannot decrease.

7.5. Stationary Black Holes

A black hole formed as a result of gravitational collapse will initially be a nonstationary object. Owing to its strong gravitational field it can attract surrounding matter. The gravitational waves emitted in the last phases of the collapse can be scattered and interact with the newly formed black hole for a long time. As follows from Price's analysis, if we wait long enough, the space-time outside the black hole will settle down to a stationary state. The matter liable to be attracted by the hole will eventually be accreted, and gravitational waves and other perturbations of the curvature will either be damped or radiated to infinity. Of course this picture applies only to small disturbances. If we are to believe in the cosmic censorship hypothesis, however, even strong perturbations will not be able to destroy the horizon, and we can still expect that as time runs on the whole system will asymptotically approach a stationary state.

It is worth-while to give some thought to the possibility that the cosmic censorship hypothesis is not satisfied. What would happen then? In some situations the process of gravitational collapse might then give rise to a singularity which would not be hidden inside a horizon. All information from the neighbourhood of such a naked singularity would be able to reach observers at infinity. Examples of a space-time with a naked singularity are furnished by the Reissner–Nordström solution for $\frac{Ge^2}{c^4} > \left(\frac{GM}{c^2}\right)^2$ and the Kerr solution for $a^2 > \left(\frac{GM}{c^2}\right)^2$. It is sometimes maintained that the existence of a naked singularity would mean a breakdown of all the laws of physics. The author does not share this pessimistic point of view. In the whole space-time outside the singular region the laws of physics hold and the causal principle remains firm. A naked singularity, if one exists, will have to behave so as not to violate those laws. The initial singularity in the hot model of the Universe is still "visible", which does not prevent us from discovering and formulating new causal laws of physics. Hereinafter we shall assume, however, that the cosmic censor exists, and therefore after a sufficiently long time any black hole will "settle down" and continue in a stationary state.

We shall now look at the properties of a stationary black hole. In an asymptotically flat space only one black hole can be in such a state. If we admit the possibility of the existence of charged black holes, there may be infinitely many of them†. They will attract each other gravitationally and repel each other by electromagnetic interactions. In the expanding Universe there may also be arbitrarily many black holes.‡ For simplicity we shall assume that the space-time to be considered contains only one black hole.

We call a black hole stationary if the asymptotically flat space-time containing it has a timelike Killing vector in the neigbourhood of $\mathscr{I}^+$ and $\mathscr{I}^-$. In general, this Killing vector will not be timelike on the horizon. The existence of the Killing vector implies that the area of a spacelike cross-section of the horizon of a stationary black hole is independent of time. The null geodesic generators of the horizon do not converge, hence $\theta=0$. The rate of change of θ with respect to the affine parameter is given by the equation

$$\frac{d\theta}{d\lambda}=-\theta^2-\sigma\bar{\sigma}+(\epsilon+\bar{\epsilon})\theta-R_{\mu\nu}l^\mu l^\nu, \tag{7.49}$$

where σ is the shear parameter and $\epsilon+\epsilon$ measures the changes of the tangent vector l^μ relative to the vector transported parallelly. The parameter ε vanishes if l^μ is transported parallelly. Equation (7.49) implies that on the horizon $\sigma=0$ and $R_{\mu\nu}l^\mu l^\nu=0$. The rate of change of the shear σ along the null geodesics is given by the equation

$$\frac{d\sigma}{d\lambda}=-2\theta\sigma+(3\epsilon-\bar{\epsilon})\sigma+C_{\mu\nu\rho\sigma}l^\mu m^\nu l^\rho m^\sigma. \tag{7.50}$$

It follows that on the horizon $C_{\mu\nu\rho\sigma}l^\mu m^\nu l^\rho m^\sigma=0$. Here $m^\mu=\frac{1}{\sqrt{2}}(r^\mu+is^\mu)$, where r^μ and s^μ are orthogonal spacelike unit vectors generating 2-dimensional spacelike subspaces orthogonal to l^μ, and $C_{\mu\nu\rho\sigma}$ is the traceless part of the Riemann curvature tensor, the Weyl tensor

$$C_{\mu\nu\rho\sigma}=R_{\mu\nu\rho\sigma}+g_{\mu[\sigma}R_{\rho]\nu}+g_{\nu[\rho}R_{\sigma]\mu}+\tfrac{1}{3}Rg_{\mu[\rho}g_{\sigma]\nu}. \tag{7.51}$$

The fact that $R_{\mu\nu}l^\mu l^\nu=0$ and $C_{\mu\nu\rho\sigma}l^\mu m^\nu l^\rho m^\sigma=0$ on the horizon means that no matter or gravitational radiation may cross the horizon of a stationary black hole.

The vanishing of θ implies that the horizon of a stationary black hole is a marginally trapped surface. Hence in the stationary case the apparent horizon coincides with the event horizon. We have seen before that any apparent horizon is topologically equivalent to $R^1\times S^2$, we may thus conclude that in the stationary case, spacelike cross-sections of the horizon are topologically 2-spheres.

The following very important theorem about the properties of stationary black holes was proved by Hawking. First, if near $\mathscr{I}^+$ and $\mathscr{I}^-$ the space-time is stationary but not static, the Killing vector ξ^μ, which is timelike at infinity, is spacelike near the horizon.

† Such a system, however, is not stable.

‡ The notion of a black hole can be generalized to the case of cosmological space-times. This would require a more precise definition of points at infinity, a problem somewhat outside the scope of this book.

The space-time region outside the horizon where the vector ξ^μ is spacelike is called the ergosphere. Second, if the ergosphere intersects the horizon, there exists one more spacelike Killing vector η^μ with closed trajectories. This implies that a stationary nonstatic black hole must be axisymmetric. The two Killing vectors commute. This fact imposes quite strong restrictions on the properties of the space-time. For if we choose the coordinates so that $\xi^\mu = \frac{\partial}{c\partial t}$ and $\eta^\mu = \frac{\partial}{\partial \varphi}$, the metric, as we shall see later, must be invariant under the simultaneous transformations $t \to -t$ and $\varphi \to -\varphi$. We are now faced with the very important problem of finding what metric should be used for the description of stationary black holes. The class of stationary, axisymmetric space-times is very wide, and it seemed at first that a black hole might have a very complex structure, so that a general metric with many parameters would be needed. However, it will be seen later that the existence of a regular horizon is a very strong restriction, which makes it possible to give the metric of a stationary black hole in an explicit form.

Considering the static case, Israel has shown that among all asymptotically flat, static space-times satisfying the vacuum Einstein equations and having equipotential surfaces ($\xi^\mu \xi_\mu = \text{const}$, $t = \text{const}$, where $\xi_\mu = \alpha t_{,\mu}$ is the timelike Killing vector) topologically equivalent to a 2-sphere, *only the Schwarzschild space-time has its singularities surrounded by a regular horizon.* The idea of the proof is as follows. The staticity of the space-time implies the existence of a family of spacelike hypersurfaces $t = \text{const}$. On a hypersurface $t = \text{const}$ one can consider the subspaces $\xi_\mu \xi^\mu = \xi^2 = \text{const}$. Einstein's equations can be reduced to a system of elliptical equations describing the internal geometry of the equipotential surfaces, their embedding in the hypersurface $t = \text{const}$ and the internal geometry of the hypersurface $t = \text{const}$. The imposed boundary conditions and the assumed topological conditions imply the vanishing of certain integrals which control the external geometries of the equipotential surfaces. This in turn implies the spherical symmetry of the problem. In Chapter 1 we showed that the only asymptotically flat, spherically symmetric vacuum solution of Einstein's equations is the Schwarzschild solution. We thus come to a very important conclusion: a static black hole must be spherically symmetric, and therefore it is described by the Schwarzschild metric. Thus a static black hole is fully characterized by its mass.

Israel has generalized his theorem to the case of a space-time satisfying the source-free Einstein–Maxwell equations. The only static, asymptotically flat solution with a regular horizon is then the Reissner–Nordström metric

$$ds^2 = \left(1 - \frac{2GM}{c^2 r} + \frac{Ge^2}{c^4 r^2}\right) c^2 dt^2 - \frac{dr^2}{1 - \frac{2GM}{c^2 r} + \frac{Ge^2}{c^4 r^2}} - r^2 (d\theta^2 + \sin^2\theta \, d\varphi^2), \tag{7.52}$$

with $\left(\frac{GM}{c^2}\right)^2 > \frac{Ge^2}{c^4}$. If a magnetic monopole were discovered, the Reissner–Nordström solution could easily be generalized to apply to gravitational fields produced by spherically symmetric distributions of mass, electric charge and magnetic monopole. In this case a spherically symmetric black hole would be fully characterized by its mass and electric and magnetic charge.

The case of a stationary but nonstatic black hole has proved much more difficult. Carter has obtained a number of partial results indicating that the Kerr solutions are probably the only uncharged stationary black hole solutions. As follows from Hawking's theorem, any stationary black hole is axisymmetric. In addition to the timelike Killing vector ξ^μ there exists a spacelike Killing vector η^μ, which commutes with ξ^μ. Now consider the family of local 2-subspaces generated by these two vectors. We can describe it by means of the bivector $\xi_{[\alpha}\eta_{\beta]}=S_{\alpha\beta}$. Carter has shown that if

$$S_{\alpha\beta}R^\beta{}_\gamma\epsilon^{\gamma\delta\sigma\tau}S_{\sigma\tau}=0 \tag{7.53}$$

on a connected open subset U of the space-time, and if

$$S_{\alpha\beta}=0 \tag{7.54}$$

at some point of U, then

$$S_{[\alpha\beta;\gamma}S_{\sigma]\tau}=0 \tag{7.55}$$

on U.

If the subset U contains at least one point lying on the axis of symmetry, condition (7.54) will be satisfied since η^μ vanishes on the axis of symmetry. Condition (7.53) is plainly satisfied in empty space and when the energy-momentum tensor is that of a source-free electromagnetic field. Equality (7.55) implies that there exists locally a family of 2-dimensional spacelike hypersurfaces which are orthogonal to the timelike 2-surfaces generated by $\xi^{[\mu}\eta^{\nu]}$. Therefore the space-time outside a stationary black hole can be decomposed into the subspaces of transitivity determined by the Killing vectors ξ^μ and η^μ and the family of 2-dimensional spacelike subspaces which are orthogonal to them, which means that the space-time admits the discrete symmetry $t\to -t$, $\varphi\to -\varphi$. The square of the bivector $S_{\alpha\beta}$, $S^2=S_{\alpha\beta}S^{\alpha\beta}$, is zero on the horizon and on the rotation axis, but $S_{\alpha\beta}=0$ on the axis only. Outside the horizon $S>0$, except on the rotation axis. This means that everywhere outside the horizon and off the axis, there is some linear combination of the Killing vectors which is timelike.

The most important result of Carter was to show that empty stationary asymptotically flat space-times which

1° have no closed timelike curves,

2° are axisymmetric

and in which

3° the past horizon $\mathscr{H}^-$ intersects the future horizon $\mathscr{H}^+$ in a compact connected 2-surface,

fall into disjoint continuous families depending on at most two parameters: the mass M and the angular momentum per unit mass a. One such family, depending continuously on the parameter a, was known, namely the Kerr solutions for $\left(\frac{GM}{c^2}\right)^2 \geqslant a^2$. That this is in fact the only family of solutions satisfying the assumptions of Carter's theorem remained a conjecture for four years to be finally proved by Robinson. In this way we now know that any uncharged stationary black hole is described by the Kerr metric. Recently Mazur has shown that any charged stationary black hole is described be the

Kerr–Newman metric

$$ds^2 = \frac{\Delta(cdt - a\sin^2\theta\, d\varphi)^2 - \sin^2\theta\,[acdt - (r^2+a^2)\,d\varphi]^2}{\Sigma} - \Sigma\left(\frac{dr^2}{\Delta} + d\theta^2\right),$$

$$\Sigma = r^2 + a^2\cos^2\theta, \quad \Delta = r^2 - \frac{2GMr}{c^2} + a^2 + \frac{Ge^2}{c^4}, \tag{7.56}$$

where M is the mass of the black hole, $a = J/Mc$ the angular momentum per unit of mass and e the total charge. The potential of the electromagnetic field is given by

$$\mathscr{A} = A_\mu dx^\mu = \frac{er(cdt - a\sin^2\theta d\varphi)}{r^2 + a^2\cos^2\theta}. \tag{7.57}$$

Just as in the case of the Reissner–Nordström solution, there is room here for a magnetic monopole. Thus, any stationary black hole is fully characterized by three parameters: mass, angular momentum and charge. *Black holes have no hair*, says John Wheeler succinctly, emphasizing their primitive structure.

7.6. Kerr–Newman Black Holes

It follows from the Mazur theorem that charged stationary black holes are described by the Kerr–Newman metric. We shall now examine the properties of such black holes more closely. The Kerr–Newman metric has the form

$$ds^2 = \frac{\left(r^2 - \frac{2GM}{c^2}r + a^2 + e^2G/c^4\right)(cdt - a\sin^2\theta d\varphi)^2 - \sin^2\theta\,[acdt - (r^2+a^2)\,d\varphi]^2}{r^2 + a^2\cos^2\theta}$$

$$-(r^2 + a^2\cos^2\theta)\left(\frac{dr^2}{r^2 - 2GMr/c^2 + a^2 + Ge^2/c^4} + d\theta^2\right). \tag{7.58}$$

The horizon is the surface

$$r = r_H = \frac{GM}{c^2} + \sqrt{\left(\frac{GM}{c^2}\right)^2 - a^2 - \frac{Ge^2}{c^4}}. \tag{7.59}$$

It exists only when $\left(\frac{GM}{c^2}\right)^2 \geqslant a^2 + \frac{Ge^2}{c^4}$. The stationary limit surface can be determined by the condition $g_{00} = 0$ from which we find

$$r = r_E = \frac{GM}{c^2} + \sqrt{\left(\frac{GM}{c^2}\right)^2 - a^2\cos^2\theta - \frac{Ge^2}{c^4}}. \tag{7.60}$$

The part of the space-time contained between r_H and r_E is called the ergosphere. A comparison of formulae (7.59) and (7.60) shows that the stationary limit surface is tangent to the horizon at two points which lie on the rotation axis. The relative position of the two surfaces is illustrated in Fig. 7.9.

The Killing vector $\xi^\mu = \frac{\partial}{c\partial t}$, which is timelike at infinity, becomes a null vector on the outer boundary of the ergosphere and is spacelike inside it. As shown by Carter, it is possible to find a linear combination of the Killing vectors which is timelike near the horizon and null on the horizon. This combination is the vector $k^\mu = \xi^\mu + \frac{a}{r_H^2 + a^2}\eta^\mu$; the quantity $\Omega_H = \frac{ca}{r_H^2 + a^2}$ is called the angular velocity of the black hole. Ω_H is constant on the horizon.

Although the Killing vector ξ^μ is spacelike inside the ergosphere, the space-time is stationary not only outside the ergosphere but in the whole region outside the horizon. The reason is the existence in the ergosphere of the timelike Killing vector k.† However, the ergosphere differs in an essential way from the space extending outside to infinity. If $r > r_E$, the curve $r = r_0 = \text{const} > r_E$, $\theta = \text{const}$, $\varphi = \text{const}$ is timelike and, though it is not a geodesic, it may be the world-line of an observer travelling on a rocket. In contrast, inside the ergosphere the curve $r = r_0 < r_E$, $\theta = \text{const}$, $\varphi = \text{const}$ is spacelike, and no observer can remain at a fixed point of the space. In the ergosphere every observer must be in motion.

Now imagine the following "Penrose process" (Penrose 1969, Christodoulu 1970). We send a composite particle from infinity into the ergosphere. There it splits into two

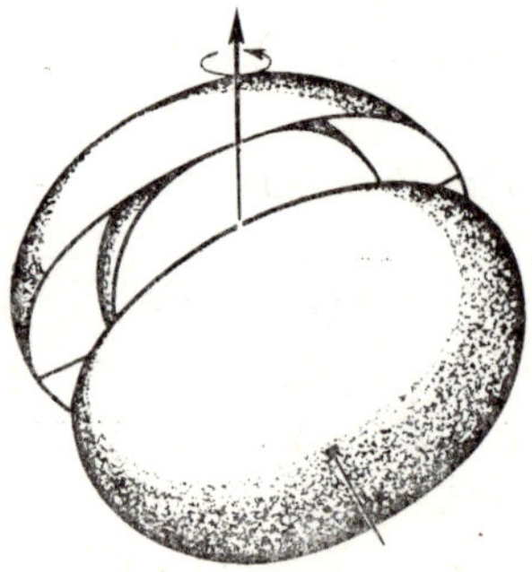

FIG. 7.9. Ergosphere of Kerr geometry, the region between the horizon (the inner surface) and the stationary limit surface (the outer surface). The latter touches the horizon at two points lying on the rotation axis (adapted from Rees, Ruffini and Wheeler, 1974).

particles, one of which, B, falls into the black hole, and the other, A, returns to infinity. Let the energy of the original particle be $E/c = \xi^\alpha p_\alpha$ and its angular momentum component along the z-axis $J = \eta^\alpha p_\alpha$, where p_α is the four-momentum of the particle. E and J are constant along the particle world-line. Inside the ergosphere the particle splits, and according to the conservation laws for momentum, energy and angular momentum we have

† However, there is no continuous Killing vector field timelike near both infinity and the horizon.

$p^\alpha = p^\alpha_A + p^\alpha_B$, $E = E_A + E_B$ and $J = J_A + J_B$. Since ξ^α is spacelike in the ergosphere, we can choose p_B so that $E_B = \xi^\alpha p_{B\alpha} < 0$.† This is possible because the scalar product of a spacelike vector and a timelike vector can be negative. The particle B cannot escape to infinity,

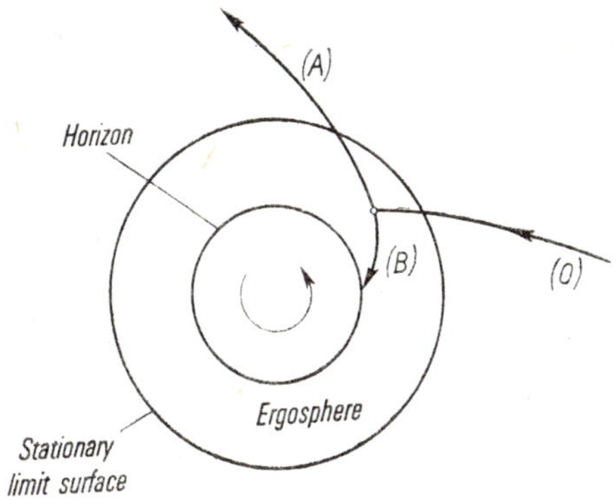

FIG. 7.10. The Penrose process. A composite particle decays in the ergosphere. One of the components falls through the horizon; the other escapes to infinity carrying away more energy than was brought by the original particle.

it falls into the black hole. The particle A can escape to infinity where its energy $E_A = E - E_B > E$ will be greater than that of the original particle. The process is illustrated in Fig. 7.10. Let us examine it in more detail. To this end, consider the motion of a test particle of mass m and charge q in Kerr–Newman space-time (Carter, 1968). The velocity four-vector of the particle satisfies the equation

$$mc\frac{Du^\alpha}{ds} = \frac{q}{c}F^\alpha{}_\beta u^\beta . \tag{7.61}$$

The motion of the particle can be found directly from equation (7.61) or by using the Hamilton–Jacobi equation

$$g^{\alpha\beta}\left(S_{,\alpha} + \frac{q}{c}A_\alpha\right)\left(S_{,\beta} + \frac{q}{c}A_\beta\right) = m^2c^2 . \tag{7.62}$$

As there exist two Killing vectors ξ^α and η^α, the total energy of the particle and the axial component of its angular momentum are conserved. Hence

$$\begin{aligned} \frac{E}{c} &= \xi^\alpha\left(mcu_\alpha + \frac{q}{c}A_\alpha\right) = \text{const}, \\ J_z &= \eta^\alpha\left(mcu_\alpha + \frac{q}{c}A_\alpha\right) = \text{const}. \end{aligned} \tag{7.63}$$

† Recall that E_B is the energy measured by an observer at rest at infinity. For a local observer the energy of the particle B would be positive. This situation is possible because the Killing vector connected with translations in time, which is timelike at infinity, is spacelike in the ergosphere. On the other hand, there is a Killing vector which is timelike in the ergosphere and which at infinity, in the neighbourhood of $\mathscr{I}^+$, is spacelike.

The velocity four-vector is a unit vector; using (7.63), one can write the condition $u^\alpha u_\alpha = 1$ in the form

$$\alpha E^2 - 2c\beta E + c^2\gamma = 0, \tag{7.64}$$

where

$$\alpha = (r^2+a^2)^2 - \Delta a^2 \sin^2\theta > 0 \quad \text{for} \quad r > r_H,$$

$$\beta = (J_z a + eqr)(r^2+a^2) - J_z a\Delta, \tag{7.65}$$

$$\gamma = \left(\frac{J_z a}{c} + \frac{eqr}{c^2}\right)^2 - \Delta\left(\frac{J_z}{c\sin\theta}\right)^2 - m^2\Delta(r^2+a^2\cos^2\theta) - (r^2+a^2\cos^2\theta)\left[(p^r)^2 + \Delta(p^\theta)^2\right].$$

Here $\Delta = r^2 - \frac{2GM}{c^2} r + a^2 + \frac{Ge^2}{c^2}$. For the energy of the particle we obtain the expression

$$\frac{E}{c} = \frac{\beta + \sqrt{\beta^2 - \alpha\gamma}}{\alpha}, \tag{7.66}$$

(having chosen the root corresponding to $p^t > 0$). The energy of the particle will be negative when $\beta < 0$ and $\gamma > 0$ (α is always positive outside the horizon). These conditions will be satisfied if at least one of the inequalities $J_z a < 0$, $eq < 0$ holds. Falling into the black hole, such a particle will reduce the angular momentum of the hole, or its charge, or both. From the point of view of an observer at infinity, this way of extracting energy from the black hole does not violate the law of energy conservation. The energy of the whole system, i.e. the black hole and the test particle, is unchanged. The particle returning to the observer at infinity has gained energy at the expense of the rotational or the electrostatic energy of the black hole. The process can only be applied to a rotating or charged black hole, since it is only then that there exists an ergosphere or a space-time region of similar properties.

One of the most important properties of black holes is the fact that the area of the horizon cannot decrease, and in the case of stationary black holes it does not change with time at all. Let us calculate the area of the surface of a Kerr–Newman black hole. The horizon is given by the equation $r = r_H$. The line element (7.58) is singular on the horizon. However, it is merely a coordinate singularity, and introducing, for example, the retarded time u instead of the time coordinate ct, we obtain a new form of the metric

$$ds^2 = \frac{\Delta - a^2\sin^2\theta}{\Sigma} du^2 - 2dr\,du + 2a\frac{r^2+a^2-\Delta}{\Sigma}\sin^2\theta\,du\,d\varphi + 2a\sin^2\theta\,dr\,d\varphi - \Sigma\,d\theta^2$$
$$- \frac{(r^2+a^2)^2 - \Delta a^2\sin^2\theta}{\Sigma}\sin^2\theta\,d\varphi^2, \tag{7.67}$$

which is regular on the horizon. Here $\Delta = r^2 - \frac{2GM}{c^2} r + a^2 + \frac{Ge^2}{c^4}$, $\Sigma = r^2 + a^2\cos^2\theta$. The area of the horizon is understood to be the area of a 2-dimensional cross-section of the horizon with a spacelike or null surface. The existence of the Killing vector k^μ on the horizon implies that the areas of all such cross-sections are equal. Inserting $u = \text{const}$

and $r=r_H$ into the line element (7.67), we get

$$\mathrm{d}s^2=-(r_H^2+a^2\cos^2\theta)\,\mathrm{d}\theta^2-\frac{(r_H^2+a^2)^2}{r_H^2+a^2\cos^2\theta}\sin^2\theta\,\mathrm{d}\varphi^2. \tag{7.68}$$

Knowing the metric, we can calculate the surface area:

$$A=4\pi(r_H^2+a^2)\,. \tag{7.69}$$

Inserting r_H given by (7.59) and introducing the angular momentum $a=J/Mc$ gives

$$A=4\pi\left[2\left(\frac{GM}{c^2}\right)^2+2\left(\left(\frac{GM}{c^2}\right)^4-\frac{J^2G^2}{c^6}-\frac{G^3e^2M^2}{c^8}\right)^{\frac{1}{2}}-\frac{Ge^2}{c^4}\right]. \tag{7.70}$$

For a static uncharged black hole a simpler formula is found by putting $J=e=0$ in (7.70)

$$A=16\pi\left(\frac{GM}{c^2}\right)^2=4\pi r_g^2\,. \tag{7.71}$$

Using (7.70), we can determine the dependence of the mass on A, J and e,

$$M=\frac{c^2}{G}\left(\frac{A}{16\pi}+\frac{\pi}{A}\left(\frac{Ge^2}{c^4}\right)^2+\frac{1}{2}\frac{Ge^2}{c^4}+\frac{4\pi}{A}\frac{J^2G^2}{c^6}\right)^{\frac{1}{2}}. \tag{7.72}$$

Christodoulou and independently Hawking have introduced the notion of the "irreducible mass" of a black hole. The irreducible mass M_{ir} is related to the area of the horizon by

$$A=16\pi\left(\frac{GM_{\text{ir}}}{c^2}\right)^2, \tag{7.73}$$

It follows that for a Schwarzschild black hole $M_{\text{ir}}=M$, and for a Kerr–Newman black hole

$$\left(\frac{GM_{\text{ir}}}{c^2}\right)^2=\frac{1}{4}\left[2\left(\frac{GM}{c^2}\right)^2+2\left(\left(\frac{GM}{c^2}\right)^4-\frac{J^2G^2}{c^6}-\frac{G^3e^2M^2}{c^8}\right)^{\frac{1}{2}}-\frac{Ge^2}{c^4}\right]. \tag{7.74}$$

The mass of a Kerr–Newman black hole can be expressed in terms of the irreducible mass, electric charge and angular momentum. This relationship can be written as

$$\left(\frac{GM}{c^2}\right)^2=\left(\frac{GM_{\text{ir}}}{c^2}+\frac{e^2}{4M_{\text{ir}}c^2}\right)^2+\frac{1}{4}\frac{J^2}{M_{\text{ir}}{}^2c^2}. \tag{7.75}$$

Owing to its simple relation to the area of the horizon, the irreducible mass plays an important part in black hole dynamics. All classical processes involving a black hole proceed in such a way that the area of the horizon does not decrease. Hence the irreducible mass does not decrease either. If during a process the irreducible mass of the black hole remains unchanged, the process is said to be reversible. If the irreducible mass increases, the process is called irreversible. It is seen from formula (7.74) that a process is reversible if the increase in the mass of the black hole is compensated by a change in the angular momentum and charge. As an example, let us find under what conditions the Penrose process of extracting energy from a black hole is reversible. If a particle with energy E, angular momentum with respect to the rotation axis J_z, mass m and charge q falls through the horizon, the mass of the black hole will change by $\delta M = E/c^2$, its angular momentum by $\delta J = J_z$ and its charge by $\delta e = q$. Equation (7.66) implies that for a fixed J_z the energy of the particle at a given point of the space will be minimum when $m = p^r = p^\theta = 0$. As seen by a distant observer, the energy gap† between the positive and the negative energy states vanishes on the horizon. Using (7.66) and (7.65), we find that the energy is then equal to

$$E = \frac{cJ_z a + eqr_H}{r_H^2 + a^2}. \tag{7.76}$$

The change in the mass of the black hole δM cannot be less than

$$\frac{ca\delta J + er_H \delta e}{c^2(r_H^2 + a^2)}; \tag{7.77}$$

this corresponds to $\delta M_{ir} \geqslant 0$. The irreducible mass is unchanged only when

$$\delta M = \frac{ca\delta J + er_H \delta e}{c^2(r_H^2 + a^2)}, \tag{7.78}$$

i.e. when relation (7.76) is satisfied. For the process of the particle capture by the black hole to be reversible, the mass of the particle should be very small, $m \approx 0$, and the particle should move in such a way that $p^r = p^\theta = 0$ near the horizon. These conditions can be satisfied when the particle moves on a circular orbit near the horizon. In the case of an extreme Kerr–Newman black hole, where M, a and e are such that $\frac{G^2M^2}{c^4} - a^2 - \frac{Ge^2}{c^4} = 0$, the position of the horizon is given by

$$r = r_H = \frac{GM}{c^2}. \tag{7.79}$$

Christodoulou and Ruffini have shown that in the case of an extreme Kerr–Newman black hole the radius of the innermost marginally stable circular orbit of a test particle approaches r_H when $J_z a > 0$. The transformation of the hole due to the capture of a test particle from such an orbit is reversible only when

$$\left|\frac{q\sqrt{G}}{m}\right| \to \infty, \quad \frac{e\sqrt{G}}{M} \to 0, \quad \text{and} \quad \frac{Geq}{mM} \to -\infty. \tag{7.80}$$

Reversible processes constitute the most effective method of extracting energy from a black hole. A reversible process is reversible in the sense that there exists another process by which the black hole can be restored to its initial state.

Now let us estimate the maximum amount of energy that can be extracted from a rotating black hole by a reversible process. If the hole is uncharged and rotates with the

† The positive and the negative energy states correspond to two different roots of equation (7.64), E_+ and E_-. As is easy to verify, this equation has one double root $E_+ = E_-$ only on the horizon. In general $E_+ \neq E_-$ and so there is an energy gap of width $E_+ - E_-$.

maximum angular velocity possible, so that $\left(\frac{GM}{c^2}\right)^2 = a^2$, then $M_{ir} = \frac{\sqrt{2}}{2} M$, whence we find $\frac{\Delta M}{M} = 1 - \frac{\sqrt{2}}{2} \approx 0.29$. Thus a Kerr black hole can give off about 29% of its total energy. Still greater amounts of energy can be obtained from charged rotating black holes. A similar calculation leads to the conclusion that in this case one can extract as much as 50% of the energy of the hole. It should be remembered, hovewer, that the above estimate apply only to reversible processes. In real situation a reversible process will hardly be realizable at all and thus the extracted amounts of energy will be much smaller.

7.7. The Four Laws of Black Hole Mechanics

So far we have considered black holes in an empty space or in a space endowed with an electromagnetic field whose sources were hidden below the horizon. In the latter case the most general solution of Einstein's equations describing the properties of the space-time outside the horizon is the Kerr–Newman solution. A stationary black hole may also exist in a space with matter, provided the matter is distributed axisymmetrically and rotates stationarily about the hole. An example is a rotating black hole surrounded by a thin disc rotating in the equatorial plane. Exact solutions describing such a configuration are not known. However, by using the general properties of black holes and the fact that the space-time is stationary and axisymmetric outside the horizon, it is possible to derive a general expression for the total mass and the total angular momentum which an observer at infinity would assign to a given configuration.

In any stationary axisymmetric asymptotically flat space-time there exist two Killing vectors, ξ^μ and η^μ, of which the first is timelike near spacelike and null infinities and the second is spacelike and has closed orbits. The vectors ξ^μ and η^μ commute. Using Killing's equation $\xi_{(\alpha;\mu)} = 0$ and the fact that for any vector field ζ^α

$$\zeta^\alpha{}_{;\mu;\nu} - \zeta^\alpha{}_{;\nu;\mu} = -\zeta^\beta R^\alpha{}_{\beta\mu\nu}, \tag{7.81}$$

one gets

$$\xi^{\alpha;\beta}{}_{;\beta} = -R^\alpha{}_\beta \xi^\beta \tag{7.82}$$

and, similarly,

$$\eta^{\alpha;\beta}{}_{;\beta} = -R^\alpha{}_\beta \eta^\beta . \tag{7.83}$$

Integrating equation (7.82) or (7.83) over a spacelike hypersurface Σ, remembering that $\zeta^{\alpha;\beta}$ and $\eta^{\alpha;\beta}$ are antisymmetric, and using the Gauss theorem, one obtains

$$\int_{\partial\Sigma} \zeta^{\alpha;\beta} d\Sigma_{\alpha\beta} = -\int_\Sigma R^\alpha{}_\beta \zeta^\beta d\Sigma_\alpha, \tag{7.84}$$

where $\partial\Sigma$ is the boundary of Σ, and $d\Sigma_{\alpha\beta}$ and $d\Sigma_\alpha$ are appropriately oriented surface elements on $\partial\Sigma$ and Σ. If the surface Σ is chosen so as to be asymptotically flat, to be tangent to the Killing vector η^μ, and to intersect the horizon along a 2-surface ∂B, then $\partial\Sigma$ will be composed of a surface $\partial\Sigma_\infty$ at infinity and of ∂B. If a space-time with a bounded distribution of matter is asymptotically flat and admits a timelike Killing vector ξ_α, the total mass M

assigned to the system by the observer at infinity will be

$$M = -\frac{1}{4\pi}\frac{c^2}{G}\int_{\partial\Sigma\infty} \xi^{\alpha;\beta}\mathrm{d}\Sigma_{\alpha\beta}. \tag{7.85}$$

Substituting ξ^α for ζ^α in (7.84) and using (7.85) and the field equations $R^\alpha{}_\beta - \frac{1}{2}\delta^\alpha{}_\beta R = \frac{8\pi G}{c^4} T^\alpha{}_\beta$, one obtains

$$M = \frac{1}{c^2}\int_\Sigma (2T^\alpha{}_\beta - \delta^\alpha{}_\beta T)\xi^\beta \mathrm{d}\Sigma_\alpha + \frac{1}{4\pi}\frac{c^2}{G}\int_{\partial B} \xi^{\alpha;\beta}\mathrm{d}\Sigma_{\alpha\beta}. \tag{7.86}$$

Similarly, from relation (7.84) with η^α instead of ζ^α one finds that the total angular momentum of the system can be represented as

$$J = -\frac{1}{c}\int_\Sigma T^\alpha{}_\beta \eta^\beta \mathrm{d}\Sigma_\alpha - \frac{1}{8\pi}\frac{c^3}{G}\int_{\partial B} \eta^{\alpha;\beta}\mathrm{d}\Sigma_{\alpha\beta}. \tag{7.87}$$

The linear combination of the Killing vectors ξ^μ and η^μ given by $k^\alpha = \xi^\alpha + (\Omega_H/c)\eta^\alpha$, where $\Omega_H = \frac{ca}{r_H^2 + a^2} = \text{const}$, is a null vector on the horizon. Using the vector k^α, we can transform formula (7.86) to

$$M = \frac{1}{c^2}\int_\Sigma (2T^\alpha{}_\beta - \delta^\alpha{}_\beta T)\xi^\beta \mathrm{d}\Sigma_\alpha + \frac{2}{c^2}\Omega_H J_H + \frac{1}{4\pi}\frac{c^2}{G}\int_{\partial B} k^{\alpha;\beta}\mathrm{d}\Sigma_{\alpha\beta}, \tag{7.88}$$

where

$$J_H = -\frac{1}{8\pi}\frac{c^3}{G}\int_{\partial B} \eta^{\alpha;\beta}\mathrm{d}\Sigma_{\alpha\beta} \tag{7.89}$$

is the angular momentum of the black hole. The surface element $\mathrm{d}\Sigma_{\alpha\beta}$ of the horizon can be expressed by the bivector $k_{[\alpha}l_{\beta]}$, where l_β is also a null vector such that $k^\alpha l_\alpha = 1$, and by the scalar surface element $\mathrm{d}A$: $\mathrm{d}\Sigma_{\alpha\beta} = 2k_{[\alpha}l_{\beta]}\mathrm{d}A$. Now consider the expression $2k^{\alpha;\beta}k_{[\alpha}l_{\beta]}$; it is easy to verify that it is equal to $-k_{\alpha;\beta}l^\alpha k^\beta$. Put $\kappa = -k_{\alpha;\beta}l^\alpha k^\beta$. The scalar κ measures the deviation of the parameter of the trajectory of the vector k^α from the affine parameter. To find the physical meaning of κ consider a particle moving outside the horizon on a circular orbit with the angular velocity Ω_H. Its velocity four-vector is $u^\alpha = u^0(\xi^\alpha + (\Omega_H/c)\eta^\alpha)$ and the acceleration four-vector is $\dot{u}^\alpha = u^\alpha{}_{;\beta}u^\beta$. The absolute value of the acceleration divided by u^0 (so as to measure the velocity change with respect to the time coordinate t and not the proper time) tends to κ as the particle approaches the horizon. Following Hawking we call κ the surface gravity of the black hole. Similarly to Ω_H, κ is constant on the surface of the black hole (see "the zero law" below).

The formula for the mass of the system can now be written

$$M = \frac{1}{c^2}\int_\Sigma (2T_\alpha^\beta - \delta_\alpha{}^\beta T)\xi^\alpha \mathrm{d}\Sigma_\beta + \frac{2}{c^2}\Omega_H J_H + \frac{\kappa}{4\pi G}A, \tag{7.90}$$

where A is the area of the horizon. If the space outside the horizon is empty, for a Kerr black hole we have

$$\Omega_H = \frac{ca}{r_H^2 + a^2}, \tag{7.91}$$

$$\kappa = \frac{c^2 r_H - GM}{r_H^2 + a^2}, \tag{7.92}$$

$$A = 4\pi(r_H^2 + a^2). \tag{7.93}$$

Note that the surface gravity of a Kerr black hole decreases to zero as $\left(\frac{GM}{c^2}\right)^2 \to a^2$.

Now consider two neighbouring configurations of a black hole surrounded by a stationary axisymmetric distribution of matter. The difference between the masses of the two systems as seen by an observer at infinity can be found by calculating variation in the equilibrium mass (7.90). Assuming that the distribution of matter is described by the hydrodynamic energy-momentum tensor, one obtains after some calculations

$$\delta Mc^2 = \int \Omega \delta \mathrm{d}J + \int (v_\alpha v^\alpha)^{\frac{1}{2}} \mu \delta \mathrm{d}N + \int (v_\alpha v^\alpha)^{\frac{1}{2}} T \delta \mathrm{d}S + \Omega_H \delta J_H + \frac{\kappa c^2}{8\pi G}\delta A, \tag{7.94}$$

where Ω is the angular velocity of the fluid, $v^\alpha = \xi^\alpha + (\Omega/c)\eta^\alpha$ is the tangent vector to the particle trajectories (it is not a unit vector), μ is the chemical potential and T the temperature. $\delta \mathrm{d}J$, $\delta \mathrm{d}N$ and $\delta \mathrm{d}S$ are the changes in the angular momentum, particle number and entropy of the matter flowing through the surface element $\mathrm{d}\Sigma_\alpha$.

The general properties of black holes can be brought together as the following four laws of black hole dynamics, first formulated as such by Bardeen, Carter and Hawking.

The zeroth law: *The surface gravity κ of a stationary black hole is constant over the horizon.*

To show that this is true, let m^α and $\bar{m}^\alpha$ be a complex conjugate pair of vectors generating the tangent space to ∂B. They are orthogonal to the vectors k^α and l^α introduced before. Let us calculate the change of κ in the direction of the vector m^α:

$$\kappa_{;\alpha} m^\alpha = -(k_{\alpha;\beta} l^\alpha k^\beta)_{;\gamma} m^\gamma = -R_{\alpha\beta\mu\nu} k^\alpha m^\beta k^\mu l^\nu; \tag{7.95}$$

we have used the fact that the null generators of the horizon do not diverge and are shear-free; so $\theta = k_{\alpha;\beta} m^\alpha \bar{m}^\beta = 0$ and $\sigma = k_{\alpha;\beta} m^\alpha m^\beta = 0$. Owing to the vanishing of θ and σ on the horizon, from the equation $(k_{\alpha;\beta} m^\alpha \bar{m}^\beta)_{;\mu} m^\mu = 0$ we obtain the relation $R_{\alpha\beta} k^\alpha m^\beta - R_{\alpha\beta\mu\nu} k^\alpha m^\beta k^\mu l^\nu = 0$. To show that $R_{\alpha\beta} k^\alpha m^\beta = 0$, we have to resort to the constraints set on the energy-momentum tensor.

The energy-momentum tensor is required to satisfy the inequality $T_{\alpha\beta} u^\alpha u^\beta \geqslant \frac{1}{2} T^\alpha_\alpha$ for any non-spacelike u^α. Physically, this means that for every observer the local energy density is non-negative and the local energy density flux vector is not spacelike. Therefore the vector $T_{\alpha\beta} k^\beta$ cannot be spacelike and, moreover, since on the horizon $T_{\alpha\beta} k^\alpha k^\beta = 0$, $T_{\alpha\beta} k^\beta$ must be either zero or proportional to k_α. It follows that $T_{\alpha\beta} k^\alpha m^\beta = 0$ on the horizon. We have thus shown that $\kappa_{,\alpha} m^\alpha = 0$, whence $\kappa = \mathrm{const}$ on the horizon.

The first law: *Any two neighbouring stationary axisymmetric solutions for a perfect fluid with circular flow and a central black hole are related by*

$$\delta Mc^2 = \frac{\kappa c^2}{8\pi G}\delta A + \Omega_H \delta J_H + \int \Omega \delta \, \mathrm{d}J + \int (v_\alpha v^\alpha)^{\frac{1}{2}} \mu \delta \, \mathrm{d}N + \int (v_\alpha v^\alpha)^{\frac{1}{2}} T \delta \, \mathrm{d}S . \quad (7.96)$$

The second law: *The area of the horizon of a black hole cannot decrease.*

In fact, if some number of black holes merge together, the area of the horizon of the resulting black hole A is not less than the sum of the areas of the initial horizons,

$$A \geqslant A_1 + A_2 + \dots \quad (7.97)$$

The only other additive physical quantity that cannot decrease is entropy. This suggests that the area of the horizon can be regarded as a quantity determining the entropy of the black hole. Relation (7.96) then implies that κ should be an analogue of temperature. These analogies and important conclusions which follow from them will be considered in the last section of the present chapter.

The last, third law still lacks a formal proof. We present it for completeness.

The third law: *No finite sequence of processes can reduce the surface gravity κ to zero.*

For a Kerr–Newman black hole, κ is zero only when $\left(\frac{GM}{c^2}\right)^2 - a^2 - \frac{Ge^2}{c^4} = 0$, i.e. when the hole is extreme. A further increase of the angular momentum or charge of the hole would result in the formation of a naked singularity, a drastic change in the global properties of the solution. In all real situations, where changes of state of a black hole are caused by capture of individual particles or by accretion of gas, the closer the state of the hole is to the extreme state, the more difficult it is to change it (Wald, 1975). In other words, the third law states that it is impossible to make the horizon disappear by any process. It reflects the control that the cosmic censor exercises over black holes.

7.8. Stability of Schwarzschild and Kerr Solutions

The proof of the uniqueness of the Schwarzschild solution and the Kerr solution provides very strong evidence for the existence of black holes. However, this is not an argument that can be regarded as complete. Black holes are formed as a result of gravitational collapse of stars or stellar systems with sufficiently large masses. The initial state of the system may be neither spherically symmetric nor even axisymmetric. Price has shown that, if a horizon is formed, dynamic perturbations decay with time. Still, it is conceivable that in some situation initially small perturbations may grow as the collapse progresses and the final state may be dynamic, with no horizon formed at all. Such situations, however, will be impossible if the Schwarzschild solution and the Kerr solution are stable.

As we already mentioned in Chapter 4 while examining stability of spherically symmetric stars, there are no satisfactory general methods by which to investigate the stability of complex physical systems, except observing the evolution of a given system under small perturbations. Investigations of the stability of the Schwarzschild solution were initiated by Regge and Wheeler nearly twenty years ago. In their now classical work they developed a general formalism for the description of small perturbations in the Schwarzschild space-time. If $g_{\mu\nu}$ is the metric of the unperturbed space-time, we represent perturbations as small corrections $h_{\mu\nu}$ to the metric $g_{\mu\nu}$. Thus the metric of the perturbed space-time is of the form $\tilde{g}_{\mu\nu}=g_{\mu\nu}+h_{\mu\nu}$. We use $\tilde{g}_{\mu\nu}$ to calculate the Ricci tensor. To within terms linear in $h_{\mu\nu}$ we get

$$R_{\mu\nu}(g+h)\approx R_{\mu\nu}(g)+\delta R_{\mu\nu}(h), \tag{7.98}$$

where

$$\delta R_{\mu\nu}=-\delta\Gamma^{\beta}{}_{\mu\nu;\beta}+\delta\Gamma^{\beta}{}_{\mu\beta;\nu}, \tag{7.99}$$

and

$$\delta\Gamma^{\alpha}{}_{\beta\gamma}=\tfrac{1}{2}g^{\alpha\nu}(h_{\beta\nu;\gamma}+h_{\gamma\nu;\beta}-h_{\beta\gamma;\nu}); \tag{7.100}$$

the indices in $h_{\mu\nu}$ are raised or lowered with the aid of $g_{\mu\nu}$. If the original space contains no matter, the field equations take the simple form $R_{\mu\nu}(g)=0$. There are two kinds of perturbations one can consider those which are purely gravitational, and those related to matter. For matter generates perturbations, $h_{\mu\nu}$ satisfies the equation

$$\delta G_{\mu\nu}(h)=\frac{8\pi G}{c^4}\,\delta T_{\mu\nu}, \tag{7.101}$$

where $\delta G_{\mu\nu}(h)$ denotes the linearized part of the Einstein tensor and $\delta T_{\mu\nu}$ describes the distribution of matter. Equation (7.101) written explicitly reduces to

$$\delta R_{\mu\nu}-\tfrac{1}{2}h_{\mu\nu}R^{\alpha}{}_{\alpha}-\tfrac{1}{2}g_{\mu\nu}h_{\alpha\beta}R^{\alpha\beta}-\tfrac{1}{2}g_{\mu\nu}g_{\alpha\beta}\,\delta R^{\alpha\beta}=\frac{8\pi G}{c^4}\,\delta T_{\mu\nu}. \tag{7.102}$$

In the following we shall be interested in purely gravitational perturbations. In this case the equation for small corrections reduces to

$$\delta R_{\mu\nu}(h)=0. \tag{7.103}$$

Before we go any further, let us make one more general remark. The corrections $h_{\mu\nu}$ are not determined uniquely. If we perform a coordinate transformation

$$x^{\mu}\to x^{\mu}+\xi^{\mu}(x^{\alpha}), \tag{7.104}$$

where $\xi^{\mu}(x)$ are small coordinate deformations, $h_{\mu\nu}$ will change according to

$$h_{\mu\nu}\to h_{\mu\nu}+\xi_{\mu;\nu}+\xi_{\nu;\mu}. \tag{7.105}$$

It is straightforward to check that these changes do not affect the form of eqn. (7.103). The functions ξ_{μ} will get cancelled. In other words, if $h_{\mu\nu}$ is a solution of eqn. (7.103), then $h_{\mu\nu}+\xi_{\mu;\nu}+\xi_{\nu;\mu}$ is also a solution for arbitrary ξ_{μ}. This freedom should be borne in mind. Physically significant results should be independent of the choice of ξ_{μ}.

In the case of the Schwarzschild metric, owing to its spherical symmetry, the perturbations $h_{\mu\nu}$ can be expanded into spherical harmonics $Y_l^m(\theta,\varphi)$. The perturbations corresponding to given l and m are axisymmetric, so without loss of generality we can restrict our attention to axisymmetric perturbations. They fall into two disjoint classes according as they are odd or even. By suitable gauging, $h_{\mu\nu}$ can always be reduced to the canonical form. For odd perturbations we have

$$h_{\mu\nu}=\begin{bmatrix} 0 & 0 & 0 & h_0(r) \\ 0 & 0 & 0 & h_1(r) \\ 0 & 0 & 0 & 0 \\ h_0(r) & h_1(r) & 0 & 0 \end{bmatrix} e^{-i\omega t}\sin\theta\frac{\partial}{\partial\theta}P_l(\cos\theta), \tag{7.106}$$

and for the even ones

$$h_{\mu\nu}=\begin{bmatrix} H_0(1-r_g/r) & H_1 & 0 & 0 \\ H_1 & H_2(1-r_g/r)^{-1} & 0 & 0 \\ 0 & 0 & r^2K & 0 \\ 0 & 0 & 0 & r^2K\sin^2\theta \end{bmatrix} e^{-i\omega t}P_l(\cos\theta). \tag{7.107}$$

$P_l(\cos\theta)$ are the Legendre polynomials. Inserting $h_{\mu\nu}$ into equation (7.103), we obtain a system of equations for the radial components of the perturbations. In order to discuss these equations it is convenient to introduce a new radial coordinate r^* defined by

$$r^*=r+r_g\ln(r/r_g-1). \tag{7.108}$$

The equations for odd perturbations then reduce to one second-order equation

$$\frac{d^2Q}{dr^{*2}}+(\omega^2-V_{\rm eff})Q=0, \tag{7.109}$$

where

$$Q=\frac{h_1(r)}{r}(1-r_g/r), \tag{7.110}$$

and

$$V_{\rm eff}=(1-r_g/r)\left(\frac{l(l+1)}{r^2}-\frac{3r_g}{r^3}\right); \tag{7.111}$$

$h_0(r)$ is related to Q by the equation

$$h_0(r)=\frac{i}{\omega}\frac{d}{dr^*}(rQ). \tag{7.112}$$

The new radial coordinate r^* ranges from $-\infty$ to $+\infty$, with $r^*\to\infty$ as $r\to\infty$ and $r^*\to-\infty$ as $r\to r_g$. The dependence of the effective potential on r was shown earlier in Fig. 7.5. The potential attains a maximum at $r\approx\frac{3}{2}r_g$ and decays very quickly when $r^*\to\pm\infty$.

We now need to specify the boundary conditions for equation (7.109) at infinity and on the horizon. The Schwarzschild coordinates are singular on the horizon. Therefore, we must first pass to the Kruskal coordinates, use them to define the boundary conditions near the horizon, and then return to the Schwarzschild coordinates. Owing to the fact that the effective potential vanishes at infinity and on the horizon, we have

$$Q_\infty\sim e^{\pm i\omega r}, \quad Q_H\sim e^{\pm i\omega r^*}. \tag{7.113}$$

Now consider the case where ω is purely imaginary, i.e. the case of perturbations growing with time. If $\omega=i\sigma$, perturbations dying out at infinity will be of the form $Q_\infty\sim e^{-\sigma r}$.

If Q is positive, equation (7.109) implies that $\frac{d^2Q}{dr^{*2}}$ is non-negative over the entire range of r^* from $-\infty$ to $+\infty$. It follows that a solution vanishing at infinity cannot be matched to a solution vanishing near the horizon. Thus near the horizon the solution regular at infinity behaves as $Q_H \sim e^{-\sigma r^*}$. The reason that in the Schwarzschild coordinates the perturbations grow unboundedly near the horizon is not enough to reject such perturbations as non-physical. We have to pass to Kruskal coordinates. The component h_{03} is transformed most easily:

$$h_{03}^K = 2r_g(U^2 - V^2)^{-1}[Uh_{03} - V(1-r_g/r)h_{13}]. \tag{7.114}$$

The relationship between the Schwarzschild and the Kruskal coordinates was given in Chapter 1 (see formulae (1.77)–(1.78)). From formula (7.112) we find $h_0(r) \approx -r_g e^{-\sigma r^*}$, and from (7.110) $h_1(r) \approx \frac{r}{1-r_g/r} e^{-\sigma r^*}$. Inserting these results into (7.114) and omitting the angular variables, we obtain $h_{03}^K \approx -2r_g^2(U^2-V^2)^{-1}(U+V)e^{-\sigma r^*}e^{\sigma t}$. Using formula (1.78), we can eliminate t and r^*, to obtain finally $h_{03}^K \approx -2r_g^2(U-V)^{-2\sigma/r_g}$. Since at $t=0$ and near the horizon $V=0$, $h_{03}^K \approx -2r_g^2 U^{-2\sigma/r_g}$. If u is small enough, h_{03}^K can be arbitrarily large, which contradicts our assumption that the perturbation is small compared to the background. Thus we come to a very important conclusion: perturbations which grow with time are non-physical, for they do not satisfy, even at the initial moment, the condition of regularity over the whole space.

Perturbations with real frequencies represent gravitational waves; they are regular gravitational perturbations. Near the horizon and at infinity both Q_H and $Q_\infty \approx e^{\pm i\omega r^*}$. The properties of black holes discussed before imply that only ingoing waves are admissible near the horizon, hence in this case

$$Q_H \approx e^{-i\omega r^*}, \qquad Q_\infty \approx e^{\pm i\omega r^*}. \tag{7.115}$$

As previously, we must check whether the real-frequency perturbations are regular in Kruskal coordinates. It turns out that the amplitude of the monochromatic waves is divergent, for near the horizon we have $h_{03}^K \approx 2r_g^2(U+V)^{-1}(U+V)^{-2i\omega r_g}$. This time, however, the singularity is not serious and we can eliminate it by passing to wave packets. These will be regular and physically acceptable.

The case of even perturbations is more complicated. For $l > 1$ the field equations impose a further condition on H_0 and H_2, namely $H_0 = H_2 = H$. Thus there remain only three independent radial functions. The field equations reduce to one differential equation for the function $S = \frac{H_1(r)}{r}$ and relations between H and K. The equation is rather complicated and we shall consider only its asymptotic forms. Introducing the dimensionless quantities $x = r/r_g$, $x^* = r^*/r_g$ and $\omega^* = \omega r_g$, Vishveshwara obtained the equations

$$\frac{d^2S_\infty}{dx^2} + \omega^{*2}S_\infty = 0 \quad \text{for} \quad x \to \infty \tag{7.116}$$

and

$$\frac{d^2S_H}{dx^{*2}}+2\frac{dS_H}{dx^*}+(\omega^{*2}+1)S_H=0 \quad \text{for} \quad x\to 1. \tag{7.117}$$

The asymptotic solutions for imaginary frequencies $\omega^*=i\alpha$ have the form

$$S_\infty \approx e^{\pm\alpha x}, \quad S_H \approx e^{-(1\pm\alpha)x^*}. \tag{7.118}$$

To ensure the regularity of the solution at infinity, we assume $S_\infty \approx e^{-\alpha x}$. If $\alpha>1$, the regular solution near the horizon is $S_H \approx e^{(\alpha-1)x^*}$. As in the case of odd perturbations, by using the equation for the function S it can be shown that a solution which is regular at infinity cannot be matched to a solution which is regular near the horizon. A solution which is divergent near the horizon in the Schwarzschild coordinates is also divergent in the Kruskal coordinates. It is a non-physical perturbation.

The situation is similar when $\alpha=1$ and when $0<\alpha<1$. We thus come to the important conclusion that perturbations which grow with time and are regular at infinity are unbounded near the horizon. Conversely, it can also be shown that perturbations which are regular near the horizon are unbounded at infinity. In other words, the *Schwarzschild metric does not admit small perturbations with an amplitude growing with time.*

In the case of odd perturbations the above results can be made more precise. The differential operator

$$D=-\frac{d^2}{dr^{*2}}+V_{\text{eff}} \tag{7.119}$$

has a self-adjoint extension and is positive-definite. The spectrum of a positive self-adjoint operator contains no negative eigenvalues. From the properties of the operator D alone it follows that every regular perturbation can be resolved into odd perturbations of real frequencies.

Although no one has yet managed to obtain similar results for even perturbations at this level of generality, the asymptotic properties of the latter and certain general similarities with odd perturbations lead us to expect that an analogous theorem holds for them too. The crucial missing link is the proof that the even real-frequency solutions form a complete set of functions.

To examine stationary perturbations of Schwarzschild space-time it is enough to set $\omega=0$ in the equations for perturbations (7.109). The odd perturbations with $l=1$ are given by the equation

$$\frac{d^2h_0}{dr^2}=\frac{2h_0}{r^2}. \tag{7.120}$$

The solution regular at infinity is

$$h_0=\frac{b}{r}, \tag{7.121}$$

where $b=\text{const}$, and therefore

$$h_{03}=\frac{b}{r}\sin^2\theta. \tag{7.122}$$

Comparing this result with the weak-field form (3.50), we can interpret it as a rotational perturbation. It is regular near the horizon. If we expanded the Kerr metric in the parameter a and retained only linear terms then, regarding these terms as small perturbations of the Schwarzschild metric, we would obtain $h_{03}=\frac{ar_g}{r}\sin^2\theta$, which shows how b is related to the angular momentum.

Analysing stationary odd perturbations with $l>1$, Regge and Wheeler found that they cannot be regular. As in the case of time-dependent perturbations, those perturbations which are regular at infinity are singular near the horizon, and conversely: those regular near the horizon are singular at infinity.

Stationary even perturbations with $l=0$, $l=1$ correspond to a small change in the mass and to a small shift of the centre of symmetry, respectively. Regular perturbations with $l>1$ do not exist. These results tell us that a stationary black hole is fully characterized by its mass and angular momentum, and all the other quantities are uniquely determined by these two.

Stability of the Kerr solution is a conjecture which is still waiting for a complete proof. At first it seemed that the problem was hopelessly difficult. That investigations of stability of the Schwarzschild solution led to satisfactory results so quickly was due to the fact that the original space-time admitted a four-parameter symmetry group and it was known right from the start that the equations for perturbations would be separable. Kerr space-time admits only a two-parameter symmetry group, and finding separable equations could, in principle, prove impossible. In 1972 Teukolsky noticed, however, that the equations for certain combinations of the components of the perturbed curvature tensor can be separated. It turned out that in the case of purely gravitational or electromagnetic perturbations those combinations carry full information about the properties of the perturbed field. If we write the Kerr metric in Boyer–Lindquist coordinates, it will take the form

$$ds^2=\frac{\Delta-a^2\sin^2\theta}{\Sigma}c^2dt^2+\frac{2r_g ar\sin^2\theta}{\Sigma}c\,dt\,d\varphi-\frac{\Sigma}{\Delta}dr^2-\Sigma d\theta^2$$

$$-\sin^2\theta\left[r^2+a^2+\frac{a^2r_g r\sin^2\theta}{\Sigma}\right]d\varphi^2, \qquad (7.123)$$

$$\Delta=r-r_g r+a^2, \quad \Sigma=r^2+a^2\cos^2\theta, \quad r_g=\frac{2GM}{c^2}.$$

Let

$$l^\mu=\left(\frac{r^2+a^2}{\Delta},1,0,\frac{a}{\Delta}\right), \quad m^\mu=\frac{1}{\sqrt{2}(r+ia\cos\theta)}\left(ia\sin\theta,0,1,\frac{i}{\sin\theta}\right)$$

and

$$n^\mu=\frac{1}{2\Sigma}(r^2+a^2,-\Delta,0,a).$$

Separable equations can be obtained for

$$\psi_0=\delta R_{\alpha\beta\gamma\delta}l^\alpha m^\beta l^\gamma m^\delta \quad \text{and} \quad \psi_4=\delta R_{\alpha\beta\gamma\delta}n^\alpha\bar{m}^\beta n^\gamma\bar{m}^\delta \qquad (7.124)$$

in the case of gravitational perturbations, and for

$$\phi_0=\delta F_{\mu\nu}l^\mu m^\nu \quad \text{and} \quad \phi_2=\delta F_{\mu\nu}\bar{m}^\mu n^\nu \tag{7.125}$$

in the case of electromagnetic perturbations. The peturbations satisfy the homogeneous, separable wave equation

$$\mathscr{F}\psi(t,r,\theta,\varphi)=0, \tag{7.126}$$

where

$$\mathscr{F}=\left[\frac{(r^2+a^2)^2}{\Delta}-a^2\sin^2\theta\right]\frac{\partial^2}{c^2\partial t^2}+\frac{2r_g ar}{\Delta}\frac{\partial}{c\partial t}\frac{\partial}{\partial\varphi}+\left(\frac{a^2}{\Delta}-\frac{1}{\sin^2\theta}\right)\frac{\partial^2}{\partial\varphi^2}$$
$$-\Delta^{-s}\frac{\partial}{\partial r}\left(\Delta^{s+1}\frac{\partial}{\partial r}\right)-\frac{1}{\sin\theta}\frac{\partial}{\partial\theta}\left(\sin\theta\frac{\partial}{\partial\theta}\right)-2s\left[\frac{a(r-\frac{1}{2}r_g)}{\Delta}+\frac{i\cos\theta}{\sin^2\theta}\right]\frac{\partial}{\partial\varphi}$$
$$-2s\left[\frac{\frac{1}{2}r_g(r^2-a^2)}{\Delta}-r-ia\cos\theta\right]\frac{\partial}{c\partial t}+(s^2\cot^2\theta-s); \tag{7.127}$$

$s=0$ corresponds to scalar perturbations, $s=\pm1$ to electromagnetic perturbations, and $s=\pm2$ gravitational perturbations.

In the following we are interested in gravitational perturbations only, and so we take $s=\pm2$. Assuming that ψ is of the form

$$\psi=\mathrm{e}^{-i\omega t}\mathrm{e}^{im\phi}S(\theta)R(r), \tag{7.128}$$

we obtain from (7.126) the system of two equations

$$\Delta^{-s}\frac{\mathrm{d}}{\mathrm{d}r}\left(\Delta^{s+1}\frac{\mathrm{d}R}{\mathrm{d}r}\right)+\left(\frac{K^2-2is(r-\frac{1}{2}r_g)K}{\Delta}+4is\omega r-\lambda\right)R=0, \tag{7.129}$$

$$\frac{1}{\sin\theta}\frac{\mathrm{d}}{\mathrm{d}\theta}\left(\sin\theta\frac{\mathrm{d}S}{\mathrm{d}\theta}\right)+\left(a^2\omega^2\cos^2\theta-\frac{m}{\sin^2\theta}-2a\omega s\cos\theta\right.$$
$$\left.-\frac{2ms\cos\theta}{\sin^2\theta}-s^2\cot^2\theta+s+A\right)S=0, \tag{7.130}$$

where $K=(r^2+a^2)\,\omega-am$, $\lambda=A+a^2\omega^2-2am\omega$ and $A=$const. Equation (7.130) together with the boundary conditions ensuring the regularity of the solution for $\theta=0$ and $\theta=\pi/2$ can be regarded as a Sturm–Liouville eigenvalue problem for the separation constant $A={}_sA_l^m(a\omega)$. The theory of Sturm–Liouville equations implies that for every fixed m, s and $a\omega$ the eigenfunctions ${}_sS_l^m$ form a complete, orthogonal system of functions.

In order to examine the asymptotic behaviour of the radial function R, introduce the variable r^* defined by

$$\frac{\mathrm{d}r^*}{\mathrm{d}r}=\frac{r^2+a^2}{\Delta}, \tag{7.131}$$

and replace R by

$$Y=\Delta^{s/2}(r^2+a^2)^{1/2}R. \tag{7.132}$$

Equation (7.129) will then take the form

$$\frac{d^2Y}{dr^{*2}}+\left\{\frac{K^2-2is(r-\frac{1}{2}r_g)K+\Delta(4ir\omega s-\lambda)}{(r^2+a^2)^2}-G^2-\frac{dG}{dr^*}\right\}Y=0, \tag{7.133}$$

where $G=\frac{s(r-\frac{1}{2}r_g)}{r^2+a^2}+\frac{r\Delta}{(r^2+a^2)^2}$. It is now easy to find the asymptotic form of the solution. As $r\to\infty$ and $r^*\to\infty$, we obtain $Y\sim r^{\pm s}\,e^{\pm i\omega r^*}$, so $R\sim\frac{1}{r}e^{-i\omega r^*}$ or $\frac{1}{r^{2s+1}}e^{i\omega r^*}$ (see the extensive discussion by Chandrasekhar, 1979 and 1983). Physically, we can interpret the functions

$$\psi_4\sim\frac{1}{r}e^{i\omega r^*},\qquad \psi_0\sim\frac{1}{r^5}e^{i\omega r^*} \tag{7.134}$$

as representing outgoing waves, and the functions

$$\psi_0\sim\frac{1}{r}e^{-i\omega r^*},\qquad \psi_4\sim\frac{1}{r^5}e^{-i\omega r^*} \tag{7.135}$$

as representing ingoing waves. Near the horizon, when $r\to r_H$, or $r^*\to-\infty$, we get $Y\sim\Delta^{\pm s/2}\,e^{\pm i\alpha r^*}$, where $\alpha=\omega-m\frac{a}{r_g r_H}$, which leads to $R\sim e^{i\alpha r^*}$ or $R\sim\Delta^{-s}e^{-i\alpha r^*}$. Of these two possibilities we have to choose that which ensures that no information can escape from inside the horizon. Considering the behaviour of the perturbations in the local frame of an observer near the horizon, we come to the conclusion that the solution which is regular for $r\to r_H$ has asymptotically the form

$$R\sim\Delta^{-s}e^{-i\alpha r^*}. \tag{7.136}$$

Using this information, Press and Teukolsky, and independently Detweiler and Ipser, carried out a stability analysis for the Kerr solution. The Kerr solution becomes a Schwarzschild solution when $a\to0$. An instability in the Kerr solution, if it can occur at all, should appear for large $a\left(a^2\to\left(\frac{GM}{c^2}\right)^2\right)$. Consider a perturbation of a given frequency ω. In general, ω may be complex, and if it lies in the lower half of the complex plane, i.e. if Im $\omega<0$, the configuration is stable, while if Im $\omega>0$, the configuration is unstable. Transition from stability to instability occurs when ω crosses the real axis. Press and Teukolsky restricted their attention to perturbations of real frequencies. They solved numerically the equation for the radial function $R(r)$ subject to the boundary conditions

$$R\sim\Delta^{-s}e^{-i\alpha r^*}\quad\text{when } r\to r_H,\qquad R\sim\frac{1}{r^{2s+1}}e^{i\alpha r^*}\quad\text{when } r\to\infty. \tag{7.137}$$

They further examined the radial functions for $l\leqslant3$ and $0<a<\frac{GM}{c^2}$ and found no configurations at the stability limit for any real frequency. They integrated the radial equation numerically, constructing the solution step by step from the horizon upward. Asymptotically, for large R, they decomposed the solution into parts corresponding to the ingoing and the outgoing waves

$$R = Z_{in}(\omega)\frac{1}{r}e^{-i\omega r^*} + Z_{out}(\omega)\frac{1}{r^{2s+1}}e^{i\omega r^*} \tag{7.138}$$

and examined the behaviour of $Z_{out}(\omega)/Z_{in}(\omega)$ in relation to ω. Configurations at the stability limit correspond to those values of ω for which $Z_{in}=0$. As an illustration, Fig. 7.11 and Fig. 7.12 show the dependence of the function $Z(\omega)=Z_{out}(\omega)/Z_{in}(\omega)$ on frequency for $l=2$, $m=\pm 1$ and for $l=3$, $m=\pm 2$. The solutions are regular not only as functions of ω but also as functions of the parameter a. Although these results are not yet enough to state with complete certainty that the Kerr solution is stable, they provide very strong evidence in favour of this hypothesis.

As in the case of Schwarzschild space-time, regular stationary perturbations of Kerr space-time involve small changes in the mass and angular momentum, and shifts of the centre. All other stationary perturbations are singular either at infinity or near the horizon.

Stability of the Schwarzschild solution has been established by a rigorous proof; everything seems to indicate that the Kerr solution is also stable. Thus, if the cosmic censorship hypothesis is true, gravitational collapse of stars whose masses are somewhat greater than the mass of the Sun should lead to the formation of black holes. The confirmation of the existence of black holes is a matter of astronomical observations. Discussing

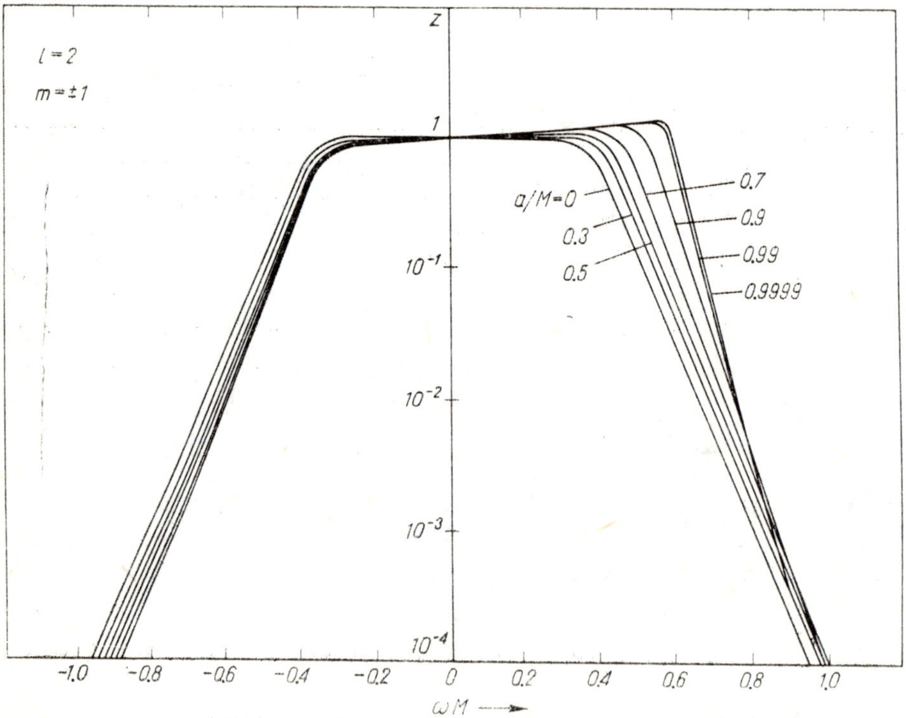

FIG. 7.11. The outgoing to ingoing wave amplitude ratio vs. ωM for $l=2$, $m=\pm 1$ measured asymptotically far from a Kerr black hole (adapted from Press and Teukolsky, 1973).

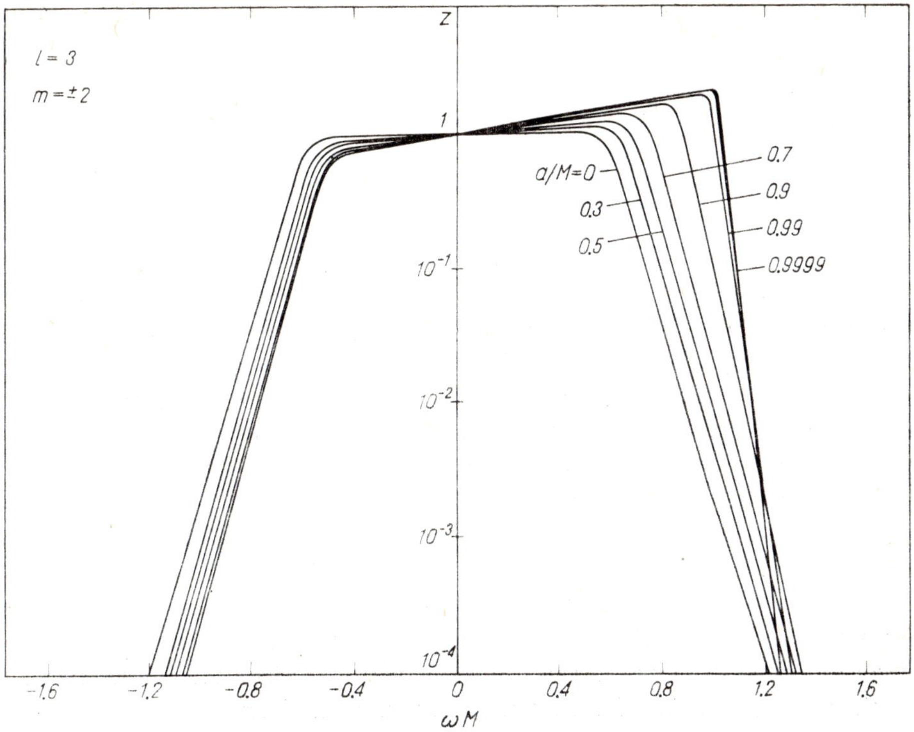

FIG. 7.12. The outgoing to ingoing wave amplitude ratio vs. ωM for $l=3$, $m=\pm 2$ measured asymptotically far from a Kerr black hole (adapted from Press and Teukolsky, 1973).

the process of gravitational collapse, we came to the conclusion that there was no possibility of such a process ever being observed. Nor can we directly observe a black hole. Therefore, we shall have to investigate the processes that may occur near black holes and so provide a basis for astronomical search for black holes. We shall take up this problem in Chapter 8.

7.9. Particle Creation by Black Holes. The Hawking Process

In 1974, analysing the behaviour of a scalar field in an asymptotically flat space-time containing a black hole, Hawking reached the conclusion that a black hole emits particles as if it were a body of temperature $T=\dfrac{\kappa h}{2\pi k c}$, where κ is the surface gravity of the hole, c the speed of light and k the Boltzmann constant. Before we present Hawking's arguments in more detail, it is worthwhile to recall the successive steps of the reasoning which lead to this very important discovery.

A Schwarzschild spherically symmetric black hole is a very simple object. Its external gravitational field is static, and it is only through this field that it can interact with its surroundings. Stationary Kerr black holes are more interesting. As shown by Penrose, owing to the existence of the ergosphere it is possible to extract, to some extent at least, their rotational energy. Similarly, energy can be drawn from charged static black holes. The Penrose process consists in sending a composite particle into the ergosphere, where it splits into two parts, one of which, whose energy is negative from a distant observer's point of view, falls into the black hole, while the other escapes to infinity carrying more energy than the original particle. We discussed this process in greater detail in Section 6.

Zeldovich pointed out, and Starobinsky confirmed it by direct calculation, that rotational energy can be drawn from a black hole by scattering on it electromagnetic or gravitational waves. It turned out that depending on the frequency, an incident wave may be absorbed, or it may undergo superradiant scattering in which case it extracts energy from a rotating black hole. Superradiant scattering occurs when the angular frequency of the incident wave satisfies the inequality $\omega < m\Omega_H$, where m is the projection of the angular momentum of the wave on the rotation axis of the black hole, and Ω_H is the angular velocity of the hole.† The maximum amplification of an electromagnetic wave is 4.4% and occurs when $l=m=1$, and that of a gravitational wave is 138%, for $l=m=2$. Zeldovich also advanced the hypothesis of spontaneous emission of particles by a rotating black hole. It is to be noted that both the superradiant scattering and spontaneous emission of particles take place at the expense of the rotational energy of the black hole and that both effects disappear when $\Omega_H=0$. At the same time, Bekenstein proposed a method by which to assign entropy to black holes. His considerations were based on a formal similarity between the first law of black hole mechanics and the first law of thermodynamics. The former can be written in the form

$$\delta(Mc^2) = \frac{\kappa c^2}{8\pi G}\delta A + \Omega_H \delta J_H, \tag{7.139}$$

which resembles the first law of thermodynamics

$$dU = TdS + pdV, \tag{7.140}$$

with $\delta A \geqslant 0$ and $dS \geqslant 0$. According to Bekenstein, the entropy of a black hole should be proportional to the surface area of the hole. If this were so, the black hole should have a certain temperature assigned to it. Bekenstein's arguments were based not only on that formal similarity. The entropy of a black hole should depend only on the fundamental parameters characterizing the hole, i.e. on mass and angular momentum, and should possess two properties:

(a) the entropy should increase whenever particles or radiation fall into the black hole,

(b) when black holes merge together, the entropy of the resulting black hole should not be less than the sum of the entropies of the original components.

A quantity satisfying these conditions would have to be a monotonic function of the surface area, $S(A)$, such that $\frac{d^2S}{dA^2} \geqslant 0$. Bekenstein suggested the simplest possibility $S(A)$

† If the incident wave is charged, superradiant scattering by a charged rotating black hole occurs when $\omega < m\Omega_H + e\phi_H$, where ϕ_H is the Coulomb potential on the horizon and e the charge of the scattered field. In this case it is possible to extract electrostatic energy from the hole.

$=\gamma\dfrac{c^3A}{Gh}$, but did not succeed in determining the correct value of γ ($\gamma=\frac{1}{4}$;see an excellent review by Sciama, 1976).

Hawking adopted a different approach to the problem. He investigated the properties of a scalar field in the space-time of a spherically symmteric collapsing star. The space-time metric outside a static black hole is given by the Schwarzschild line element. The wave equation in this space-time takes the form

$$\frac{1}{1-(r_g/r)}\frac{\partial^2\phi}{c^2\partial t^2}-\frac{1}{r^2}\left[r^2(1-r_g/r)\frac{\partial\phi}{\partial r}\right]_{,r}-\frac{1}{r^2}\left[\frac{1}{\sin\theta}\left(\sin\theta\frac{\partial\phi}{\partial\theta}\right)_{,\theta}+\frac{1}{\sin^2\theta}\frac{\partial^2\phi}{\partial\varphi^2}\right]=0. \quad (7.141)$$

By spherical symmetry we can seek a particular solution of this equation of the form

$$\phi=\frac{1}{r\sqrt{2\pi\omega}}e^{-i\omega t}R_{\omega lm}(r)\,Y_l^m(\theta,\varphi), \quad (7.142)$$

where $Y_l^m(\theta,\varphi)$ are spherical harmonic functions, and $R_{\omega lm}(r)$ satisfies the equation

$$\frac{d^2R}{dr^{*2}}+\left[\left(\frac{\omega}{c}\right)^2-\frac{r_g(r-r_g)}{r^4}-\frac{l(l+1)(r-r_g)}{r^3}\right]R=0, \quad (7.143)$$

where r^* is given by (7.108).

In flat Minkowski space-time a physical interpretation of solutions (7.142) presents no difficulty. Owing to the fact that the Minkowski space-time is invariant under the Poincaré group of transformations, and basically because it admits a global timelike Killing vector field, it is possible to introduce invariant notions of particle, antiparticle and vacuum states. (Recall that if ξ^μ is a timelike Killing vector, then a fundamental solution – a spherical or plane wave with wave vector k_μ($k_\mu k^\mu=0$ from the wave equation) – is called a positive-frequency solution if $\xi^\mu k_\mu>0$. The wave packets formed from positive-frequency fundamental solutions represent particles. Complex conjugates of the positive-frequency fundamental solution give fundamental solutions for antiparticles).

In Schwarzschild space-time there is no global timelike Killing vector field. The space-time region $r<r_g$ is non-stationary and admits no timelike Killing vector. The method of distinguishing between particles and antiparticles proposed above loses sense and so does the notion of a particle. Still, the situation is not hopeless, because a timelike Killing vector does exist near future and past null infinities $\mathscr{I}^+$ and $\mathscr{I}^-$ (De Witt, 1975). In his work, Hawking uses this fact to analyse the behaviour of a massless scalar field in the space-time of a collapsing star. The history of the collapse process is illustrated in Fig. 7.13. The solution of the wave equation in the region between $\mathscr{I}^-$, $\mathscr{H}^+$ and $\mathscr{I}^+$ is determined uniquely by the boundary conditions on $\mathscr{I}^-$, or on $\mathscr{I}^+$ and $\mathscr{H}^+$, if the star is not the source of the scalar field. Otherwise we have to add information about the behaviour of the scalar field at the surface of the star. The initial conditions given on $\mathscr{I}^-$ and $\mathscr{I}^+$ have a simple physical interpretation. The usual assumption is that there are no outgoing particles on $\mathscr{I}^-$ and no ingoing particles on $\mathscr{I}^+$. The initial conditions given on $\mathscr{H}^+$ cannot be interpreted so simply.

In Minkowski space-time, for a source-free wave equation, all the particles ingoing through $\mathscr{I}^-$ are eventually found on $\mathscr{I}^+$, and there will be exactly as many of them on $\mathscr{I}^+$ as there were on $\mathscr{I}^-$. Hawking has shown that in space-time containing a black hole more particles reach $\mathscr{I}^+$ than there were originally on $\mathscr{I}^-$. He interprets this effect as

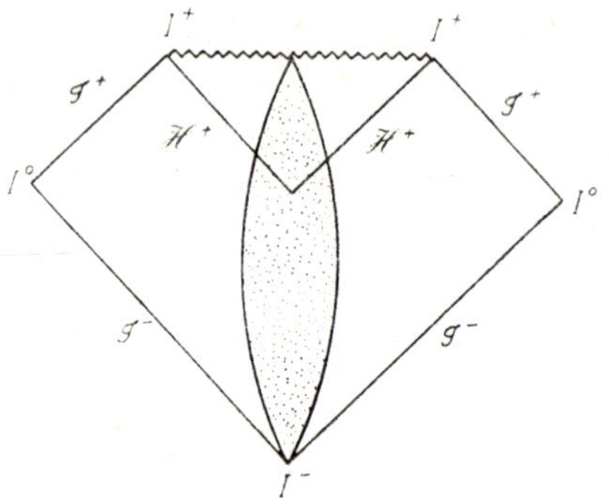

FIG. 7.13. Asymptotically flat space-time with a spherically symmetric collapsing star.

creation of particles. The probability of such emission of a particle and the probability of absorption of a particle in the same state are related exactly as in the theory of thermal radiation. Hawking's original calculations are very instructive but they are somewhat complicated. Here let us present a simpler, though more formal, derivation proposed by Damour and Ruffini.

Equation (7.143) implies that near the horizon, for $r \to r_g$, solution (7.142) becomes

$$\phi^{\text{in}} \sim e^{-i\omega(t+r^*/c)} = e^{-i\omega v}, \tag{7.144}$$

$$\phi^{\text{out}} \sim e^{-i\omega(t-r^*/c)} = e^{2i\omega r^*/c} e^{-i\omega v} = (r-r_g)^{2i\omega r_g/c} e^{-i\omega v}. \tag{7.145}$$

The ingoing wave function is regular near the horizon and can be extended to the region $r<r_g$, whereas the outgoing wave function is singular near the horizon and cannot be straightforwardly extended inside. It is known from the quantum field theory, however, that a wave function can be analytically continued to the complex manifold $z=x+iy$, where x is an arbitrary vector and y is a timelike vector such that $y^0<0$. The regularized extension of ϕ^{out}, resolved into a wave outgoing from the horizon and a wave infalling on the singularity, is

$$\phi^{\text{out}}_{\text{reg}} = N[H(r-r_g)\phi^{\text{out}}(r-r_g) + e^{2\pi\omega r_g/c} H(r_g-r)\phi^{\text{out}}(r_g-r)]. \tag{7.146}$$

Here $H(x)$ is the step function

$$H(x) = \begin{cases} 0, & x<0, \\ 1, & x>0. \end{cases} \tag{7.147}$$

The normalization factor N has been chosen so that

$$\langle \phi^{\text{out}}_{\text{reg}\,\omega_1}, \phi^{\text{out}}_{\text{reg}\,\omega_2} \rangle = -\delta(\omega_1-\omega_2)\,\delta_{l_1 l_2}\,\delta_{m_1 m_2}; \tag{7.148}$$

hence

$$|N|^2 = (e^{4\pi\omega r_g/c} - 1)^{-1}. \tag{7.149}$$

Thus the probability of the emission of a particle by the black hole per unit of time and unit of surface area is $|N|^2/2\pi$. Not all the particles emitted by the black hole will escape to infinity. They meet on their way a potential barrier, which, as follows from equation (7.143), is produced by the centrifugal term and the curvature of the space. The situation is presented schematically in Fig. 7.14. Denote by Γ_ω the transmission coefficient of the barrier. The particle flux per unit of time and unit of frequency which will be transmitted to infinity has the intensity

$$\frac{\Gamma_\omega}{2\pi}(e^{4\pi\omega r_g/c}-1)^{-1}. \tag{7.150}$$

Comparing this result with Planck's formula, we conclude that a *spherical black hole*

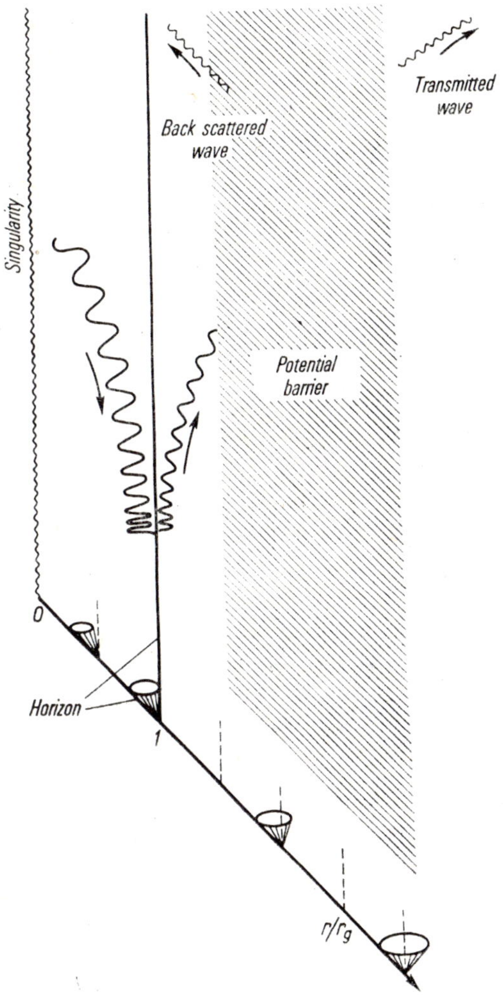

FIG. 7.14. Particle creation near the horizon. Some particles are backscattered by the effective potential and will not reach the observer at infinity (adapted from Damour and Ruffini, 1976).

emits particles as if it were a hot body of temperature

$$T_H = \frac{hc}{4\pi r_g k} = \frac{\kappa h}{2\pi kc}. \tag{7.151}$$

Note that the rate of the particle emission by the black hole is independent of the history of the hole, hence of the details of the collapse, and is fully determined by the parameters of the black hole. The fundamental role in the process of emission is to be attributed to the presence of the future horizon. Therefore, particle emission by a black hole also occurs when the hole is static. This quantum emission of particles by black holes is known as the "Hawking process".

The expression "particle creation by a black hole", which we often use in this context, requires a few words of comment. As has already been mentioned, the standard notion of a particle becomes meaningless near the horizon, and therefore the functions (7.144) and (7.145) cannot be interpreted as particle wave functions. For this reason one cannot say that the created particles are sent from the surface of the black hole. *Creation of particles is a non-local process*; it takes place in the strong gravitational field of the black hole.

Using formula (7.151), one can calculate the temperature of a black hole. On inserting the numerical values one gets

$$T_H = 6{\cdot}2 \cdot 10^{-8} \frac{M_\odot}{M} K, \tag{7.152}$$

which implies that the temperature of typical black holes of stellar size is very low, and therefore the process of particle creation by such holes can in practice be neglected. The Hawking process plays an important part in the evolution of black holes of small mass. Small black holes could have been formed only at very early stages of the evolution of the universe. A black hole with a mass of 10^{15} g would have had enough time by now to "evaporate" through emission of particles. As the mass of a black hole decreases, its temperature rises and, according to Hawking's estimate, during the last 0.1 s of its life the hole emits energy equal to 10^{30} erg. However, these estimates should be viewed with scepticism as nothing is known about the last stages of the "evaporation" of black holes.

The Hawking process is a very important step towards a full understanding of the connection between quantum phenomena, thermodynamics and the relativistic theory of gravitation. Although the estimates quoted above imply that the Hawking process will hardly be of any relevance to black hole astrophysics, it may have great cosmological significance.

Literature

AMES, W. and THORNE, K. S. (1968) *Astrophys. J.* **151**, 659.

ARNETT, W. D. (1979) *Physical Aspects of Collapse*, in Proceedings of Ninth Texas Symposium on Relativistic Astrophysics, Annals New York Acad. Sci.

BARDEEN, J., CARTER, B. and HAWKING, S. (1973) *Comm. Math. Phys.* **31**, 161.

BEKENSTEIN, J. D. (1974) *Phys. Rev.* **D9**, 3292.

BEKENSTEIN, J. D. (1980) *Physics Today*, p. 24, January.

BOYER, R. and LINDQUIST, R. (1967) *J. Math. Phys.* **8**, 265

CARTER, B. (1968) *Phys. Rev.* **174**, 1559.

CARTER, B. (1973) *Black Hole Equilibrium States*, in *Black Holes*, ed.: DeWitt, C., Gordon and Breach, N. Y.

CARTER, B. (1979) *The General Theory of the Mechanical, Electromagnetic and Thermodynamic Properties of Black Holes*, in *General Relativity, an Einstein Centenary Survey*, eds.: Hawking, S. W., Israel, W., Cambridge University Press, Cambridge.

CARTER, B., GIBBONS, G. W., LIN, D. N. C. and PERRY, M. J. (1976) *Astron. Astrophys.* **52**, 327.

CHANDRASEKHAR, S. (1979) *An Introduction to the Theory of the Kerr Metric and Its Perturbations*, in *General Relativity, an Einstein Centenary Survey*, eds.: Hawking, S. W., Israel, W., Cambridge University Press, Cambridge.

CHANDRASEKHAR, S. (1983) *The Mathematical Theory of Black Holes*, Clarendon Press, Oxford.

CHRISTODOULOU, D. (1970) *Phys. Rev. Lett.* **25**, 1596.

CHRISTODOULOU, D. and RUFFINI, R. (1971) *Phys. Rev.* **D4**, 3552.

DAMOUR, T. and RUFFINI, R. (1976) *Phys. Rev.* **D14**, 331.

DAVIES, P. C. W. (1976) *Proc. Roy. Soc. London*, **A351**, 129.

DETWEILER, S. L. and IPSER, J. R. (1973) *Astrophys. J.* **185**, 675.

DE WITT, B. S. (1975) *Physics Reports* **19**, 295.

HARTLE, J. B. (1973) *Relativistic Stars, Gravitational Collapse and Black Holes*, in *Relativity, Astrophysics and Cosmology*, ed.: Israel, W., D. Reidel Publ. Comp., Dordrecht.

HAWKING, S. W. (1973) *The Event Horizon*, in *Black Holes*, eds.: De Witt, C., Gordon and Breach, N. Y.

HAWKING, S. W. (1976) *Comm. Math. Phys.* **43**, 199.

HAWKING, S. W. (1977) *Scientific American*, **236**, p. 34, January.

HAWKING, S. W. and ELLIS, G. F. R. (1973) *The Large Scale Structure of Space-Time*, Cambridge University Press, Cambridge.

HAWKING, S. W. and PENROSE, R. (1969) *Proc. Roy. Soc. London* **A314**.

MAY, M. and WHITE, R. (1966) *Phys. Rev.* **141**, 1232.

MAZUR, P. (1982) *J. Phys.* **A15**, 3173.

MISNER, C. and SHARP, D. (1964) *Phys. Rev.* **B136**, 571.

MISNER, C., THORNE, K. and WHEELER, J. (1973) *Gravitation*, W. Freeman, San Francisco.

OPPENHEIMER, R. and SNYDER, H. (1939) *Phys. Rev.* **56**, 455.

PARKER, L. (1975) *Phys. Rev.* **D12**, 1519.

PENROSE, R. (1969) *Rivista del Nuovo Cimento, Supplemento Speciale* **1**, 252.

PRESS, W. and TEUKOLSKY, S. (1973) *Astrophys. J.* **185**, 649.

PRICE, R. (1972) *Phys. Rev.* **D5**, 2419.

REES, M., RUFFINI, R. and WHEELER, J. A. (1974) *Black Holes, Gravitational Waves and Cosmology*, Gordon and Breach, N. Y.

REGGE, T. and WHEELER, J. (1957) *Phys. Rev.* **108**, 1063.

SCIAMA, D. W. (1976) *Vistas Astron.* **19**, 385.

SEXL, R. V. (1975) *Acta Phys. Austriaca* **B42**, 303.

STAROBINSKY, A. A. (1973) *Soviet Phys.-JETP*, **37**, 28.

STEWART, J. and WALKER, M. (1973) *Black Holes – the Outside Story*, Springer Tracts in Modern Physics, No. 69, Springer-Verlag, Berlin.

TEUKOLSKY, S. (1973) *Astrophys. J.* **185**, 635.

TEUKOLSKY, S. and PRESS, W. (1974) *Astrophys. J.* **193**, 443.

UNRUH, W. G. (1977) *Phys. Rev.* **D15**, 365.

VISHVESHWARA, C. (1970) *Phys. Rev.* **D1**, 2870.

WALD, R. M. (1975) *Comm. Math. Phys.* **45**, 9.

ZELDOVICH, YA. B. (1971) *Soviet Phys.-JETP Letters* **15**, 180.

ZELDOVICH, YA. B. (1972) *Soviet Phys.-JETP*, **35**, 1085.

CHAPTER 8

ASTROPHYSICS OF NEUTRON STARS AND BLACK HOLES

8.1. Spherically Symmetric Accretion of Matter Onto Neutron Stars and Black Holes

Neutron stars come into existence as a result of the dramatic explosions of supernovae. Part of the stellar mass together with the envelope is thrown off; the hot central part contracts and, depending on its mass, forms a neutron star or a black hole. The temperature in the core of a newly formed neutron star is very high, reaching 10^8 K. Owing to the high thermal conductivity of the degenerate neutron and electron gases, a temperature gradient will occur only in a thin layer near the surface. The star gradually cools down, emitting neutrinos and photons. Initially, the cooling progresses fast, and

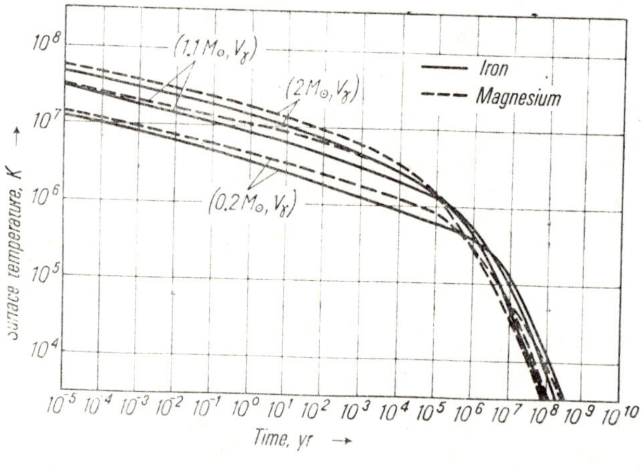

FIG. 8.1. The rate of cooling of a neutron star for different models of the star and different chemical compositions of the stellar surface. The differences are insignificant, and irrespective of the model and the chemical composition the temperature of the star falls to below 10^6 K after about 10^4 years (adapted from Tsuruta and Cameron, 1966).

after about 10^4 years the temperature at the surface of the star drops to 10^6 K. An analysis of the cooling rate by Tsuruta and Cameron (1966) shows that further temperature fall is much slower (Fig. 8.1). If we regard a hot neutron star as a perfect black body, it radiates mainly the X-rays while optically the star will be very weak. At first, before pulsars were discovered, astronomer set their hopes on the possibility of discovering such weak X-ray sources.

Apart from thermal radiation various other possibilities were considered. It can be safely assumed that a supernova explosion excites all the vibrational modes of the star. Thorne showed that nonradial oscillations die out very quickly. The characteristic decay period depends on the mass of the configuration and ranges from several seconds to tenths of a second. Radial oscillations are not damped so fast. A considerable part of the energy of the explosion can be stored in the form of oscillations. Zeldovich and Novikov estimate that the oscillation energy is of the order

$$E_{osc} \approx 10^{53} \frac{M}{M_\odot} \left(\frac{\delta R}{R}\right)^2 \text{ erg}, \tag{8.1}$$

where $\delta R/R$ is the relative amplitude of oscillations. Oscillations of the star cause changes in the density distribution within it. If the central density exceeds 10^{14} g/cm^3, there may occur a URCA process. During compression the dominant reaction is

$$n+n \to n+p+e+\tilde{\nu}_e, \tag{8.2}$$

whereas during expansion

$$n+p+e \to n+n+\nu_e \tag{8.3}$$

dominates. In this way, the oscillation energy is eventually dissipated. The characteristic decay period of radial oscillations is estimated to be at most a few hundred years. Owing to dissipative processes, a newly created neutron star will very quickly reach a state of stationary equilibrium. There is hardly a chance, therefore, for the initial active period to be observed. Radial oscillations decay because of the neutrino emission, and the non-radial ones are damped by the emission of gravitational waves.

The prospects of detection of a solitary black hole also appear quite hopeless. Massive black holes emit no radiation, and the only means by which they interact with the surrounding matter is the gravitational field.

In real situations, both neutron stars and black holes are surrounded by clouds of gas. The problem of accretion of gas onto a spherically symmetric body in the Newtonian theory of gravitation was investigated by Bondi (1952). He considered the process of adiabatic accretion of gas satisfying the equation of state $p \sim \rho^\Gamma$. Bondi's considerations were later generalized to the relativistic case by Salpeter (1964).

Far from the black hole or the neutron star the particles of gas move freely. Denote by T_∞ the temperature of the gas and by v_s the speed of sound in that region. The kinetic energy of the disordered thermal motion of the particles is greater than their potential energy. The pull of the hole or the neutron star is hardly felt by the particles which are outside a certain critical radius r_a. At the border of this region the escape velocity is comparable to the velocity of the thermal motion. If M is the mass of the central body, then

$$r_a = \frac{2GM}{v_s^2} = 10^{14} \left(\frac{M}{M_\odot}\right) \left(\frac{T_\infty}{10^4 \text{K}}\right)^{-1} \text{ cm}. \tag{8.4}$$

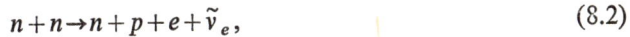

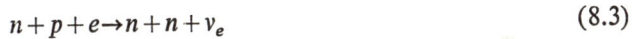

For $r < r_a$, the particles are falling approximately freely on the centre. If the density of the gas for $r > r_a$ is ρ_∞, the amount of matter falling on the centre can be found from

$$\frac{dM}{dt} = \frac{4\pi G^2 M^2 \rho_\infty}{v_s^2} = 5 \cdot 10^{10} \left(\frac{M}{M_\odot}\right)^2 \frac{\rho_\infty}{10^{-24}\ \mathrm{g/cm^3}} \left(\frac{T_\infty}{10^4 \mathrm{K}}\right)^{-\frac{3}{2}} \mathrm{g/s}. \tag{8.5}$$

Assuming that the infall of gas is adiabatic, one can determine how its density and temperature change with distance from the centre. The density increases as $r^{-3/2}$ and temperature as $r^{-3(\Gamma-1)/2}$. The particles are accelerated in the gravitational field and at a distance $r = r_s$ from the centre their velocity becomes equal to the speed of sound. For the sonic radius r_s we get the estimate

$$r_s = \frac{(5-3\Gamma)}{4} \frac{GM}{v_s^2}. \tag{8.6}$$

In the case of spherically symmetric accretion the surface $r = r_s$ is not distinguished by anything particular. The temperature of the gas near the surface of the black hole has been estimated by Eardley and Press to be

$$kT \sim m_p c^2 \left(\frac{m_p c^2}{kT_\infty}\right)^{\frac{3\Gamma-5}{2}} \tag{8.7}$$

and is very high only when $\Gamma \approx 5/3$. For $\Gamma < 5/3$ the gas temperature near the horizon is comparatively low.

This simple picture of accretion can be made more realistic by taking into account various possible cooling processes. This will reduce the value of Γ. It is even conceivable that the gas temperature falls as contraction progresses. Spherically symmetric accretion is an inefficient converter of rest-mass energy into radiation. Allowing for a motion of the gas cloud relative to the centre makes no substantial difference. If the velocity of the motion of a neutron star or black hole is greater than the speed of sound in the gas, a shock wave will develop in front of the star. It will be a weak shock, and the efficiency of the transformation of the particle kinetic energy into radiation will be very low. It is estimated that the process of spherical accretion onto a neutron star may account for an energy release per unit of time a few orders of magnitude less than the luminosity of the Sun.

It became necessary to find a different, more efficient model of accretion. Attention was drawn to close binary systems. Before we present a model of accretion of gas in a binary system, let us recall the basic properties of such systems.

8.2. Binary Star Systems

In spite of appearances, stars very often occur in systems composed of at least two stars. It is estimated that about 60% of stars are members of multiple stellar systems. Here we shall consider the most common kind, namely systems composed of two stars. The motion of such a system is well known from theoretical mechanics. Let us recall some of its properties and introduce notation used in observational astronomy.

Binary systems can be divided into visual, spectroscopic and eclipsing binaries. In a vsiual binary system, the two companion stars are sufficiently far apart to be observed independently. If the distance between the stars is comparable with their dimensions, there occur strong tidal effects, and there is a possibility of mass transfer and strong mutual irradiation.

The velocities of the orbital motion of stars in such systems are very high, causing Doppler shift of spectral lines. Such systems are called spectroscopic binaries. If the inclination of the orbit to our line of sight is sufficiently small, both stars may partially or completely eclipse one another, which will result in periodic changes in the luminosity. In this case we call the system eclipsing.

Close binary systems, in which matter can flow from one star to the other, are of particular interest to us. Before we examine this process in more detail, let us investigate the dynamics of a binary system. If the interaction between the components is not too strong, in analysing the dynamics of the system we can regard them as material points.

In a noninertial reference frame bound to one of the stars, the velocity of a test particle satisfies the equation

$$\frac{d\boldsymbol{v}}{dt} = -2\boldsymbol{\Omega}\times\boldsymbol{v} - \nabla\phi, \tag{8.8}$$

where

$$\phi = -\frac{GM_A}{|\boldsymbol{r}-\boldsymbol{r}_A|} - \frac{GM_B}{|\boldsymbol{r}-\boldsymbol{r}_B|} - \tfrac{1}{2}|\boldsymbol{\Omega}\times\boldsymbol{r}|^2 \tag{8.9}$$

M_A and M_B are masses of the stars, $\boldsymbol{r}_A$ and $\boldsymbol{r}_B$ – their positions, and Ω is the angular velocity of the stars about each other. If we move radially away from one of the stars towards the other, the potential ϕ is initially dominated by the gravitational potential of the first star. The equipotential surfaces are nearly spherical. As we move further, the attraction of

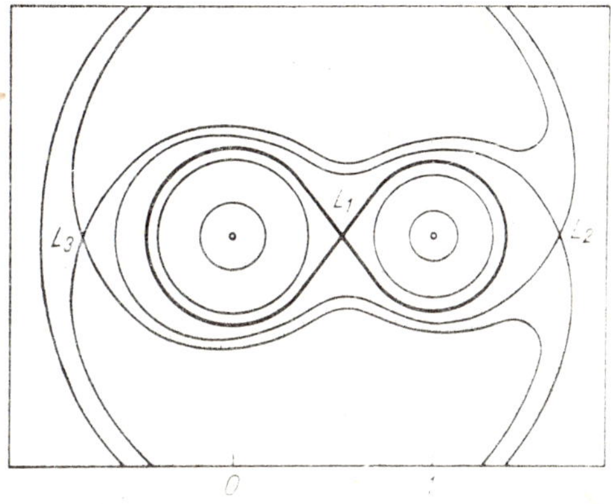

FIG. 8.2. Lagrange points and Roche lobe.

the first star decreases while the gravitational pull of the second star begins to play an increasing role. At some distance the two forces will exactly balance each other. We call this point the first Lagrange point. The shape of the equipotential surfaces in the orbital plane is shown in Fig. 8.2. The Lagrange point is the common point of two equipotentials of which one surrounds star *A* and the other surrounds star *B*. The innermost common equipotential surface is called the Roche lobe. It is through the Lagrange point that matter can flow from one star to the other. Having noted this posibility, to be examined later, let us now review the information that can be obtained about the components of a spectroscopic binary system.

Analysis of the spectrum of a spectroscopic binary system permits us to determine the radial velocity curve (Fig. 8.3). It is a curve representing the time-dependence of the radial

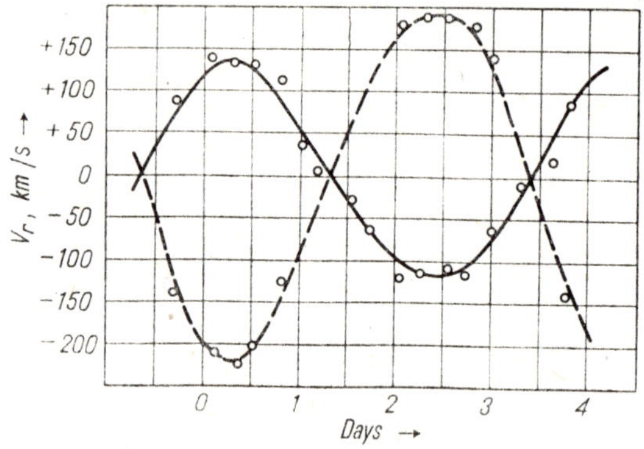

FIG. 8.3. Radial velocity curves for the components of a binary system.

velocity projected on the line of sight. We can calculate from it the following elements of the orbit:

(a) the orbital period P,
(b) the time t_0 of periastron passage,
(c) the eccentricity e,
(d) the distance ω between the periastron and the node,
(e) the product of the semi-major axis a by the sine of the inclination angle.

Denote by K_A and K_B the semi-amplitudes of the radial velocity curves of the components A and B. By Kepler's second law we have

$$a_1 \sin i = 13751\,(1-e^2)^{\frac{1}{2}} K_A P, \tag{8.10}$$

where K_A is given in km·s^{-1}, P in mean solar days and a_1 in kilometers. If the spectrum of the system contains lines of both stars, we can determine the radial velocity curve for either of them. We then have another independent relationship

$$a_2 \sin i = 13751(1-e^2)^{\frac{1}{2}} K_B P. \tag{8.11}$$

Dividing (8.10) by (8.11) and using Kepler's third law gives

$$\frac{K_A}{K_B} = \frac{M_B}{M_A}, \tag{8.12}$$

where M_A and M_B are the masses of the components A and B. Adding relations (8.10) and (8.11) and denoting the sum $K_A + K_B$ by K, we get

$$a \sin i = 13751\,(1-e^2)^{\frac{1}{2}} P \cdot K, \tag{8.13}$$

where $a = a_1 + a_2$ is the semi-major axis of the relative orbit. Kepler's third law can be written in the form

$$\frac{a^3}{P^2(M_A + M_B)} = 2.5 \cdot 10^{19}, \tag{8.14}$$

where a is expressed in km, P in mean solar days and M_A and M_B in solar mass units. Eliminating a from (8.13) and (8.14), we obtain

$$(M_A + M_B)\sin^3 i = 1.036 \cdot 10^{-7}(1 - e^2)^{3/2} K^3 P. \tag{8.15}$$

Using (8.12), we can calculate the product of the mass of each component with the cube of the sine of the orbit inclination angle:

$$M_A \sin^3 i = 1.036 \cdot 10^{-7}(1 - e^2)^{3/2} K^2 K_B P, \tag{8.16}$$

$$M_B \sin^3 i = 1.036 \cdot 10^{-7}(1 - e^2)^{3/2} K^2 K_A P. \tag{8.17}$$

If the spectrum of our spectroscopic binary contains lines of only one component, we can determine only K_A and P. Eliminating K from (8.15) and (8.17) we then obtain

$$\frac{M_B^3 \sin^3 i}{(M_A + M_B)^2} = 1.036 \cdot 10^{-7}(1 - e^2)^{3/2} K_A^3 P. \tag{8.18}$$

The right-hand side of this relation involves quantities which can be determined from observation; the left-hand side is the so-called *mass function*

$$f(M) = \frac{M_B^3 \sin^3 i}{(M_A + M_B)^2}. \tag{8.19}$$

In order to find the masses of a spectroscopic binary one first has to determine the mass of one of the components or the mass ratio from some other considerations; assuming that $i = 90°$ one can then use the mass function to calculate the minimum value of the mass of the second component (for more details see Heintz, 1978).

In early attempts to show that spectroscopic binaries may contain a black hole as one of their components, the tendency was to investigate binary systems with large mass functions. The mass of the luminous component was estimated on the basis of the spectral type, and then the mass function was used to determine the minimum mass of the second component. If it proved greater than $1.4 M_\odot$, it could, in principle, be a black hole. The spectral type, however, does not allow the mass of the star to be determined precisely, and therefore the list of binary systems suspected of containing a black hole was viewed with scepticism. It was also doubted whether a black hole could be formed within a binary system at all. As we know, neutron stars and black holes arise at the final stages of the evolution of massive stars, as a result of supernova explosions. It was believed that a supernova explosion in a close binary would pull the system apart. Now we know that this is not so. Indeed, there exist pulsars, and hence neutron stars, which are members of binary systems. It is a fact, therefore that a close binary can remain a bound system after the explosion of one of its components, and there is no reason, it seems, why one of the stars could not subsequently become a black hole as a result of gravitational collapse.

In such a system the matter flowing through the Lagrange point will be captured by the black hole. The process of accretion will differ from that occurring in the spherically symmetric case. The falling matter will have non-zero angular momentum. It will form a disc, gradually lose its angular momentum owing to dissipative processes, and approach the horizon, to be eventually drawn in by the black hole. Accretion of matter onto a black hole or a neutron star in a close binary system is discussed in greater detail in the next section.

8.3. Accretion of Matter Onto a Black Hole or Neutron Star in a Binary System

A black hole, a neutron star or a white dwarf in a binary system will have a long-lasting source of matter in its companion. Depending on the evolutional stage of the accompanying normal star, its radius may be less than or equal to the radius of its Roche lobe (sometimes it may even be greater). In the first case, the source of matter is the stellar wind – a stream of particles blown off the upper strata of the atmosphere of the star by radiation pressure and accelerated to very high velocities. In the second case, a substantial mass stream can flow through the inner Lagrange point.

We shall now present in some detail a qualitative picture of accretion, basing ourselves on the works of Shakura and Sunyaev (1973), Novikov and Thorne (1973), and Pringle and Rees (1972). Assume that the distance between the centres of mass of the stars in the system under consideration is within a factor of 2 or so of the radius of the normal star. It can also be assumed without loss of generality that as a result of the long-lasting interaction between the stars their orbits are circular and that the normal star corotates with its companion.

The flow of gas from the normal star to its compact companion is most conveniently described in a rotating frame bound to the centre of the compact star. In this frame

$$\frac{d\boldsymbol{v}}{dt} = -\nabla\phi - 2\boldsymbol{\Omega}\times\boldsymbol{v} - \frac{1}{\rho_0}\nabla p + \frac{2}{\rho_0}\nabla(\eta\cdot\sigma), \tag{8.20}$$

where $\boldsymbol{v}$ is the gas velocity relative to the rotating frame, ϕ is the effective gravitational potential given by formula (8.9), ρ_0 and p are the rest-mass density and the pressure of the gas, η is the viscosity and σ the shear.

The gas particles flowing through the Lagrange point are accelerated in the gravitational field of the compact star by Coriolis forces and gravitational attraction, and, depending on their angular momentum, either escape from the system or begin to move in closed orbits determined by conservation of angular momentum. In this way a disc is formed around the compact star. Typical trajectories of the particles are shown in Fig. 8.4. The incoming particles interact viscously with the gas of the disc. Some of them lose angular momentum and join the disc. Others pick up some angular momentum from the disc and are ejected out of it and back to the normal star or are scattered into interstellar space.

The gas particles which form the disc move on roughly circular orbits with the velocity

$$v_\phi = \sqrt{\frac{GM_A}{r}}, \tag{8.21}$$

where M_A is the mass of the compact star. The disc structure is very sensitive to any chalt ges in the dissipation rate. The principal dissipative factor is viscosity; it is very difficuo however, to take account of all the processes responsible for it. The dominant sources c viscosity are turbulent motions and chaotic magnetic fields in small regions of the disc

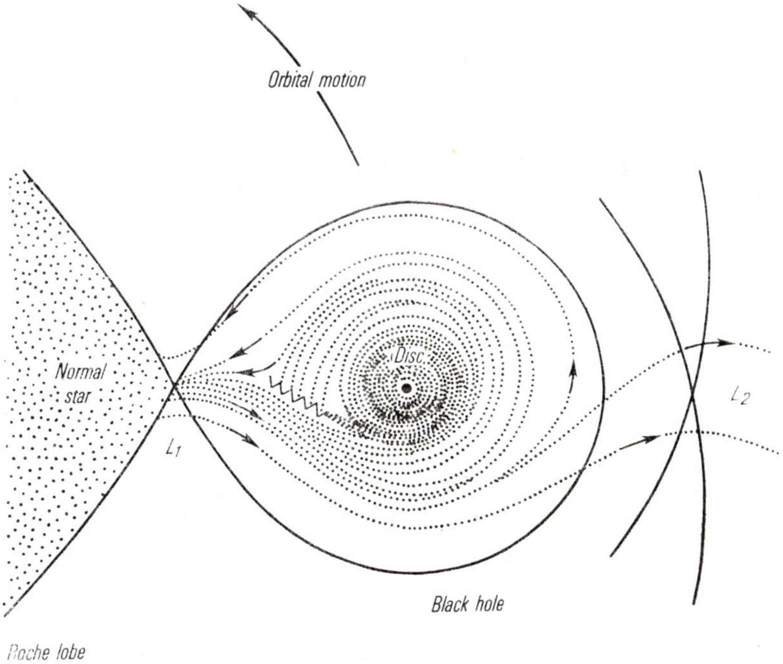

FIG. 8.4. Typical trajectories of particles flowing through the Lagrange point L_1. Some of the particles lose enough angular momentum to get deposited in the disc around the black hole, some return to the normal star, and others are scattered into interstellar space (adapted from Eardley and Press, 1975).

the so-called magnetic cells. We shall assume that viscous stresses are proportional to the shear tensor. Thus, using (2.135), we get

$$\tau_{r\varphi}=2\eta\sigma_{r\varphi}. \tag{8.22}$$

Using formula (2.52) for the shear tensor $\sigma_{\alpha\beta}$, and formula (8.21) we find

$$\sigma_{r\varphi}=\tfrac{3}{4}\left(\frac{GM_A}{r^3}\right)^{\frac{1}{2}}. \tag{8.23}$$

Let $2h$ be the thickness of the disc; then $\Sigma=2h\rho_0$ is the surface density .Some information about the stationary structure of the disc can be derived from conservation laws. At a distance r from the centre the mass flow should be equal to the mass deposited in the disc in a unit of time; hence

$$-2\pi r\Sigma v^r=\dot{M}, \tag{8.24}$$

where v^r is the radial velocity of the gas. The rate of change of the angular momentum of the inflowing gas must equal the rate of change due to viscosity and the gravitational pull of the compact star, and therefore

$$\dot{M}\left(\frac{GM_A}{r}\right)^{\frac{1}{2}}\cdot r=2\pi r\cdot 2h\cdot \tau_{r\varphi}\cdot r+\dot{J}_A, \tag{8.25}$$

where $\dot{J}_A$ is the rate of change of the angular momentum of the compact star.

The change in the angular momentum of the compact star cannot exceed, in order of magnitude, the angular momentum of the gas at the inner edge of the disc r_i; hence the restriction

$$\dot{J}_A=\beta\dot{M}(GM_A r_i)^{\frac{1}{2}} \tag{8.26}$$

for some $|\beta|\leqslant 1$. From (8.25) and (8.26) we now find

$$2h\tau_{r\varphi}=\frac{\dot{M}}{2\pi r^2}[(GM_A r)^{\frac{1}{2}}-\beta(GM_A r_i)^{\frac{1}{2}}]; \tag{8.27}$$

for $r\gg r_i$ this becomes

$$2h\tau_{r\varphi}=\frac{\dot{M}(GM_A r)^{\frac{1}{2}}}{2\pi r^2}. \tag{8.28}$$

In the stationary case the product $2h\tau_{r\varphi}$ is determined uniquely by the accretion rate and the mass of the central body M_A. This allows self-regulation of the disc structure. If, at some moment, $2h\tau_{r\varphi}$ is smaller than required by (8.28), mass will begin to accumulate in this region because the viscous stresses will not be sufficient to dispose of the additional amount of angular momentum. As a result, either the disc thickness or viscosity will increase, and the disc will regain its equilibrium. A similar regulation of the disc structure takes place when $2h\tau_{r\varphi}$ exceeds the equilibrium value.

The presence of viscosity causes a release of heat. The entire difference between the gravitational potential energy of the particles at the inner and the outer edges of the disc is converted into thermal energy. The amount of heat released in the disc per unit volume is given by (cf. (2.137))

$$\varepsilon=4\eta\sigma^2=\tau_{r\varphi}\sigma_{r\varphi}, \tag{8.29}$$

and therefore, if we remember that the disc has two faces, the heat generated per unit area is

$$2h\cdot 2\varepsilon=2\sigma_{r\varphi}\cdot 2h\tau_{r\varphi}=\frac{3\dot{M}}{4\pi r^2}\frac{GM_A}{r}\left[1-\beta\left(\frac{r_i}{r}\right)^{\frac{1}{2}}\right]. \tag{8.30}$$

The amount of heat released in the disc region between radii r_1 and r_2 is

$$\int_{r_1}^{r_2} 2h2\varepsilon 2\pi r dr=\tfrac{3}{2}\dot{M}GM_A\left(\frac{1}{r_1}-\frac{1}{r_2}\right),\qquad r_2>r_1\gg r_i. \tag{8.31}$$

It is worth noting here that the thermal energy released in the region $r_1<r<r_2$ is greater than the released gravitational potential energy. According to the virial theorem, only half of the gravitational potential energy can be converted into heat. We thus have to make clear where the remaining part of the heat comes from. The viscous stresses not only trans-

port angular momentum from the internal regions of the disc outward, but also transmit energy. The rate at which energy is transported across the cylinder of radius r is

$$\dot{E}=\Omega\cdot\dot{J}=\dot{M}\frac{GM_A}{r}\left[1-\beta\left(\frac{r_i}{r}\right)^{\frac{1}{2}}\right]. \tag{8.32}$$

It is only the sum of half of the released gravitational potential energy and the energy transported by the viscous stresses that gives the total heating rate (8.31).

The released energy is almost immediately radiated away. In a stationary state the amount of energy released in an element of the disc per unit time is equal to the energy radiated away. If we denote by $2F$ the total flux of radiation per unit area, from (8.30) we get

$$F=\frac{3\dot{M}}{8\pi r^2}\frac{GM_A}{r}\left[1-\beta\left(\frac{r_i}{r}\right)^{\frac{1}{2}}\right]. \tag{8.33}$$

The total luminosity L of the disc is now readily calculated as

$$L=\int_{r_i}^{\infty}2F\cdot 2\pi r dr=(\tfrac{3}{2}-\beta)\dot{M}\frac{GM_A}{r_i}. \tag{8.34}$$

In order to determine the spectrum of the radiation emitted by the disc, we shall assume that locally it is blackbody. Denoting by $T_e(r)$ the effective temperature of the disc at radius r, we have the relation

$$F(r)=\sigma T_e^4(r), \tag{8.35}$$

where $\sigma=5.67\cdot 10^{-5}$ erg K^{-4} cm^{-2} s^{-1} is the Stefan–Boltzmann constant. Comparing (8.35) with (8.33), we find the dependence of the effective temperature on the radius

$$T_e(r)=T_e(r_i)\left(\frac{r_i}{r}\right)^{\frac{3}{4}}\left(\frac{1-\beta\left(\frac{r_i}{r}\right)^{\frac{1}{2}}}{1-\beta}\right)^{\frac{1}{4}}, \tag{8.36}$$

where $T_e(r_i)$ is the temperature at the inner edge of the disc. For a given frequency ν the luminosity of the disc can now be found from

$$J_\nu=4\pi^2\int_{r_i}^{r_0}B_\nu(T)\,r dr=\frac{8\pi^2 h\nu^3}{c^2}\int_{r_i}^{r_0}\frac{r dr}{e^{\frac{h\nu}{kT_e}}-1}; \tag{8.37}$$

here r_0 denotes the outer radius of the disc. Inserting (8.36) and integrating over three different regions, we find that

(a) when $h\nu\ll kT_e(r_0)$,

$$J_\nu\sim r_0^2\,T_e(r_0)\,\nu^2, \tag{8.38}$$

(b) when $kT_e(r_0) \ll h\nu \ll kT_e(r_i)$,

$$J_\nu \sim r_i^2 T_e^{8/3}(r_i) \cdot \nu^{1/3}, \tag{8.39}$$

(c) when $h\nu \gg kT_e(r_i)$,

$$J_\nu \sim r_i^2 e^{-\frac{h\nu}{kT_e(r_i)}}. \tag{8.40}$$

The dependence of the luminosity of the disc on frequency is very similar to the spectrum of optically thin plasma, which absorbs low-frequency radiation. A more detailed analysis of disc radiation carried out by Shakura and Sunyaev (1973) led to the spectrum shown in Fig. 8.5, where only the observationally interesting frequency range is included.

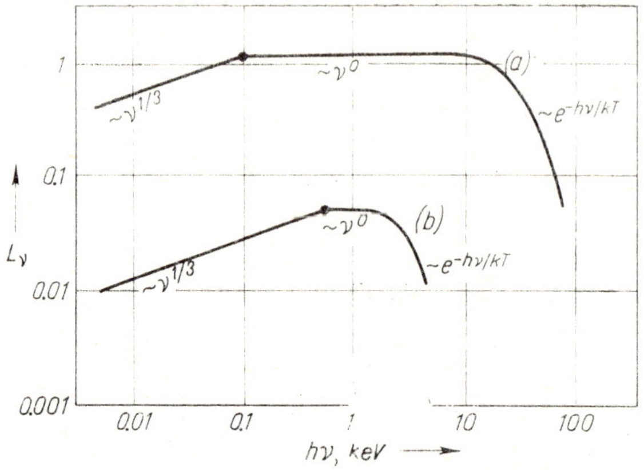

Fig. 8.5. The predicted spectrum of radiation emitted by a disc around a black hole in a binary system for different accretion rates and viscosities inside the disc. It is assumed that the black hole has the mass $M = M_\odot$. In model (a) the accretion rate is $10^{-8}\ M_\odot$/yr and viscosity is small; in model (b) $\dot{M} = 10^{-6} M_\odot$/yr and the viscosity is an order of magnitude greater than in model (a) (adapted from Shakura and Sunyaev, 1973).

The luminosity of the disc can be estimated without any complex calculations. As we have seen, the energy released is to order of magnitude equal to the change in gravitational potential energy. If the central body is a white dwarf or neutron star, we must calculate the change in the gravitational potential energy of a test particle with a mass m approaching the surface of the star from infinity. Measuring the energy changes in units of mc^2, we get

$$E = \frac{GM_A}{2c^2R} = \begin{cases} 10^{-4} & \text{for white dwarf,} \\ 10^{-2} & \text{for neutron star.} \end{cases} \tag{8.41}$$

If the central body is a black hole, the inner edge of the disc may reach the innermost stable circular orbit, and then

$$E \sim 1 - \sqrt{8/9} \approx 0.057 \tag{8.42}$$

when the hole is static, and

$$E \sim 1 - \frac{1}{\sqrt{3}} \approx 0.42 \tag{8.43}$$

when it rotates with the maximum possible angular velocity. Thus, the luminosity of the disc around a white dwarf is

$$L \sim 10^{-4}\dot{M}c^2 \sim 10^{34}\left(\frac{\dot{M}}{10^{-9}M_\odot/\text{yr}}\right)\text{erg/s}. \tag{8.44}$$

and that for a black hole or neutron star is

$$L \sim 0.1\dot{M}c^2 \sim 10^{37}\left(\frac{\dot{M}}{10^{-9}M_\odot/\text{yr}}\right)\text{erg/s}. \tag{8.45}$$

It follows from these estimates that even at a low accretion rate of merely $10^{-9}M_\odot$ per year *the luminosity of the disc around a neutron star or black hole may be four orders of magnitude greater than the luminosity of the Sun.*

The qualitative structure of the disc discussed above is very sensitive to changes in the coefficient of viscosity. It is usually assumed that the viscous stresses are proportional to the total pressure:

$$\tau_{r\varphi} = \alpha p, \tag{8.46}$$

where α is a phenomenological coefficient lying between 10^{-3} and 1, and p is the total pressure in the disc. Such models are called α models of the disc. In the outer regions of the disc, far from the black hole, where the temperature is still low, the dominant form of pressure is gas pressure. Near the black hole, it is radiation pressure that dominates. Lightman and Eardley (1974) have pointed out that the region dominated by radiation pressure is liable to instabilities. Dropping the assumption of stationarity, they examined a thin disc structure and showed that the disc tends to break up into rings of size $\Delta r \sim h$, and that the process of fragmentation occurs on a characteristic time scale $\left(\frac{\Delta r}{r}\right)^2 \cdot \tau$, where τ is the characteristic drifting time of a given element of the disc towards the centre. It seems that this local instability has no major effect on the overall picture of accretion. Indeed, the thickening of the disc in the region where instabilities of this kind may occur inhibits their occurrence. In spite of this controlling factor, however, the innermost regions of the disc may suffer fluctuations characterized by very rapid time-variation.

Spherical and disc accretion in different astrophysical situations has been studied intensively in recent years. Accretion onto magnetized white dwarfs and neutron stars in binary systems has been investigated by Lynden-Bell and Pringle (1974), Fabian *et al.* (1976), Sunyaev (1978), and Lamb (1979), among others. More realistic disc models have also been proposed; see, for example, Abramowicz *et al.* (1978), Kozłowski *et al.* (1978), Carter (1979a), Blandford (1979), Blandford and Thorne (1979), Jaroszyński *et al.* (1979), Paczyński and Wiita (1979). Disc accretion onto supermassive black holes has also been investigated (Carter, 1979b; Gunn, 1979; McCray, 1979).

8.4. Compact X-Ray Sources and Observational Data

When in 1962 a rocket-borne X-ray detector was for the first time carried beyond the atmosphere, not only was X-ray background radiation observed, but also the first compact X-ray source was discovered. It was designated Sco X-1. During the following eight years several more discrete sources were discovered and their approximate positions were determined. A turning point was the launching in December 1970 of the UHURU satellite, the first satellite designed especially for observations of X-ray sources. It was equipped with a system of detectors sensitive to X-rays of energy 2keV to 20keV. By means of the UHURU satellite more than 90 new compact X-ray sources were discovered. As follows from the spatial distribution shown in Fig. 8.6, most of them are clustered along the galactic plane. Sources lying far from this plane are distributed isotropically, and for the most part they are extragalactic. Some of them have been successfully identified with external galaxies or clusters of galaxies.

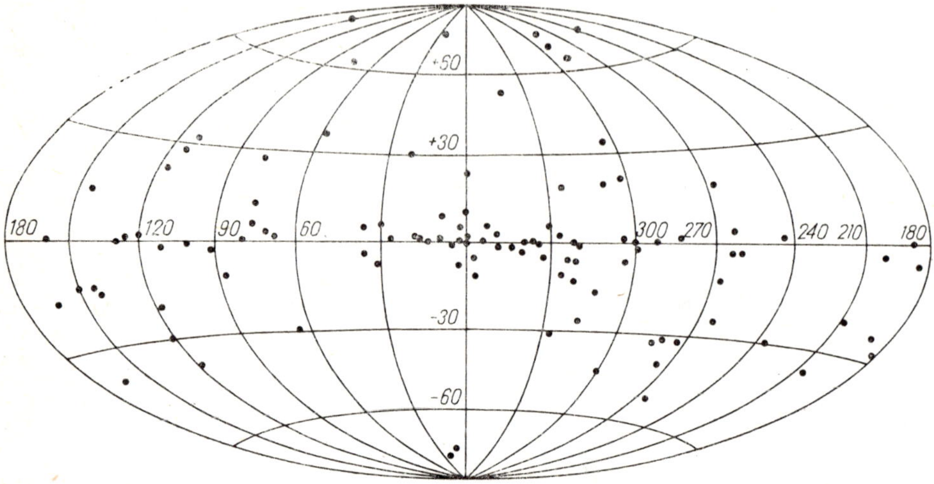

Fig. 8.6. Spatial distribution of X-ray sources in galactic coordinates (adapted from Gursky, 1973).

Properties of the galactic X-ray sources have been discussed in detail by Gursky (1973), Tananbaum and Tucker (1974), Blumenthal and Tucker (1974) and Sanford *et al.* (1982). The luminosity of galactic sources can be estimated by the following simple argument. Five sources were found in the Magellanic Clouds. Knowing the distance to the Magellanic Clouds and the flux density of the radiation that reaches the Earth, one can calculate the luminosity of these sources. Their intensity is very high, they send about 10^{38} erg/s, exceeding the luminosity of the Sun by five orders of magnitude. If it is assumed that the mechanism of X-ray emission is the same for all sources, the luminosity of the sources that exist in our own galaxy should also be of the order of 10^{38} erg/s. This is the upper bound. On the other hand, examining the distribution of galactic sources in relation to the intensity of radiation, one notices the absence of weak sources. The estimate for the minimum luminosity derived from this information is 10^{36} erg/s.

If absorption effects are neglected, the flux density of X-rays is observed to be a monotonically decreasing function of energy. X-ray sources radiate most of their energy in the range 2–10keV. This fact may be an internal property of the sources, but it is also possible that it is due to absorption, particularly for low-energy X-rays.

Generally, the following four statements can be made about bright galactic X-ray sources:

(a) Their luminosity is within the range 10^{36}–10^{38} erg/s.

(b) The number of X-ray sources in the Galaxy is unexpectedly small; about 100 of them are known.

(c) The radiation mechanism produces mainly photons of energies between 2 and 10keV.

(d) Sources in the outer regions of the Galaxy seem to confirm its spiral structure.

To provide more information about the properties of X-ray sources, let us discuss a few of the best-studied objects.

(a) *Sco X-1*. As has already been mentioned, this was the first discrete X-ray source observed. Four years after its discovery, it was correctly identified with an optical object. This identification has since been confirmed by observations of simultaneous luminosity variations in the optical, radio and X-ray parts of the spectrum.

The X-ray emission in the 1–10keV energy range can be described very well by a black-body spectrum with a temperature $T \sim 5 \cdot 10^7$K. At energies greater than 20keV there are departures from this pattern, which should be attributed to non-thermal processes.

Optically, Sco X-1 is an object of variable luminosity between the 12th and 13th stellar magnitude. The intensity of its radiation varies by a few percent over short periods of the order of minutes and may change by a factor of two on a time scale of hours. The optical spectrum is continuous with a few high-excitation emission lines. No periodic variation of the spectrum that would signify the presence of a companion star is observed. Also, there is no evidence that Sco X-1 contains an ordinary star.

In the radio range, Sco X-1 is a very weak, variable source. Two stationary radio sources are present in its vicinity.

Simultaneous observations in the optical, radio and X-ray regions of the spectrum have revealed that Sco X-1 has two well-defined states. In the quiescent state, which may last a few days, the intensity throughout the spectrum undergoes small variations, not exceeding 20%. The active state begins abruptly. Both in the optical and in the X-ray regions of the spectrum, and also in radio, there occur continuous changes of the luminosity by a factor of two or more over a time scale of less than one hour. The average luminosity in this period is about twice as high as in the quiescent state.

These changes have an effect on the spectrum as well. In the active state, the X-ray spectrum indicates a rise in the temperature of the source and the maximum optical intensity is blue-shifted.

Both the quiescent and the active states last a few days. Transition from one state to another occurs at random. No correlation between the two states have been observed.

It is generally accepted that the X-ray emission of Sco X-1 is generated by thermal bremsstrahlung in a region of the size of 10^3–10^4 km. The optical emission is produced outside this region and the radio emission in the outermost layer of plasma. If we assume that the luminosity of Sco X-1 is 10^{36} erg/s, its distance from the Earth would be at least 200 pc. At this distance, its estimated optical luminosity is 10^{33} erg/s, about that of the Sun. If Sco X-1 was a normal star, its luminosity would be much smaller. Although so much is already known about Sco X-1, fundamental information on its mass and dimensions is still missing.

Despite numerous attempts, no emission lines have yet been discovered in the X-ray region of the spectrum.

(b) *Cyg X-2*. In most of its properties Cyg X-2 resembles Sco X-1. Its luminosity (about 40 times less than that of Sco X-1) and the apparent lack of a quiescent state (characteristic of Sco X-1) are the basic differences. In both optical and X-ray regions of the spectrum, the luminosity of the source undergoes violent changes, increasing or decreasing by a factor of two within a few hours. The X-ray spectrum corresponds to blackbody radiation with $T=4\cdot 10^7$ K.

In the optical spectrum Cyg X-2 differs from Sco X-1 considerably. The spectral lines exhibit shifts which imply motion of the source with a radial velocity of the order of several hundred km/s. However, no periodicity has been observed which would provide evidence that Cyg X-2 is a binary system. The optical properties of the source can be accounted for by the presence of a small white dwarf of type *G*. If this is the case, the distance to Cyg X-2 is estimated to be 500–700 pc. The X-ray luminosity is then 10^{36} erg/s and the optical luminosity is 100 times lower.

Several other X-ray sources have properties similar to those of Sco X-1 and Cyg X-2. However, our knowledge of those sources is much poorer as none of them has been identified optically. If, as in the case of Sco X-1 and Cyg X-2, their optical luminosity is 100–1000 times less than their X-ray luminosity, they would appear as very weak stars of at most 17 m. The regions of uncertainty within which these X-ray sources are located are too large for their positive optical identification.

(c) *Centaurus X*-3. This was the first X-ray source observed to pulse and found to be a binary. The observed X-ray emission has a component pulsing periodically with a period of 4.84239 s. Centaurus X-3 is thus an X-ray pulsar. At the same time it is a binary system, for the X-ray emission is eclipsed every 2.08707 days. The period of the pulsar is also observed to vary periodically with a 2-day period. This variation in period can be explained by the motion of the source around a central star at a linear velocity of 415 km/s.

The pulsed portion of the X-ray emission sometimes disappears for a period of several days. The transitions occur at random and show no correlation with any other properties of the object. During these periods and during the eclipse, continuous X-ray emisson is observed whose intensity is about 1/10 of the maximum intensity of the pulsed portion. Its presence during the eclipse implies that it is produced by some mechanism other than the pulsed emission.

The eclipses of the pulsing source are not abrupt. Intensity decays slowly, the spectrum changes, and its low-energy part fades out. The effect may be explained by the screening of the source by the atmosphere of the central star.

The credit for the optical identification of Centaurus X-3 is due to a Polish astronomer, W. Krzemiński. Cen X-3 has turned out to be a weak, spectroscopic binary star.

The fact that Centaurus X-3 is a binary system and the observed periodicity of the variations of its properties permit us to estimate its radius to be about $8R_\odot$. Thus, Cen X-3 is a close binary. From the variation of radial velocity we find the value of the mass function to be 15 $M_\odot$. Assuming that one of the stars fills its Roche lobe, we can estimate the masses of both components. The result for the X-ray source is a value between $0.2M_\odot$ and $0.7M_\odot$. (The latest estimate is 0.3–$1.7M_\odot$ (Rappaport, 1979).)

Centaurus X-3 is believed to be a rotating magnetic neutron star. Its X-ray emisson is probably produced when accreting matter funneled by a strong magnetic field falls onto the neutron star near the magnetic poles (Gursky and van den Heuvel, 1975; Rappaport, 1979). If this is the case, the radiation should be beamed.

(d) *Hercules X-1*. This is another pulsating X-ray source. Its properties are very similar to those of Cen X-3. The pulse period of the X-ray source is considerably shorter and equals 1.24 s. The source is eclipsed periodically every 1.70 days; the actual duration of the eclipse is 5 hours. Like Cen X-3, Her X-1 is a binary system. In this case, however, the periodic component of the X-ray emisson disappears for much longer periods of time. They recur approximately every 35 days. During this time the pulsating source is on for only about 12 days.

The average intensity of the pulsating source shows a characteristic variation. The source appears abruptly, its average intensity increases to a maximum after about three days, and then slowly declines.

Her X-1 is optically identified with an irregular variable star of 13.5 m, HZ Hercules. No periodic changes corresponding to the 35-day X-ray cycle have been observed in its optical spectrum.

From the observed variation of the radial velocity and the eclipse duration it is possible to determine a few important parameters of the system. The mass function is found to be $1M_{\odot}$, the diameter of the orbit of the X-ray source is $8 \cdot 10^{11}$ cm and the diameter of the central star is $3 \cdot 10^{11}$ cm. Knowing the mass function and assuming that the central star fills its Roche lobe, one estimates the mass of the X-ray source to lie between 0.6 $M_{\odot}$ and $2M_{\odot}$. Her X-1 is believed to be a rotating magnetic neutron star. The short pulse period excludes the possibility of a white dwarf. The double pulse shape resembles the signals of the pulsar PSR 0532+21. At the present time we know altogether 16 X-ray pulsars. For a review of their properties see Rappaport (1979).

Recently, sources of yet another kind have been discovered. They are bursting sources: the burst of X-rays repeat quasi-periodically. Some of them have been identified with globular clusters. The bursting X-ray sources are believed to be neutron stars in binary systems. Their peculiar properties are probably connected with thermonuclear flashes triggered by the matter accreting onto a magnetized neutron star, but may also result from instabilities in the accretion flow. The properties of bursting X-ray sources have been reviewed by Lewin (1979) and Lewin and Clark (1980).

The number of known X-ray sources increased dramatically after the launching in November 1978 of the HEAO-2 satellite, the most sophisticated high-energy astronomical observatory, now called the Einstein Observatory. HEAO-2 is equipped with an X-ray telescope with spatial and spectral resolution comparable to optical telescopes. Its sensitivity is also much higher than that of any earlier X-ray detector. The Einstein Observatory has opened a new era of X-ray astronomy by allowing us to observe and study in detail very weak X-ray sources. Its technical design and first observational results are reviewed by Giacconi (1980), who started work on focussing X-rays in 1959 and has been involved in all the advances of X-ray astronomy. (See also reviews by Hartline (1979) and Giacconi (1981)).

Galactic X-ray sources can be roughly divided into four groups: X-ray stars, X-ray pulsars, bursting sources and the group that contains Sco X-1 and other sources with similar properties that show no periodic changes. The sources of the second and third group are believed to be close binary systems containing neutron stars. Characteristics of a few X-ray sources are summarized in Table 8.1.

The literature on observational properties and models of compact X-ray sources has grown vast in the last decade. We mention the works of Bradt and Giacconi (1973),

TABLE 8.1. *Characteristics of X-ray sources*

Source	Fast-variation characteristics	Slow-variation characteristics	Optical object	Distance kpc	Radio-source
Her X-1	pulse 1.24 s	eclipse period 1.7 days, periodically variable every 35 days	Hz Her $P = 1.7$ day binary	2–6	No
Cen X-3	pulse 4.8 s	eclipse of the X-ray source, period 2.1 days	Cen-3 binary	5–10	No
Cyg X-1	irregular variation 1 ms	No clear characteristics	HDE 226868 binary $P = 5.6$ d	2.5	Irreg. source
Sco X-1	Irregular variation min.	flares 10 min–1 hour	No positive identification		Yes
Cyg X-3	Irregular variation min.	variation with period 4.8 h	None	10	Yes

Blumenthal and Tucker (1974), Gursky and Ruffini (1975), Lamb and Pines (1979), Baity and Peterson (1979), and Blandford and Thorne (1979).

We now turn our attention to the first X-ray source believed to contain a black hole as one of its components.

8.5. Cyg X-1 as Observational Confirmation of the Existence of a Black Hole

The X-ray source which has been investigated most extensively is Cyg X-1. It was the first source observed to have variable intensity. The data provided by the UHURU satellite implies that Cyg X-1 is a rapidly varying X-ray source. Its intensity undergoes significant fluctuations on a very short time scale of the order of 50 msec. This restricts the size of the active region to be less than 15000 km. Precise positional data permitted astronomers to find that Cyg X-1 is also a variable radio source and, more importantly, made its optical identification possible. The position of Cyg X-1 coincides with that of the supergiant BO HDE 226868. The spectrum of this star changes periodically with a 5.6 day period and shows no irregularities. The only peculiarity was the discovery of H_β emission lines undergoing random shifts corresponding to radial velocities of about 100 km/s, indicative of mass flow. However, the presence of mass streaming or the fact that the system is binary are nothing unusual for stars of this type. Observations of the X-ray emission of Cyg X-1 have not revealed any periodic variations with a 5.6 day period.

Twice between December 1970 and December 1975 an abrupt change in the intensity of the X-ray emission was observed. The first change was reported in March 1971 (Tananbaum *et al.*, 1972), when the intensity of the source in the 2–6 keV energy range dropped by a factor of four and that in 10–20 keV by a factor of two. At the same time weak radio signals were received. The second transition was observed in early May 1975 (Hjellming, Gibson, and Owen, 1975), when the X-ray emission in the 1–2 keV energy range increased tenfold, with much lower rises in other ranges and no changes at all at energies exceeding 8 keV. The radio intensity also changed: it doubled at 8 MHz and fell as much as seven times at 2.7 MHz. The spectral index was also affected. From March 1971 till the end of April 1975 the radio spectrum was flat and the spectral index was null. During the active period in May 1975 the radio source emitted various amounts of energy at different frequencies. The dependence of the emitted energy on frequency corresponded to a spectral index of 0.31. If the radio emission was produced by thermal processes, this meant that the capacity of the source increased. If the emission mechanism did not involve thermal processes, such a change of the spectral index would have implied the appearance of self-absorption. After about two weeks the emission of the source both in the X-ray and in the radio regions of the spectrum returned to the previous level.

Random observations of Cyg X-1 in the optical part of the spectrum have also shown significant variations. In October 1973 and August 1974 the optical spectrum of Cyg X-1 was analysed for correlations between 1 ms counts. No systematic changes were noticed. The observations were repeated in June 1975 and then a strong pulsed component was discovered with a very short pulse period of 83.531 ± 0.008 ms (Auriemma *et al.*, 1976). The active period lasted about 10 minutes, after which the pulse died out and no further regularities were observed. A similar active period, about one hour long, was observed on 18th July, 1975. There again appeared strong periodic pulsations, with a period of approximately 83 ms. For the pulse to be seen against the bright central star, its intensity could not be less than 10^{35} erg/s.

Soon after Cyg X-1 was optically identified with a BOI supergiant and observational evidence for its binary nature was obtained, an emission line He II 4686 was discovered in its spectrum. The line was found to undergo periodic shifts which are out of phase with the periodic variations in the radial velocity of the primary component. The conclusion was drawn that the region of emission of the He II must be placed on the other side of the centre of mass than the main star. If the emission region is associated with the second object, then, provided that the primary star is a typical supergiant, it is possible to determine the masses of both components. For more information on the existing optical and X-ray data see a detailed review by Oda (1977).

The mass normally associated with a supergiant is 20 $M_\odot$, and in any case it should not be less than 10 $M_\odot$. The velocity of the supergiant, determined on the basis of the spectral changes, is 74 km/s, and that of the second component, based on the He II line, is 100 km/s. It follows from (8.12) that the ratio of the two masses is 1.35. Therefore the mass of the unseen object should be between 11 $M_\odot$ and 8 $M_\odot$. If the X-ray emission is produced by gas accretion onto the small star, the mass estimates for this object imply that it must be a black hole. The mass of this component is much greater than the maximum stable mass of a neutron star. The weakest point in this reasoning is the assumption that the region of emission of the He II line moves with the secondary. However, even if we determine the mass function using the primary star spectrum alone and assume the mass of the primary to be 12 $M_\odot$, the minimum mass we get for its companion is 3.35 $M_\odot$. This mass is also above the maximum mass of a stable white dwarf or a neutron star. Thus, providing we exclude exotic models of the X-ray source, the simplest way to interpret the observed properties of Cyg X-1 is to accept that this is a binary system with a black hole as one of its components.

A question that suggests itself is why no X-ray eclipse of the system is observed. From the optical data, the distance of Cyg X-1 is estimated to be 2000 pc. If we assume that the region emitting the He II line is associated with the unseen component, then, knowing the masses of both components, we can determine from (8.15) the inclination of the orbit; it is 26°. The radius of a typical BOI supergiant is estimated to be 16 $R_\odot$ and the separation of the centres of mass of the two components is about 30 $R_\odot$. The geometry of the system thus implies that an eclipse would be possible only if the orbit inclination angle were at least 60°.

As we have already mentioned, the optical data about the central star provide no evidence for the presence of an X-ray source in its vicinity. There is nothing surprising in this fact, for even if the X-ray source radiates the same fraction of the total energy in the visible spectrum as does the brightest source, Sco X-1, such an amount of light would at most be a mere 10^{-3} of the brightness of the supergiant.

The observed variations of the Cyg X-1 radio and X-ray emissions and the discovery of rapidly-varying periodic fluctuations in the visible part of the spectrum confirm the disc model of accretion. Thorne and Price (1975) have examined various types of instability in the accretion disc. They found that physical conditions present in different parts of the disc may lead to instabilities and fluctuations with very different time scales. Until the present time, all considerations have concerned the stationary state or small perturbations of it. More precise information may be provided only by investigations of the time-dependence of various processes that can occur in the disc without breaking it up. At present, the stationary disc model shows good qualitative agreement with observational data. The interpretation it affords of the X-ray emission of Cyg X-1 inclines us to believe that this system contains a black hole. (For a critical review of all the arguments leading to this conclusion see Eardley *et al.*, 1978.) Whether or not this is true will be shown by more precise observational data.

8.6. Black Holes in Spherical Clusters and Galactic Nuclei

Spherical clusters contain from several hundred to a few million stars, and galaxies many more. Giant elliptic galaxies may consist of about 10^{12} stars. Such large numbers make it natural to use statistical mechanics. Gravitational interactions are long-range interactions and, since all masses are positive, the screening effect does not occur. The statistical sum is divergent, both for small and for large distances. As compared with a system of charged particles interacting electromagnetically, both these divergencies present serious difficulties. In the case of plasma, it is possible, owing to the screening, to introduce an effective potential, which vanishes quickly outside the characteristic screening radius. At small distances, on the other hand, quantum processes begin to play an important role, modifying the form of the potential. Attempts to overcome these difficulties have failed so far and no satisfactory form of the statistical sum for a system of gravitating bodies has yet been given. An alternative possibility is to adopt a statistical approach and describe the system by a distribution function satisfying the Boltzmann equation. This approach was developed by Chandrasekhar (1942, 1943) and has proved to be very useful. For detailed reviews see Lynden-Bell (1968) and Lightman and Shapiro (1978). Recently significant progress in understanding the dynamics of many-body self-gravitating systems has been made using sophisticated numerical calculations (Aarseth and Lecar, 1975; Hayli, 1975). However, observational data still remain our basic source of information (Peterson and King, 1975; Peterson, 1976; Oort, 1977).

Observation proves irrefutably that the star number density, both in stellar clusters and in galaxies, increases as the centre of the cluster or the galaxy is approached. If evolutionary processes in such a dense region give rise to a black hole, collisions with other bodies and capture of matter from the surroundings will augment its mass. Such a massive black hole will tend to shift towards the centre.

Lynden-Bell (1969) advanced the hypothesis that massive black holes are present in most galactic nuclei and in the central regions of spherical clusters. In his model of an active nucleus, the central massive black hole, whose mass may reach $10^8\ M_\odot$, is surrounded by a dense cloud of stars and scattered gas. Based on the very conservative assumption that the black hole sucks in a mass of $10^{-3}\ M_\odot$ per year, one can estimate the amount of energy released in the process of accretion. It turns out that this process can secure a luminosity for the galactic nucleus of the order of $10^9\ L_\odot$. Should such an amount of energy be released in the optical part of the spectrum, the luminosity of the galactic nucleus would increase noticeably. The main difficulty with the Lynden-Bell model is that these enormous energies produced by accretion must be emitted in the visible or radio part of the spectrum. As in the model of accretion onto a black hole in a binary system, the viscosity is main factor determining the rate of energy transport and the shape of the spectrum. However, we know very little about the dynamic viscosity of a dense cloud of stars and dust, dominated by turbulent motions and magnetic interactions between" magnetic cells". It is certainly worthwhile to investigate these processes more closely, as they may prove essential for the explanation of properties of quasars, Seyfert galaxies and active central galactic regions.

Although the Lynden-Bell model should be treated as just a working hypothesis, at least two conclusions susceptible to observational verification can be derived from it. First, the central luminosity of a galactic nucleus containing a black hole should be distinctly higher than the luminosities of normal nuclei. Secondly, a galactic nucleus contains large amounts of dust, which radiates most effectively in the infrared. Part of the energy released in the process of accretion is converted into infrared radiation. The galactic nuclei containing black holes should therefore exhibit a high excess of infrared radiation energy as compared with the nuclei of normal galaxies.

How does a black hole in the centre of a galaxy or a spherical cluster affect the stellar density distribution? It follows from the Schwarzschild solution that the gravitational field far from the black hole does not differ from the gravitational field of a spherically symmetric star. Differences can be expected only near the centre.

Let v_d be the velocity dispersion of the stars and M the mass of the central black hole. As in the case of spherically symmetric accretion, we can distinguish a critical radius

$$r_a = \frac{2GM}{v_d^2}, \tag{8.47}$$

beyond which the influence of the hole can be neglected. The density distribution in the sphere of radius r_a depends on the relative magnitude of various characteristic time scales: the dynamic time scale t_d, on which a star moving at the average velocity covers a distance equal to the characteristic size of the system; the relaxation time t_r, designating the period after which the effect of stellar collisions will make itself known; the time t_g which has elapsed since the mass of the black hole began to grow; and the time t_f measured from the moment of the formation of the hole. If t_f is much greater than any of the other characteristic times, the nucleus or the cluster has had enough time for the distribution of matter around the black hole to settle to a stationary level. This, however, is not an equilibrium distribution, for nothing that the black hole swallows can escape from it. Stars which are within the radius r_a will slowly, on a time scale of the order of the relaxation time, fall into the black hole. If the cluster is isothermal and the density of stars at its centre is n_c, then, as shown by Bahcall and Wolf (1976) and by Frank and Rees (1976), for $r_t \leqslant r \leqslant r_a$

$$n(r) \approx n_c \left(\frac{r_a}{r}\right)^{7/4}; \tag{8.48}$$

r_t is the radius at which stars are pulled apart by the tidal forces of the black hole.

The relaxation time of galactic nuclei is ordinarily much greater than the age of the universe and in their case the above description does not apply. If $t_f \gg t_d$, the evolution of the system depends on whether t_g is greater or less than t_d. If $t_g \lesssim t_d$, the growth of the black hole is very rapid, and the hole can also absorb those stars which would have had a chance to escape outside the dangerous region $r \lesssim r_a$ had the growth rate been lower. The density distribution is then (Peebles, 1972)

$$n(r) = n_c \left(1 + \frac{r_a}{r}\right)^{1/2} \quad \text{for} \quad r \lesssim r_a. \tag{8.49}$$

The rise in the central density is too small to be observed.

The case $t_g \gg t_d$, where the black hole grows slowly in the galactic nucleus or in the centre of a spherical cluster, is more realistic. Here the process is similar to adiabatic, spherically symmetric accretion. The density changes as

$$n(r) = n_c \left(\frac{r_a}{r}\right)^{3/2} \quad \text{for} \quad r < r_a. \tag{8.50}$$

A marked rise in central density should show up as a noticeable increase of the surface luminosity near the centre.

The available observational data are not too precise, but they do not exclude the possibility of the existence of black holes in galactic nuclei and spherical clusters. The problem requires further theoretical and observational studies (Blandford and Znajek, 1977; Carter, 1979a).

8.7. Interaction of a Black Hole with its Environment

Investigations of black hole interactions with various external systems, for example with a nearby star or an incident electromagnetic or gravitational wave, are of great interest for astrophysical applications. Since no exact solutions for such configurations are known, there remain approximate methods based on the assumption that the effects caused by the environment are small. In practice, therefore, one considers the linearized equations for small perturbations in a stationary Kerr background. A general method for evaluating the changes in the parameters of a black hole due to small perturbations was proposed by Hawking.

As we know, the horizon is generated by null geodesics, which form a two-parameter family. Every family of null geodesics, just like every family of timelike geodesics, can be characterized by kinematic parameters: expansion θ, shear σ and rotation ω. If the tangent vector to the null geodesics is k^α, it is always possible to construct a tetrad with k^α as one of the basis vectors. Let another vector of the tetrad be also a null vector n^α such that $n^\alpha k_\alpha = 1$, and let the remaining two vectors, a^α and b^α, be spacelike unit vectors orthogonal to each other and to k^α and n^α. From the vectors a^α and b^α we can construct the complex vector $m^\alpha = \frac{1}{\sqrt{2}}(a^\alpha + ib^\alpha)$. In the tetrad composed of the vectors k^α, n^α, m^α and $\bar{m}^\alpha$, where $\bar{m}^\alpha$ is the complex conjugate of m^α, the following orthogonality relations are satisfied:

$$k^\alpha n_\alpha = 1 = -m_\alpha \bar{m}^\alpha ; \tag{8.51}$$

the remaining scalar products are zero, in particular $m^\alpha m_\alpha = 0$. The kinematic parameters can now be determined from

$$\rho = \theta + i\omega = k_{\alpha;\beta}\, m^\alpha \bar{m}^\beta , \qquad \sigma = k_{\alpha;\beta}\, m^\alpha m^\beta . \tag{8.52}$$

Let us examine how ρ and σ change along the horizon. To this end, intersect the horizon with a spacelike surface $\Sigma(t)$. Every point of the horizon lies on the trajectory of a one-parameter group of motions, and its distance from the intersection $\mathscr{H} \cap \Sigma(t) = S(t)$ can be measured with the affine parameter v along this trajectory. The surface $S(t)$ can be parametrized with the angles θ and φ. Calculating $\mathrm{d}\rho/\mathrm{d}v$ and $\mathrm{d}\sigma/\mathrm{d}v$, we get

$$\frac{\mathrm{d}\rho}{\mathrm{d}v} = \rho^2 + \sigma\bar{\sigma} + 2\kappa\rho/c^2 + \phi_{00} , \tag{8.53}$$

$$\frac{\mathrm{d}\sigma}{\mathrm{d}v} = 2\rho\sigma + 2\kappa\sigma/c^2 + \psi_0 , \tag{8.54}$$

where

$$\begin{gathered} \kappa/c^2 = \tfrac{1}{2} k_{\alpha;\beta}\, n^\alpha k^\beta , \\ \phi_{00} = \tfrac{1}{2} R_{\alpha\beta} k^\alpha k^\beta , \qquad \psi_0 = C_{\alpha\beta\gamma\delta} k^\alpha m^\beta k^\gamma m^\delta . \end{gathered} \tag{8.55}$$

For a stationary black hole $\rho = \sigma = 0$, hence also $\phi_{00} = \psi_0 = 0$.

The definitions of ϕ_{00} and ψ_0 suggest a natural division of perturbations of a black hole into two kinds: those for which $\phi_{00} = 0$ at the horizon, to be called *gravitational perturbations*, and those for which $\phi_{00} \neq 0$ at the horizon – call them *induced perturbations.*

Let us first consider induced perturbations. Their source may be, for example, an electromagnetic, a scalar or a matter field. We treat these external fields as small perturbations. In the first approximation we can neglect quadratic terms in equation (8.53). The result is

$$\frac{d\rho}{dv} = 2\kappa\rho/c^2 + \frac{4\pi G}{c^4} T_{\alpha\beta} k^\alpha k^\beta . \qquad (8.56)$$

For perturbations of a Kerr black hole, $\kappa = \frac{r_H c^2 - GM}{r_H^2 + a^2}$. Suppose that the perturbation field is turned off after some time. The solution of (8.56) in the ensuing period is

$$\rho = -\frac{4\pi G}{c^4} \int_v^\infty e^{(2\kappa/c^2)(v-v')} T_{\alpha\beta} k^\alpha k^\beta dv' . \qquad (8.57)$$

The boundary condition used is that $\rho = \theta = 0$ at $v = \infty$, because the event horizon is tha null surface which ends up being stationary.

The rate of change of the area of the hole is related to ρ by

$$\frac{dA}{dv} = -2 \int \rho dA . \qquad (8.58)$$

Substituting ρ from (8.57) and integrating by parts, we find that the area increase of the black hole is

$$\delta A = \frac{4\pi G}{c^2 \kappa} \int T_{\alpha\beta} k^\alpha d\Sigma^\beta , \qquad (8.59)$$

where $d\Sigma^\beta = k^\beta dA dv$. The vector k^α tangent to the null geodesic generators of the horizon can be expressed as a linear combination of the Killing vectors ξ^α and η^α:

$$k^\alpha = \xi^\alpha - (\Omega_H / c) \eta^\alpha ; \qquad (8.60)$$

Ω_H is the angular velocity of the black hole. Inserting (8.60) into (8.59), we get

$$\delta A = \frac{4\pi G}{c^2 \kappa} (\delta(Mc^2) - \Omega_H \delta J) . \qquad (8.61)$$

Thus small induced perturbations turn a Kerr black hole with parameters M and J into one with parameters $M + \delta M$ and $J + \delta J$.

Small gravitiaional perturbations also affect the kinematic parameters. In the first approximation, the changes of the latter are given by the equations

$$\frac{d\rho}{dv} = \sigma\bar{\sigma} + 2\kappa\rho/c^2 , \qquad (8.62)$$

$$\frac{d\sigma}{dv} = 2\kappa\sigma/c^2 + \psi_0 . \qquad (8.63)$$

From (8.63) we get

$$\sigma = -\int_{v}^{\infty} e^{(2\kappa/c^2)(v+v')}\psi_0(v')\,dv' ; \tag{8.64}$$

inserting this into (8.62) and integrating gives

$$\rho = \int e^{(2\kappa/c^2)(v-v')}\sigma\bar{\sigma}dv' . \tag{8.65}$$

The change in the surface area of the hole can be found as in the case of induced perturbations. The result is

$$\delta A = -\frac{c^2}{\kappa}\int \sigma\bar{\sigma}dA dv . \tag{8.66}$$

Using this general formula, Hartle examined the effect exerted on a black hole by an orbiting object such as a star or a planet.

The second law of black hole mechanics implies that the surface area of the black hole will increase. In the situation considered here $\phi_{00}=0$ on the horizon, and the energy flux vanishes at the surface of the hole. The increase in the area of this surface may only be caused by changes in the internal parameters, viz. mass and angular momentum. The mass, however, must remain constant in the present case. Hence, just as the tidal friction caused by the Moon slows down the rotation of the Earth, an object orbiting a black hole will reduce the total angular momentum of the hole. Hartle has shown that if a stationary body of mass M' orbits a black hole of mass M and angular momentum $J\left(\frac{G^2M^4}{c^2} \gg J^2\right)$ at a distance r from it, the angle between the rotation axis of the hole and the radius-vector of the body being equal to θ, then the angular momentum of the black hole changes according to

$$\frac{dJ}{dt} = -\frac{2c^3J}{5GM}\left(\frac{M'}{M}\right)^2\left(\frac{GM}{c^2r}\right)^6 \sin^2\theta . \tag{8.67}$$

These changes are very slow. A peculiarity of this system is the fact that the tidal bulge always comes ahead of the position of the perturbing body, contrary to the tidal bulges on the surface of a normal star or a planet, which are always behind this position.

Hartle considered only slowly rotating black holes. Teukolsky and Press (1974) have pointed out that in the case of a rapidly rotating black hole, for which $\frac{G^2M^4}{c^2} \approx J^2$, significant new effects may occur. A body orbiting the black hole near the horizon at an angular velocity close to Ω_H would draw energy and angular momentum from the black hole. The orbiting body would also be radiating energy. If the energy it gains happens to equal the energy it loses, the body will move in a nearly circular orbit. Teukolsky and Press described such orbits as "floating". So far nobody has estimated the efficiency of this process of extracting energy from a black hole. But even if it proved high, the effect would be of little astrophysical significance, since only particles moving in a very special way could approach such floating orbits.

8.8. Equations of Motion for a Black Hole

Whenever it was previously relevant, for example when considering accretion of gas onto a black hole in a binary system, we assumed that in the first approximation the black hole moved according to Newton's laws of motion. We now propose to verify whether this is indeed the case. For simplicity, we shall restrict ourselves to a spherically symmetric black hole. As we know, it is described by the Schwarzschild metric. We shall examine the motion of the black hole in the gravitational field generated by distant masses (Demiański and Grishchuk, 1974).

Consider an isolated system of two bodies at a distance R from each other. Suppose that R is much greater than the linear dimension of either body, l_1 and l_2, but allow l_1 or l_2 or both to be comparable with the corresponding gravitational radii of the bodies.

Far from the bodies write the space-time metric in the form

$$g_{\alpha\beta}=\eta_{\alpha\beta}+h_{\alpha\beta}, \tag{8.68}$$

where $\eta_{\alpha\beta}$ is the metric of Minkowski space, and $h_{\alpha\beta}$ are small corrections. The corrections $h_{\alpha\beta}$ can be expanded in a power series with respect to a certain small parameter λ,

$$h_{\alpha\beta}=\sum_{n=1}^{\infty}\lambda^n \underset{n}{h_{\alpha\beta}}. \tag{8.69}$$

The parameter λ has only a formal meaning and serves to group terms of the same order. We shall assume that differentiation with respect to x^0 raises the order of the expression by one and that with respect to spatial variables leaves it unchanged, i.e., to order of magnitude,

$$\frac{\partial h_{\alpha\beta}}{\partial x^0}\approx\lambda\frac{\partial h_{\alpha\beta}}{\partial x^i}. \tag{8.70}$$

In the quasi-static approximation $\lambda\ll 1$, and it can be interpreted physically as the ratio of the characteristic relative velocity to the speed of light.

Formally, we can now seek vacuum solutions of Einstein's equations, assuming that $g_{\alpha\beta}$ is of form (8.68) everywhere except at singular points. In order to obtain a sensible solution, using only a finite number of terms of series (8.69), we must stay within the region where this series is convergent. It is convergent for distances much larger than the gravitational radii of the two bodies but less than the characteristic lengths of the waves emitted by the system.

By a suitable choice of the coordinate system we can always make $\underset{3}{h_{00}}$, $\underset{2}{h_{0i}}$, $\underset{1}{h_{ij}}$ and $\underset{3}{h_{ij}}$ equal to zero. Introduce the notation $\psi_{\mu\nu}=h_{\mu\nu}-\frac{1}{2}\eta_{\mu\nu}h_{\alpha\beta}\eta^{\alpha\beta}$ and $\tau=\lambda x^0$. Inserting (8.68) into the field equation $R_{00}=0$, we get in the lowest order

$$\delta^{ik}\underset{2}{\psi_{00,ik}}=0. \tag{8.71}$$

The solution of this equation which vanishes at infinity and has two monopole singularities is

$$\underset{2}{\psi_{00}}=\frac{4\phi}{c^2}=-\frac{4GM_1}{c^2 r_1}-\frac{4GM_2}{c^2 r_2}, \tag{8.72}$$

where

$$r_1^2=\delta_{ik}(x^i-\underset{(1)}{\xi^i})(x^k-\underset{(1)}{\xi^k}),\quad r_2^2=\delta_{ik}(x^i-\underset{(2)}{\xi^i})(x^k-\underset{(2)}{\xi^k}) \tag{8.73}$$

and $M_1, M_2, \underset{(1)}{\xi^i}, \underset{(2)}{\xi^i}$ are arbitrary functions of time. To lowest order, the field equations $R_{ik}=0$ lead to

$$\underset{2}{h_{ik}}=\delta_{ik}\underset{2}{h_{00}}\,.$$

From the equations $R_{0i}=0$ in the lowest order and from the equations $R_{ik}=0$ in the first-order approximation we can determine $\underset{3}{\psi_{0i}}$ and $\underset{4}{\psi_{ik}}$. The integrability conditions for these equations can be reduced to

$$\dot{M}_1=\dot{M}_2=0,$$

$$M_1\underset{(1)}{\ddot{\xi}^i}-GM_1M_2\frac{\partial}{\partial \underset{(1)}{\xi^i}}\frac{1}{R}=0,$$

$$M_2\underset{(2)}{\ddot{\xi}^i}-GM_1M_2\frac{\partial}{\partial \underset{(2)}{\xi^i}}\frac{1}{R}=0, \tag{8.74}$$

where

$$R^2=\delta_{ik}(\underset{(1)}{\xi^i}-\underset{(2)}{\xi^i})(\underset{(1)}{\xi^k}-\underset{(2)}{\xi^k}), \tag{8.75}$$

and a dot indicates differentiation with respect to τ. As follows from the form of the gravitational potential, M_1 and M_2 should be identified with the masses of the bodies.

Equations (8.74) are the Newtonian equations of motion for two bodies regarded as singularities of the field. We will show below that such a singularity, representing a source of gravity, can be replaced by a black hole.

Recall that in the case under consideration the separation of the bodies is much larger than their characteristic dimensions. In the region in which Newton's theory holds, the gravitational field produced by one of the bodies is perturbed by the field generated by the other body. For our approximation to make sense the perturbations caused by the presence of the second body should be small not only in the Newtonian region: they should remain small when extended up to the horizon.

In the Newtonian region, the perturbations of the gravitational field of a black hole due to the presence of a second body can, in the spherical coordinates, be described by the metric

$$ds^2=c^2\left(1-\frac{r_g}{r}+\frac{2\chi}{c^2}\right)dt^2-\left(1+\frac{r_g}{r}-\frac{2\chi}{c^2}\right)dr^2-r^2\left(1-\frac{2\chi}{c^2}\right)(d\theta^2+\sin^2\theta d\varphi^2), \tag{8.76}$$

where

$$\frac{2\chi}{c^2}=-\frac{2MG}{c^2\rho}\sum_{l=0}^{\infty}\left(\frac{r}{\rho}\right)^l P_l(\cos\theta), \tag{8.77}$$

$r\gg r_g$ and $\rho=\text{const}\gg r_g$. Here we are considering stationary perturbations corresponding to the instantaneous position of the bodies.

In the general case, stationary perturbations of the Schwarzschild metric can be written as

$$ds^2 = c^2\left(1-\frac{r_g}{r}\right)(1+H)\,dt^2 - \left(1-\frac{r_g}{r}\right)^{-1}(1-H)\,dr^2 - r^2(1+K)(d\theta^2+\sin^2\theta\, d\varphi^2); \tag{8.78}$$

H and K are arbitrary functions of r and θ. This general form of stationary perturbation can be derived from the even perturbations of the Schwarzschild space-time which were considered in the previous chapter. The functions H and K can be resolved into Legendre polynomials: $H=\sum_l h_l(r)P_l(\cos\theta)$, $K=\sum_l k_l(r)P_l(\cos\theta)$. Then, in the linear approximation, the field equations reduce to the system of equations

$$2(r-r_g)h' + (2r-r_g)k' - (l-1)(l+2)(h+k) = 0, \tag{8.79}$$

$$r(r-r_g)(k'+h') + h = 0, \tag{8.80}$$

where the subscript l has been omitted and a prime denotes differentiation with respect to r.

Defining w by the relation $h = r(r-r_g)w$, we get the following equation for w:

$$r(r-r_g)w'' + 3(2r-r_g)w' - (l-2)(l+3)w = 0. \tag{8.81}$$

Using the hypergeometric function, we can write the general solution of this equation in the form

$$w = c_1 F\left(2-l, l+3, 3, \frac{r}{r_g}\right) + \frac{c_2}{(r-r_g)^2 r^{l+1}} F\left(l+1, -l, 3, \frac{r}{r_g}\right), \tag{8.82}$$

where c_1 and c_2 are arbitrary constants. From equation (8.80), remembering the definition of w, one can readily find k.

Asymptotically, for $r \gg r_g$, w behaves as

$$w \to c_1 r^l + c_2 r^{-(l+1)}, \tag{8.83}$$

and therefore can be continuously matched with (8.77) if c_2 is set equal to zero. Putting $c_2 = 0$ in (8.82), we obtain a solution which for large r becomes (8.77). It is regular for small r; in particular, w is finite at $r = r_g$.

We have thus shown that the stationary perturbation of the gravitational field of a black hole due to the presence of a second body remain regular near the horizon. Hence, from the point of view of a distant observer who watches the changes in the gravitational field, a spherically symmetric black hole moves like a spherically symmetric star. (A more sophisticated derivation of this result was given by D'Eath (1975).) Basic differences are to be expected in the case where the distance between the black hole and the star or between two black holes is comparable with their dimensions. Such a system would be a strong source of gravitational waves. Equations of motion in this case are unknown.

Literature

AARSETH, S. J. (1971) *Astrophys. Space Sci.* **14**, 118.

AARSETH, S. J. and LECAR, M. (1975) *Ann. Rev. Astron. Astrophys.* **13**, 1.

ABRAMOWICZ, M. A., JAROSZYŃSKI, M. and SIKORA, M. (1978) *Astron. Astrophys.* **63**, 221.

AURIEMMA, G. *et al.* (1976) *Nature* **259**, 27.

BAHCAL, J. N. (1978) *Optical properties of binary X-ray sources* in *Physics and Astrophysics of Neutron Stars and Black Holes*, eds.: Giacconi, R., Ruffini, R., North Holland Publishing Company, Amsterdam.

BAHCALL, J. N. and WOLF, R. A. (1976) *Astrophys. J.*, **209**, 214.

BAITY, W. A. and PETERSON, L. E. (eds.) (1979) (COSPAR) *X-ray Astronomy*, Pergamon Press, Oxford.

BLANDFORD, R. D. (1979) *Accretion Discs and Black Hole Electrodynamics*, in *Active Galactic Nuclei*, eds.: Hazard, C. and Mitton, C., Cambridge University Press, Cambridge.

BLANDFORD, R. D. and THORNE, K. S. (1979) *Black Hole Astrophysics*, in *General Relativity, An Einstein Centenary Survey*, eds.: Hawking, S. W. and Israel, W., Cambridge University Press, Cambridge.

BLANDFORD, R. D. and ZNAJEK, R. L. (1977) *Mon. Not. Roy. Astr. Soc.* **179**, 433.

BONDI, H. (1952) *Mon. Not. Roy. Astr. Soc.* **112**, 195.

BLUMENTHAL, G. R. and TUCKER, W. H. (1974) *Ann. Rev. Astron. Astrophys.* **12**, 23.

BRADT, H. V. and GIACCONI, R. (eds.) (1973) *X- and Gamma Ray Astronomy*, D. Reidel Publishing Comp., Dordrecht.

CARTER, B. (1979a) *Black Holes in the Context of Galactic Nuclei: an introductory review*, in *Active Galactic Nuclei*, eds.: Hazard, C. and Mitton, S., Cambridge University Press, Cambridge.

CARTER, B. (1979b) *Perfect Fluid and Magnetic Field Conservation Laws in the Theory of Black Hole Accretion Rings*, in *Active Galactic Nuclei*, eds.: Hazard, C. and Mitton, S., Cambridge University Press, Cambridge.

CHANDRASEKHAR, S. (1942) *Principles of Stellar Dynamics*, University of Chicago Press, Chicago.

CHANDRASEKHAR, S. (1943) *Ann. N. Y. Acad. Sci.* **65**, 131.

DEMIAŃSKI, M. and GRISHCHUK, L. (1974) *Gen. Rel. Grav.* **5**, 673

D'EATH, P. D. (1975) *Phys. Rev.* **D11**, 1387; **D12**, 2183.

EARDLEY, D., LIGHTMAN, A., SHAKURA, N., SHAPIRO, S. and SUNYAEV, R. (1978) *Comments Astrophys.* 7, 151.

EARDLEY, D. and PRESS, W. (1975) *Ann. Rev. Astron. Astrophys.* **13**, 381.

FABIAN, A. C., PRINGLE, J. E. and REES, M. J. (1976) *Mon. Not. Roy. Astr. Soc.* **175**, 43.

FRANK, J. and REES, M. J. (1976) *Mon. Not. Roy. Astr. Soc.* **176**, 633.

GIACCONI, R. (1980) *Scientific American* **242**, 70, February.

GIACCONI, R. (ed.) (1981) *X-Ray Astronomy with the Einstein Satellite*, D. Reidel Publishing Company, Dordrecht.

GIACCONI, R. and RUFFINI, R. (eds.) (1978) *Physics and Astrophysics of Neutron Stars and Black Holes*, North-Holland Publishing Company, Amsterdam.

GUNN, J. E. (1979) *Feeding the Monster: Gas Discs in Elliptical Galaxies*, in *Active Galactic Nuclei*, eds.: Hazard, C. and Mitton, S., Cambridge University Press, Cambridge.

GURSKY, H. (1973) *Observations of Galactic X-Ray Sources*, in *Black Holes*, eds: De Witt, C. and De Witt, B. S., Gordon and Breach, New York.

GURSKY, H. and RUFFINI, R. (eds.) (1975) *Neutron Stars, Black Holes and Binary X-Ray Stars*, D. Reidel Publishing Comp., Dordrecht.

GURSKY, H. and SCHWARTZ, D. A. (1977) *Ann. Rev. Astron. Astrophys.* **15**, 541.

GURSKY, H. and VAN DEN HEUVEL, E. P. J. (1975) *Scientific American* **232**, 24, March.

HARTLE, J. (1973) *Phys. Rev.* **D8**, 1010.

HARTLINE, B. K. (1979) *Science* **204**, 1399, **205**, 31.

HAWKING, S. W. (1973) *The Event Horizon*, in *Black Holes*, eds.: De Witt, C. and De Witt, B. S., Gordon and Breach, New York.

HAYLI, A. (ed.) (1975) *Dynamics of Stellar Systems*, D. Reidel Publ. Comp., Dordrecht.

HEINTZ, W. D. (1978) *Double Stars*, D., Reidel Publ. Comp., Dordrecht.

HJELLMING, R., GIBSON, D. and OWEN, F. (1975) *Nature* **256**, 111.

JAROSZYŃSKI, M., ABRAMOWICZ, M. A. and PACZYŃSKI, B. (1980) *Astron. Astrophys.* **30**, 1.

KOZŁOWSKI, M., JAROSZYŃSKI, M. and ABRAMOWICZ, M. A. (1978) *Astron. Astrophys.* **63**, 209.

KYLAFIS, N. D., LAMB. D. Q., MASTERS, A. R. and WEAST, G. J. (1980)

Ann. N. Y. Acad. Sci. **336**, 520.

LAMB, F. and PINES, D. (1979) *Compact Galactic X-Ray Sources*, Physics Dept., University of Illinois at Urbana-Champaign.

LEWIN, W. M. G. (1979) *What Are X-Ray Burst Sources?*, in *Advances in Space Exploration*, Vol. 3, *X-Ray Astronomy*, eds.: Baity, W. A. and Peterson, L. E., Pergamon Press, Oxford.

LEWIN, W. M. G. and CLARK, G. (1980) *Ann. N. Y. Acad. Sci.* **335**, 451.

LIGHTMAN, A. P. and EARDLEY, D. M. (1974) *Astrophys. J.* **187**, L1.

LIGHTMAN, A. P. and SHAPIRO, S. L. (1978) *Rev. Mod. Phys.* **50**, 437.

LYNDEN-BELL, D. (1968) *Statistical Mechanics of Ste'lar Systems*, in *Astrophysics and General Relativity*, Vol. 1, eds.: Chretien, M., Deser, S., Goldstein, J., Gordon and Breach, New York.

LYNDEN-BELL, D. (1969) *Nature* **223**, 690.

LYNDEN-BELL, D. and PRINGLE, J. E. (1974) *Mon. Not. Roy. Astr. Soc.* **168**, 603.

MCCRAY, R. (1979) *Spherical Accretion onto Supermassive Black Holes*, in *Active Galactic Nuclei*, eds.: Hazard, C. and Mitton, S., Cambridge University Press, Cambridge.

NOVIKOV, I. D. and THORNE, K. S. (1973) *Black Hole Astrophysics*, in *Black Holes*, eds.: DeWitt, C. and DeWitt, B. S., Gordon and Breach, New York.

ODA, M. (1977) *Space Science Reviews* **20**, 757.

OORT, J. H. (1977) *Ann. Rev. Astron. Astrophys.* **15**, 295.

PACZYŃSKI, B. and WIITA, P. (1980) *Astron. Astrophys.* **88**, 23.

PEEBLES, P. (1972) *Gen. Rel. Grav.* **3**, 63.

PETERSON, C. J. (1976) *Astron. J.* **81**, 617.

PETERSON, C. J. and KING, I. R. (1975) *Astron. J.* **80**, 427.

PRINGLE, J. E. and REES, M. J. (1972) *Astron. Astrophys.* **21**, 1.

RAPPAPORT, S. **(1979)** *Binary X-Ray Pulsars*, a talk presented at the NATO Advanced Study Institute on Galactic X-Ray Sources, Cape Sounion, Greece, June **1979**.

SALPETER, E. (1964) *Astrophys. J.* **140**, 796.

SANFORD, P. W., LASKARIDES, P. and SALTON, J. (eds.) (1982) *Galactic X-Ray Sources*, John Wiley and Sons, New York.

SHAKURA, N. and SUNYAEV, R. A. (1973) *Astron. Astrophys.* **24**, 337.

SUNYAEV, R. A. (1978) *Theory of Accretion*, in *Physics and Astrophysics of Neutron Stars and Black Holes*, eds.: Giacconi R., and Ruffini., R., North-Holland Publishing Company, Amsterdam.

TANANBAUM, H., GURSKY, H., KELLOG, E., JONES, C. and GIACCONI, R. (1972) *Astrophys. J. Lett.* 177, L5.

TANANBAUM, H. and TUCKER, W. (1974) *Compact X-Ray Sources*, in *X-Ray Astronomy*, eds.: Giacconi, R. and Gursky, H., D. Reidel Publ. Comp., Dordrecht.

TEUKOLSKY, S. and PRESS, W. (1974) *Astrophys. J.* **193**, 443.

THORNE, K. S. (1969) *Astrophys. J.* **158**, 1.

THORNE, K. S. and PRICE, R. (1975) *Astrophys. J.* **195**, L101.

TSURUTA, S. and CAMERON, A. (1966) *Canad. J. Phys.* **44**, 1895.

ZELDOVICH, YA. B. and NOVIKOV, I. D. (1971) *Relativistic Astrophysics*, Vol. 1, *Stars and Relativity* The University of Chicago Press, Chicago.

CHAPTER 9

GRAVITATIONAL WAVES

9.1. Gravitational Waves in the Weak-Field Approximation

Contrary to Newton's theory of gravitation, the General Theory of Relativity, being a relativistic theory of gravitation, assumes that any gravitational action propagates with finite speed, which is a fundamental condition necessary for the existence of gravitational waves. Mathematically, the difference between the two theories manifests itself in the fact that the Newtonian equations of a gravitational field are differential equations of the elliptic type while the general-relativity equations are hyperbolic. To avoid the difficulty arising from the nonlinearity of the general-relativistic equations, the earliest approach was to consider small perturbations of flat Minkowski space. This approximation was proposed as early as 1916 by Einstein. As we saw in Chapter 1, in the weak-field approximation the metric of the space-time can be written

$$g_{\alpha\beta}=\eta_{\alpha\beta}+h_{\alpha\beta}\,, \tag{9.1}$$

where $\eta_{\alpha\beta}$ is the Minkowski metric and $h_{\alpha\beta}$ are small corrections. Introducing the new quantities

$$\psi_{\mu\nu}=h_{\mu\nu}-\tfrac{1}{2}\eta_{\mu\nu}\,h^{\alpha}_{\alpha}\,, \tag{9.2}$$

where the indices in $h_{\mu\nu}$ are understood to be raised and lowered with $\eta_{\mu\nu}$, and restricting the choice of the coordinate system by the additional condition

$$\psi^{\nu}_{\mu,\,\nu}=0\,, \tag{9.3}$$

we can make the field equations read

$$\psi_{\mu\nu}{}^{,\alpha}{}_{,\alpha}=-\frac{16\pi G}{c^4}\,T_{\mu\nu}\,, \tag{9.4}$$

where $T_{\mu\nu}$ describes the matter field perturbing the gravitational field. Condition (9.3) does not determine the coordinate system uniquely. The coordinate transformation

$$x'^{\mu}\rightarrow x^{\mu}+\xi^{\mu}(x^{\alpha}) \tag{9.5}$$

where

$$\xi_{\mu}{}^{,\nu}{}_{,\nu}=0 \tag{9.6}$$

is still admissible. We can now ask how many of the ten independent components of $\psi_{\mu\nu}$

are physically significant. Condition (9.3) imposes four differential relations which reduce the number of independent components to 6. The coordinate transformation (9.5) satisfying (9.6) obeys condition (9.3) and changes $\psi_{\mu\nu}$ according to

$$\psi'_{\mu\nu} \to \psi_{\mu\nu} - \xi_{\mu,\nu} - \xi_{\nu,\mu} + \eta_{\mu\nu}\xi^{\alpha}{}_{,\alpha}. \tag{9.7}$$

By a suitable choice of ξ_μ we can eliminate further 4 components of $\psi_{\mu\nu}$. Thus $\psi_{\mu\nu}$ has only two physically significant components. In fact we can choose ξ_μ so that

$$\psi^{\mu}_{\mu} = \eta^{\mu\gamma}\psi_{\mu\gamma} = 0. \tag{9.8}$$

Small perturbations in the Minkowski space-time are given by the equation

$$\psi_{\mu\nu,\alpha}{}^{;\alpha} = 0. \tag{9.9}$$

The only physically interesting solutions, however, are those which satisfy the additional conditions

$$\psi^{\nu}_{\mu,\nu} = 0 \quad \text{and} \quad \psi^{\mu}_{\mu} = 0. \tag{9.10}$$

These equations coincide with the equations of a massless field of spin 2 in the Minkowski space-time. For this reason it is conjectured that gravitons – elementary quanta of a free gravitational field – have the rest mass equal to zero and spin equal to 2. As follows from (9.7), conditions (9.10) are invariant under coordinate transformations $x'^{\mu} \to x^{\mu} + \xi^{\mu}$ provided that ξ^{μ} satisfies $\xi_{\mu,\alpha}{}^{;\alpha} = 0$ and $\xi^{\alpha}{}_{,\alpha} = 0$. It is this freedom that permits reduction of the number of the independent components of $\psi_{\mu\nu}$ to 2. Thus, gravitational waves have two independent degrees of freedom.

Every sufficiently regular solution of equation (9.9) can be represented as a superposition of plane waves. Let us therefore look at the properties of monochromatic plane waves. The general solution of equation (9.9) representing a monochromatic plane wave is of the form

$$\psi_{\mu\nu} = \mathrm{Re}(A_{\mu\nu}e^{ik_\alpha x^\alpha}), \tag{9.11}$$

where $A_{\mu\nu}$ is a constant amplitude of the wave and k_α is a constant wave vector such that

$$k^{\alpha}k_{\alpha} = 0, \quad A_{\mu\nu}k^{\nu} = 0. \tag{9.12}$$

If, in addition, condition (9.8) is imposed, then $A^{\alpha}_{\alpha} = 0$. We are still free to choose three more conditions. Usually, the amplitude of a plane gravitational wave is restricted by the requirement that for any observer moving freely in the Minkowski space-time

$$A_{\mu\nu}u^{\nu} = 0, \tag{9.13}$$

where u^{α} is the tangent vector to the observer's world-line. Under this choice of the coordinate system the amplitude matrix of the wave is traceless, and so from the point of view of the distingushed class of observers the wave is "transverse".

Consider, for example, a plane monochromatic gravitational wave propagating along the x-axis in Minkowski space parametrized by Cartesian coordinates, so that $k^{\alpha} = (\omega, \omega, 0, 0)$. Let $u^{\alpha} = (1, 0, 0, 0)$, and therefore $A_{\mu 0} = 0$ and $A_{\mu 1} = 0$, and in addition let $A^{\alpha}_{\alpha} = 0$; it follows that the wave is fully characterized by the two components $A_{23} = A_{32}$ and $A_{22} = -A_{33}$.

By a suitable choice of coordinate system, every monochromatic plane wave can be reduced to this form.

We now move on to the more general case, where the gravitational field is perturbed by a matter field. Then $\psi_{\mu\nu}$ satisfies equation (9.4); from (9.3) the integrability condition for this equation is

$$T^{\nu}_{\mu,\nu}=0. \tag{9.14}$$

Using the retarded Green function for the wave equation, we can write the general solution of (9.4) as

$$\psi_{\mu\nu}=-\frac{4G}{c^4}\int\frac{T_{\mu\nu}(r',ct')}{|\boldsymbol{r}-\boldsymbol{r}'|}\delta(|\boldsymbol{r}-\boldsymbol{r}'|-c(t-t'))\,d^3\boldsymbol{r}'c\,dt'. \tag{9.15}$$

Integration with respect to t' gives

$$\psi_{\mu\nu}=-\frac{4G}{c^4}\int\frac{T_{\mu\nu}(\boldsymbol{r}',ct-|\boldsymbol{r}-\boldsymbol{r}'|)}{|\boldsymbol{r}-\boldsymbol{r}'|}d^3\boldsymbol{r}'. \tag{9.16}$$

As in electrodynamics, if we are interested in the strength of the field far away from the source, at a distance much larger than the characteristic dimension of the source, we can expand $|\boldsymbol{r}-\boldsymbol{r}'|$ into a series

$$|\boldsymbol{r}-\boldsymbol{r}'|=r-\boldsymbol{n}\cdot\boldsymbol{r}'+\ldots, \tag{9.17}$$

where $r=|\boldsymbol{r}|$ and $\boldsymbol{n}=\boldsymbol{r}/|\boldsymbol{r}|$. In the first approximation,

$$\psi_{\mu\nu}=-\frac{4G}{c^4}\frac{1}{r}\int T_{\mu\nu}(\boldsymbol{r}\;,t-r/c)\,d^3\boldsymbol{r}'. \tag{9.18}$$

To calculate this integral we shall use condition (9.14). As we have shown in Chapter 3, for a bounded distribution of matter such that the relative velocities of its different elements are small compared with the speed of light, the integrals of T_{00} and T_{0i} over the matter-filled volume give in this approximation the total mass and the total momentum of the system. Both quantities are constant in the approximation being used. The contribution of the spatial components of the energy-momentum tensor is of interest here. Writing out relation (9.14), we get

$$T_{00,0}-T_{0i,i}=0, \qquad T_{i0,0}-T_{ik,k}=0. \tag{9.19}$$

If we multiply the first relation by x^kx^j, the second by x^j, and integrate over the whole space $t=\text{const}$, remembering that the distribution of matter is bounded, we finally obtain

$$\int T_{ik}\,d^3\boldsymbol{r}'=\tfrac{1}{2}\frac{\partial^2}{c^2\partial t^2}\int T_{00}\,x'^ix'^k d^3\boldsymbol{r}'. \tag{9.20}$$

The restrictions we have set on the distribution of matter allow us to assume that

$$T_{00}=\rho c^2, \tag{9.21}$$

and therefore

$$\int T_{ik}\,d^3\boldsymbol{r}'=\tfrac{1}{2}\frac{\partial^2}{\partial t^2}\int \rho(r',t)\,x'^ix'^k d^3\boldsymbol{r}'. \tag{9.22}$$

Put

$$I_{ik} = \int \rho(r', t) x_i' x_k' d^3 r'. \tag{9.23}$$

I_{ik} is physically interpreted as the second moment of the mass distribution. The quadrupole moment Q_{ik} introduced in Chapter 1 is related to I_{ik} by

$$Q_{ik} = 3I_{ik} - \delta_{ik} I^a_a. \tag{9.24}$$

Far away from the field source, in the wave zone, it is possible to introduce locally a coordinate system such that (9.18) will describe, up to higher-order terms, a plane wave. If the gauge is chosen so as to make $\psi_{\mu\nu}$ traceless and transverse, the only non-zero components are

$$\psi_{23} = -\frac{2G}{3c^4 r}\ddot{Q}_{23},$$

$$\psi_{22} - \psi_{33} = -\frac{2G}{3c^4 r}(\ddot{Q}_{22} - \ddot{Q}_{33}), \tag{9.25}$$

and $\psi_{22} + \psi_{33} = 0$. We can see that the radiative components of $\psi_{\mu\nu}$ depend on the second time-derivatives of the quadrupole moment.

In this approximation gravitational waves propagate with the speed of light and do not interact with themselves. However, they produce a small curving of the space-time. The curvature tensor is different from zero; in terms of the radiative components of the field it is given by

$$R_{ioko} = -\tfrac{1}{2}\ddot{\psi}_{ik} = -\tfrac{1}{2}\ddot{h}_{ik}. \tag{9.26}$$

As seen by a distant observer, small perturbations of the Minkowski space-time due to a matter field can be resolved into stationary perturbations, which depend on the total mass and the angular momentum of the system, and radiative perturbations, dependent on the second time-derivatives of the quadrupole moment.

9.2. Gravitational Waves in Curved Space

In many systems of astrophysical interest gravitational fields are strong and cannot be treated as small perturbations of Minkowski space-time. The question arises as to what should be understood by a gravitational wave in a strong gravitational field. The problem has long been a matter of considerable interest. Historically, two approaches can be distinguished: asymptotic analysis of the gravitational field generated by a bounded distribution of mass, and investigation of small perturbations of a curved space-time. In the first approach, one examines the asymptotic behaviour of the gravitational field along the light cones. It turns out that there are close analogies between the asymptotic properties of electromagnetic and gravitational fields. Owing to the works of Pirani (1957), Trautman (1958), Bondi (1962), and Sachs (1962), it has become possible to formulate general criteria which permit us to decide whether or not given space-time carries gravitational waves, and to give the relationship between the energy radiated by a system and the change in its total mass-energy. In this approach, the problem of interactions between gravitational waves and their effect on the curvature of the space-time presents serious difficulties.

The second method was proposed by Wheeler (1962) and developed by Brill and Hartle (1964), Isaacson (1968), and Thorne (1977). It consists in studying small perturbations of a curved space. If the characteristic radius of curvature of the unperturbed space is R, we are interested only in perturbations whose typical scale λ is much less than R. Perturbations affect the curvature of the space. In the following, we consider only those perturbations whose effect on the curvature of the space is small.

We write the metric of the space-time in the form

$$g_{\mu\nu}=\gamma_{\mu\nu}+h_{\mu\nu}, \tag{9.27}$$

where $\gamma_{\mu\nu}$ is the metric of the unperturbed space and $h_{\mu\nu}$ represents the perturbation. The above restriction can be translated into conditions on the derivatives of $\gamma_{\mu\nu}$ and $h_{\mu\nu}$. In the cases which interest us

$$\partial\gamma\sim\gamma/R, \qquad \partial h\sim h/\lambda, \tag{9.28}$$

with $\lambda\ll R$. The energy density of the gravitational wave is of the order of $\dfrac{c^4}{G}\left(\dfrac{h}{\lambda}\right)^2$. It follows from Einstein's equations that the curving of the space caused by the wave will be of the order of $R_{GW}^{-2}\approx(h/\lambda^2)$.

The curving of space due to the gravitational wave should be much less than the curvature of the space, $R_{GW}\ll R$,† whence $h\ll\lambda/R\ll 1$, i.e. the dimensionless amplitude of the wave should be small. Using (9.27), we can write the Ricci tensor in the form

$$R_{\alpha\beta}(\gamma_{\mu\nu}+h_{\mu\nu})=R^{(0)}_{\alpha\beta}+R^{(1)}_{\alpha\beta}+R^{(2)}_{\alpha\beta}+\dots, \tag{9.29}$$

where

$$R^{(0)}_{\alpha\beta}=R_{\alpha\beta}(\gamma_{\mu\nu}), \tag{9.30}$$

$$R^{(1)}_{\alpha\beta}=\tfrac{1}{2}\gamma^{\rho\tau}(h_{\rho\tau;\,\alpha\beta}+h_{\alpha\beta;\,\rho\tau}-h_{\tau\alpha;\,\beta\rho}-h_{\tau\beta;\,\alpha\rho}), \tag{9.31}$$

$$\begin{aligned}-2R^{(2)}_{\alpha\beta}=&\tfrac{1}{2}h^{\rho\tau}{}_{;\beta}\,h_{\rho\tau;\,\alpha}+h^{\rho\tau}(h_{\rho\tau;\,\alpha\beta}+h_{\alpha\beta;\,\tau\rho}-h_{\tau\alpha;\,\beta\rho}-h_{\tau\beta;\,\alpha\rho})+h^{\tau;\,\rho}_{\beta}(h_{\tau\alpha;\,\rho}-h_{\rho\alpha;\,\tau})\\&-(h^{\rho\tau}{}_{;\rho}-\tfrac{1}{2}h^{\rho}{}_{\rho}{}^{;\tau})(h_{\tau\alpha;\,\beta}+h_{\tau\beta;\,\alpha}-h_{\alpha\beta;\tau})\,.\end{aligned} \tag{9.32}$$

A semicolon denotes covariant differentiation with respect to the metric $\gamma_{\mu\nu}$. Indices are raised and lowered with $\gamma_{\mu\nu}$.

The unperturbed metric satisfies Einstein's equation

$$R^{(0)}_{\alpha\beta}-\tfrac{1}{2}\gamma_{\alpha\beta}R^{(0)\mu}{}_{\mu}=\frac{8\pi G}{c^4}T_{\alpha\beta}. \tag{9.33}$$

In the first approximation, linear in the amplitude of the gravitational wave, the field equations reduce to

$$R^{(1)}_{\alpha\beta}(h_{\mu\nu})=0. \tag{9.34}$$

For a unique solution, one has to specify the boundary conditions to be satisfied by $h_{\mu\nu}$. Their choice depends on the problem in hand.

† This applies to regions much larger than λ. In regions of dimensions comparable to λ, the curving of space due to gravitational waves is the major part of the total curvature.

The quantities $h_{\mu\nu}$ so found can be used to write the equations in the next-order approximation. Now we have to allow for the effect of the gravitational wave energy on the curvature of the space. It is noticeable only over space-time distances much larger than λ. Consequently, let us average the influence of the perturbations of the curvature over a region much larger than λ.† $\langle R^{(2)}_{\alpha\beta}(h)\rangle$ is the first nonvanishing correction. The field equations (9.33) now take the form

$$R^{(0)}_{\alpha\beta}-\tfrac{1}{2}\gamma_{\alpha\beta}R^{(0)\mu}{}_{\mu}+\langle R^{(2)}_{\alpha\beta}(h)\rangle-\tfrac{1}{2}\gamma_{\alpha\beta}\langle R^{(2)\mu}{}_{\mu}(h)\rangle=\frac{8\pi G}{c^4}T_{\alpha\beta}. \tag{9.35}$$

They can be written as

$$R^{(0)}_{\alpha\beta}-\tfrac{1}{2}\gamma_{\alpha\beta}R^{(0)\mu}{}_{\mu}=\frac{8\pi G}{c^4}(T_{\alpha\beta}+T^{\mathrm{GW}}_{\alpha\beta}), \tag{9.36}$$

where

$$T^{\mathrm{GW}}_{\alpha\beta}=-\frac{c^4}{8\pi G}(\langle R^{(2)}_{\alpha\beta}(h)\rangle-\tfrac{1}{2}\gamma_{\alpha\beta}\langle R^{(2)\mu}{}_{\mu}(h)\rangle) \tag{9.37}$$

is the energy-momentum tensor of the gravitational wave in the lowest approximation. The fast-varying component of the gravitational field satisfies the equations

$$R^{(1)}_{\alpha\beta}(\Delta h_{\mu\nu})+R^{(2)}_{\alpha\beta}(h_{\mu\nu})-\langle R^{(2)}_{\alpha\beta}(h_{\mu\nu})\rangle=0. \tag{9.38}$$

Here $\Delta h_{\mu\nu}$ stands for higher-order corrections to gravitational wave; they are generated by the scattering of the wave due to the curvature of the unperturbed space and from the interaction of the wave with itself.

As in the weak-field case, it is convenient to introduce new quantities $\psi_{\mu\nu}$ defined by

$$\psi_{\mu\nu}=h_{\mu\nu}-\tfrac{1}{2}\gamma_{\mu\nu}h^{\alpha}{}_{\alpha}, \tag{9.39}$$

We restrict the choice of coordinate system by imposing on $\psi_{\mu\nu}$ the additional conditions

$$\psi_{\mu}{}^{\nu}{}_{;\nu}=0, \quad \psi_{\mu}{}^{\mu}=0. \tag{9.40}$$

Expressing $h_{\mu\nu}$ by $\psi_{\mu\nu}$ and using conditions (9.40), we can rewrite equation (9.35) as

$$\psi_{\alpha\beta;\mu}{}^{;\mu}+2R^{(0)}_{\mu\alpha\nu\beta}\psi^{\mu\nu}+\psi^{\mu}{}_{\alpha}R^{(0)}_{\mu\beta}+\psi^{\mu}{}_{\beta}R^{(0)}_{\mu\alpha}=0. \tag{9.41}$$

Solutions of this generalized wave equation represent gravitational waves in the curved space. Unlike the weak-field approximation, the above equation can be applied to unbounded matter distributions and, for example, used to investigate propagation of gravitational waves in various cosmological models.

Let us now examine the energy-momentum tensor of a gravitational wave. Not being a tensor in the strict sense, it behaves like one only with respect to the coordinate transformations that leave the form of the unperturbed metric unchanged. As shown by Isaacson (1968), by averaging and using restrictions (9.40) one obtains for the energy-momentum

† The averaging is performed over a region U much larger than λ but such that every two points belonging to U can be connected by a unique geodesic. It is then possible to define a one-to-one map of the tangent space, $T_x U\to U$, $x\in U$, and an invariant averaging procedure, like is done in Minkowski space.

tensor of a gravitational wave the expression

$$T^{GW}_{\alpha\beta} = \frac{c^4}{32\pi G} \langle \psi_{\mu\nu;\alpha} \psi^{\mu\nu}{}_{;\beta} \rangle . \tag{9.42}$$

Up to higher-order terms, this tensor is traceless, $T^{GW\alpha}{}_{\alpha}=0$, and satisfies the conservation law $T^{GW\beta}{}_{\alpha;\beta}=0$.

If the metric of the unperturbed space is flat, the general case under consideration reduces to the weak-field approximation. We shall make use of this fact to calculate the energy emitted by the system in the weak-field approximation. As we saw in the preceding section, gravitational waves emitted by a bounded distribution of test matter are described in the wave zone by

$$\psi_{23} = -\frac{2G}{3c^4 R} \ddot{Q}_{23},$$
$$\psi_{22} = -\psi_{33} = -\frac{G}{3c^4 R} (\ddot{Q}_{22} - \ddot{Q}_{33}) . \tag{9.43}$$

Substituting these quantities in (9.42), we find the gravitational energy-momentum tensor. As in electrodynamics, for the energy radiated per unit of time we obtain the formula

$$\frac{dE}{dt} = -c \oint_{\Sigma} T^{0k} \eta_k \, d\Sigma , \tag{9.44}$$

where Σ is a sphere at infinity. In a similar way, we find that the angular momentum carried by gravitational waves is given by

$$\frac{dJ^k}{dt} = -c\epsilon^k{}_{lm} \oint_{\Sigma} (x^l T^{mn} - x^m T^{ln}) \, n_n \, d\Sigma . \tag{9.45}$$

Inserting the energy-momentum tensor, we finally get

$$\frac{dE}{dt} = -\frac{G}{45c^5} \langle \dddot{Q}_{ik} \dddot{Q}^{ik} \rangle \tag{9.46}$$

and

$$\frac{dJ^k}{dt} = -\frac{2G}{45c^2} \epsilon^{klm} \langle \ddot{Q}_{ln} \dddot{Q}^n{}_m \rangle . \tag{9.47}$$

We shall use these formulae when discussing sources of gravitational waves. Now let us only note that c^5/G has the dimension of power and is equal to $3.63 \cdot 10^{59}$ erg/s.

To find the propagation law for gravitational waves in curved space, we shall apply the same methods as those used in the geometrical optics approximation to investigate the propagation of electromagnetic waves. We shall assume that the solution of the wave equation (9.41) can be written in the form

$$\psi_{\mu\nu} = \mathrm{Re}(A_{\mu\nu} e^{i\phi}) , \tag{9.48}$$

where $A_{\mu\nu}$ is a slowly-varying real function, and ϕ is a real function whose gradient may be large but whose higher derivatives are slowly-varying. Let us introduce the wave propagation vector as

$$k_\alpha = \phi_{,\alpha}. \tag{9.49}$$

The additional constraints on $\psi_{\mu\nu}$ can be written as conditions on $A_{\mu\nu}$ and k_α; up to the higher-order term:

$$\gamma^{\mu\nu} A_{\mu\nu} = 0, \quad A_{\mu\nu} k^\nu = 0. \tag{9.50}$$

In the first approximation, the wave equation (9.41) leads to

$$k_\alpha k^\alpha = 0, \tag{9.51}$$

which implies that the gravitational wave propagation vector is a null vector. It follows that curves $x^\mu = x^\mu(l)$ such that

$$\frac{dx^\mu}{dl} = k^\mu, \tag{9.52}$$

are geodesics, because by virtue of (9.51) and (9.49)

$$k_{\beta;\alpha} k^\beta = k_{\alpha;\beta} k^\beta = 0. \tag{9.53}$$

Thus gravitational waves, like electromagnetic waves, propagate along null geodesics. Individual geodesics of this family are called rays.

In the next approximation, the wave equation gives

$$A_{\mu\nu;\rho} k^\rho + \tfrac{1}{2} A_{\mu\nu} k^\rho{}_{;\rho} = 0. \tag{9.54}$$

To define the amplitude of a gravitational wave, we write $A_{\mu\nu}$ in the form

$$A_{\mu\nu} = A e_{\mu\nu}, \tag{9.55}$$

where $e_{\mu\nu} = (A_{\alpha\beta} A^{\alpha\beta})^{-\frac{1}{2}} A_{\mu\nu}$ is the polarization tensor of the wave; the amplitude of the wave is $A = \sqrt{A_{\mu\nu} A^{\mu\nu}}$. Inserting (9.55) into (9.54) gives

$$(A_{,\alpha} k^\alpha + \tfrac{1}{2} A k^\alpha{}_{;\alpha}) e_{\mu\nu} + A e_{\mu\nu,\alpha} k^\alpha = 0. \tag{9.56}$$

Multiplying this equation by $e_{\mu\nu}$, we get

$$A_{,\alpha} k^\alpha + \tfrac{1}{2} A k^\alpha{}_{;\alpha} = 0, \tag{9.57}$$

which can be written in the equivalent form

$$(\ln A)_{,\alpha} k^\alpha = \frac{d \ln A}{dl} = -\tfrac{1}{2} k^\alpha{}_{;\alpha} \tag{9.58}$$

expressing the relationship between the change in the amplitude along the rays and the divergence of the rays. Furthermore, it follows from (9.56) that

$$e_{\mu\nu;\alpha} k^\alpha = 0. \tag{9.59}$$

Thus, the polarization tensor of the gravitational wave is parallel-transported along the rays, and

$$e_{\mu\nu} k^\nu = 0, \quad e_{\mu\nu} \gamma^{\mu\nu} = 0. \tag{9.60}$$

Owing to the fact that the vector k^α is tangent to the geodesic, if relations (9.60) are satisfied at an arbitrary point of the ray, they are satisfied everywhere along the ray.

It has thus turned out that in the geometrical optics approximation gravitational waves and electromagnetic waves have similar properties.

9.3. Motion of Test Particles in a Plane Gravitational Wave. Polarization

We have seen in the preceding section that a plane monochromatic gravitational wave whose length is small compared with the radius of curvature of the space can be written in the form

$$h_{\mu\nu}=A\,\mathrm{Re}(e_{\mu\nu}e^{ik_\alpha x^\alpha}), \tag{9.61}$$

where the polarization tensor $e_{\mu\nu}$ satisfies relation (9.60). We now consider the particular case of a wave propagating in the $+x$-direction in the Minkowski space; then, $k^\alpha=\omega(1, 1, 0, 0)$. Relations (9.60) and the transversality condition $u^\mu e_{\mu\nu}=0$, where $u^\mu=(1, 0, 0, 0)$, reduce the number of independent components to 2,

$$\begin{aligned} h_{22}&=-h_{33}=A\,\mathrm{Re}\,a_+\,\mathrm{e}^{i\omega(ct-x)},\\ h_{23}&=h_{32}=A\,\mathrm{Re}\,a_\times\,\mathrm{e}^{i\omega(ct-x)}, \end{aligned} \tag{9.62}$$

where a_+ and $a_\times$ represent the two independent degrees of polarization. Relations (9.62) can be written

$$\begin{aligned} h_{22}&=-h_{33}=A_+\sin[\omega(ct-x)+\varphi_+],\\ h_{23}&=h_{32}=A_\times\sin[\omega(ct-x)+\varphi_\times]. \end{aligned} \tag{9.63}$$

The alternative $A_+=0$ or $A_\times=0$ corresponds to two states of linear polariztion, and setting $A_+=A_\times$ and $\varphi_\times=\varphi_+\pm\pi/2$ gives two independent states of circular polarization.

The metric of the space-time takes the form

$$\mathrm{d}s^2=c^2\mathrm{d}t^2-\mathrm{d}x^2-(1-h_{22})\mathrm{d}y^2-(1+h_{22})\mathrm{d}z^2+2h_{23}\,\mathrm{d}y\,\mathrm{d}z. \tag{9.64}$$

To find out how the wave interacts with test particles, let us examine the motion of two neighbouring particles. We saw in Chapter 2 that the relative position vector of the particles, $n^\alpha=\delta_\perp x^\alpha$, satisfies the equation

$$\frac{\mathrm{D}^2n^\alpha}{\mathrm{d}s^2}+R^\alpha{}_{\beta\gamma\delta}u^\beta n^\gamma u^\delta=0. \tag{9.65}$$

We shall use the fact that in the linear approximation the curvature tensor of the perturbed space is

$$R_{\alpha\beta\gamma\delta}(\gamma_{\mu\nu}+h_{\mu\nu})=R^{(0)}_{\alpha\beta\gamma\delta}(\gamma_{\mu\nu})+R^{(1)}_{\alpha\beta\gamma\delta}, \tag{9.66}$$

where

$$R^{(1)}_{\alpha\beta\gamma\delta}=\tfrac{1}{2}(h_{\alpha\gamma;\,\beta\delta}+h_{\beta\delta;\,\alpha\gamma}-h_{\beta\gamma;\,\alpha\delta}-h_{\alpha\delta;\,\beta\gamma}+R^{(0)}_{\alpha\sigma\gamma\delta}h^\sigma{}_\beta-R^{(0)}_{\beta\sigma\gamma\delta}h^\sigma{}_\alpha). \tag{9.67}$$

If we assume that the particles were at rest at the initial moment, so that $u^\alpha=(1, 0, 0, 0)$ and $n^\alpha=(0, n^i)$, then only the components R_{0i0j} of the curvature tensor will occur in (9.65).

Due to the chosen gauge we have $R_{0i0j}=\frac{1}{2}h_{ij,00}$. The equation of motion (9.65) becomes

$$\frac{D^2 n^j}{ds^2}-R_{0i0j}n^i=0. \tag{9.68}$$

Taking into account that the unperturbed space is a Minkowski space and choosing the

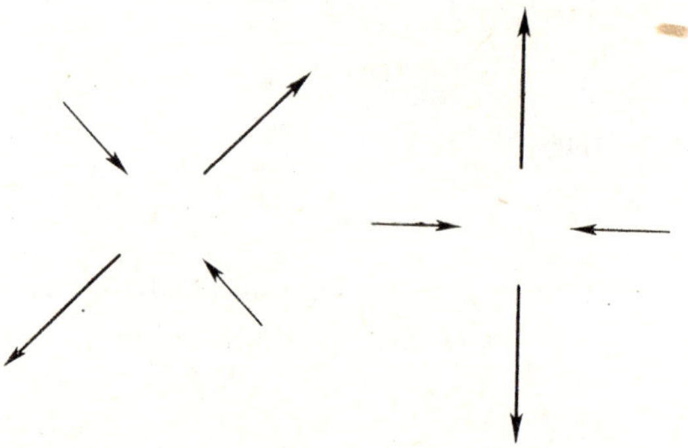

FIG. 9.1. Two independent degrees of polarization of a plane gravitational wave.

coordinate system to be bound to one of the particles, we get

$$\frac{d^2 n^j}{dt^2}=\frac{1}{2}n^i\frac{\partial^2}{\partial t^2}h_{ji}. \tag{9.69}$$

If $n^i(t=0)=n_0^i$, then

$$n^i(t)=(\delta_{ij}+\frac{1}{2}h_{ij})\,n_0^j \tag{9.70}$$

is the solution of our problem. The form of the solution implies that an incident plane wave gives rise to *relative motion of the particles in a plane perpendicular to the direction of the wave* (Fig. 9.1).

If a group of test particles are placed so as to form a circle in a plane perpendicular to the direction of the wave, their displacements relative to the centre of the circle will depend on the polarization of the wave. For linearly polarized plane waves the circle will be deformed to an ellipse. Such a system of test particles may thus serve as a primitive antenna for detecting gravitational waves.

9.4. Astrophysical Sources of Gravitational Waves

We now give certain general estimates, which provide a basis for classifying various possible sources of gravitational waves. As we have seen, the knowledge of the space-space components of the energy-momentum tensor is enough to determine the perturbations in the metric, and if the gauge is chosen so that $\psi_{\mu\nu}u^\nu=\psi_{\mu\nu}k^\nu=\psi^\mu_\mu=0$, then only the traceless part of the energy-momentum tensor is responsible for the generation of gravitational waves.

A closer analysis of the formula for the power of gravitational radiation makes it clear why no observational evidence for the existence of such radiation has yet been obtained.

Let us write this formula, after Dyson (1969), in the following way. Denote by L^{GW} the power of the gravitational radiation and let $L_0=c^5/G=3.63\cdot10^{59}$ erg/s$=3.63\cdot10^{52}$ W. Then

$$L^{\mathrm{GW}}=\frac{(L^{\mathrm{in}})^2}{L_0}, \tag{9.71}$$

where L^{in} is the internal power of the system. Let the total mass of the system be M and its characteristic dimension R. It is known from the virial theorem that the internal kinetic energy of a system is of the order of its potential energy; hence

$$L^{\mathrm{in}}\sim\frac{GM^2}{R^2}\left(\frac{GM}{R}\right)^{\frac{1}{2}}=\frac{1}{G}\left(\frac{GM}{R}\right)^{\frac{5}{2}}, \tag{9.72}$$

so finally

$$L^{\mathrm{GW}}\sim\frac{1}{G^2L_0}\left(\frac{GM}{R}\right)^5=\left(\frac{GM}{c^2R}\right)^5L_0. \tag{9.73}$$

Just as one should expect, the radiation will be strongest when the dimensions of the system are comparable to its gravitational radius; then $L^{\mathrm{GW}}\sim L_0$. In realistic situations, except for the processes involving black holes, $R\gg r_g$, and only part of the mass of the system will effectively contribute to the internal kinetic energy. For such system $L^{\mathrm{GW}}\ll L_0$. For example, for the Solar System $L^{\mathrm{GW}}\approx100$ kW.

These general estimates imply that only very special systems can be effective sources of gravitational waves.

To begin with, let us consider a close binary star system, the best-investigated source of gravitational waves. Assume that the system moves according to the Newtonian equations of motion. Place the coordinate system at the centre of mass of the binary. If m_1 and m_2 are the masses of the components, and $\boldsymbol{r}_1$ and $\boldsymbol{r}_2$ the position vectors, then

$$\boldsymbol{r}_1=\frac{m_2}{m_1+m_2}\boldsymbol{r},\qquad \boldsymbol{r}_2=-\frac{m_1}{m_1+m_2}\boldsymbol{r},\qquad \boldsymbol{r}=\boldsymbol{r}_1-\boldsymbol{r}_2. \tag{9.74}$$

As is well known, the orbit of the relative motion of the components is an ellipse, i.e.

$$\frac{a(1-e^2)}{r}=1+e\cos\varphi, \tag{9.75}$$

where a is the semi-major axis of the ellipse and $e<1$ is the eccentricity. By Newton's equations of motion, the time-dependence of $\boldsymbol{r}$ can be represented parametrically:

$$r=a(1-e\cos\xi),\qquad t=\sqrt{\frac{a^3}{G(m_1+m_2)}}(\xi-e\sin\xi), \tag{9.76}$$

and therefore

$$\dot{\varphi}=\frac{\mathrm{d}\varphi}{\mathrm{d}t}=\frac{1}{r^2}[G(m_1+m_2)a(1-e^2)]^{\frac{1}{2}}. \tag{9.77}$$

Using (9.23) and (9.24), we can calculate the quadrupole moment of the binary. The coordinate system can be chosen so that the motion will be confined to the x, y-plane. Denoting by $\mu = \dfrac{m_1 m_2}{m_1+m_2}$ the reduced mass of the binary, we have

$$\begin{aligned} Q_{xx}&=\mu r^2(3\cos^2\varphi-1), \quad & Q_{yy}&=\mu r^2(3\sin^2\varphi-1), \\ Q_{xy}&=3\mu r^2\sin\varphi\cos\varphi, \quad & Q_{zz}&=-\mu r^2. \end{aligned} \tag{9.78}$$

Substituting these values in (9.46) and using the equations of motion (9.76) and (9.77), we obtain after considerable calculations

$$-\frac{dE}{dt}=L^{GW}=\frac{8G^4m_1^2m_2^2(m_1+m_2)}{15c^5a^5(1-e^2)^5}(1+e\cos\varphi)^4[12(1+e\cos\varphi)^2+e^2\sin^2\varphi]. \tag{9.79}$$

The mean energy radiated during one period is the quantity of interest here. Averaging gives

$$\left\langle\frac{dE}{dt}\right\rangle=-\frac{32G^4m_1^2m_2^2(m_1+m_2)}{5c^5a^5}\frac{1}{(1-e^2)^{7/2}}\left(1+\tfrac{73}{24}e^2+\tfrac{37}{96}e^4\right). \tag{9.80}$$

Inserting the numerical data for the Jupiter–Sun system, we get $\langle dE/dt\rangle = -5.4$ kW. The power emitted by this system is very small. Formula (9.80) shows that $\langle dE/dt\rangle$ is very sensitive to the value of the eccentricity of the orbit. The flatter the orbit the greater the power of the radiation. Another important factor determining the power of the radiation is the semi-major axis of the ellipse a. Close binary systems are therefore of particular interest among binary star systems, hence so are spectroscopic and eclipsing binaries. Data concerning a few known binary systems and estimates for two hypothetical binaries with a very

TABLE 9.1. *Gravitational radiation from binary systems*

System	Period	Masses/$M_\odot$		Distance pc	L^{GW} erg/s	h	Flux at the Earth erg/cm²·s
Sirius	49.94 yr	2.28	0.98	2.6	$1.1\cdot10^{15}$	$1.4\cdot10^{-22}$	$1.3\cdot10^{-24}$
Fu 46	13.12 yr	0.31	0.25	6.5	$3.6\cdot10^{14}$	$8.5\cdot10^{-24}$	$7.1\cdot10^{-26}$
Per	2.867 day	4.70	0.94	30	$1.4\cdot10^{28}$	$6.8\cdot10^{-21}$	$1.3\cdot10^{-13}$
Cr B	17.36 day	2.5	0.89	22	$1.2\cdot10^{25}$	$1.7\cdot10^{-21}$	$2.1\cdot10^{-16}$
WZ Sge	81 min	0.6	0.03	100	$3.5\cdot10^{29}$	$2.4\cdot10^{-22}$	$2.9\cdot10^{-13}$
Neutron stars	12.2 s	1.0	1.0	1000	$3.25\cdot10^{41}$	$4.9\cdot10^{-20}$	$2.7\cdot10^{-3}$
	0.39 s	1.0	1.0	1000	$3.25\cdot10^{46}$	$4.9\cdot10^{-19}$	$2.7\cdot10^{-2}$

L^{GW} is the power of the radiation, h is the dimensionless amplitude of metric perturbations near the surface of the Earth. The last two rows concern hypothetical binary sources consisting of neutron stars of solar mass, at a distance of 1000 pc from the Earth. If such systems existed, their gravitational radiation would be detectable by the gravitational antennas now in operation.

short orbital period are listed in Table 9.1. For known systems, the energy flux at the surface of the Earth is very small and does not exceed 10^{-13} erg/cm²s. As we shall see later, the gravitational antennae available at present are not sensitive enough to detect gravitational waves emitted by binary systems.

Many stars go through a supernova explosion. During the explosion, part of the stellar mass is thrown off and the remaining nucleus collapses gravitationally, forming a neutron star or a black hole. Our knowledge of the dynamics of supernova explosions is very poor, and only rough estimates can be given of the energy emitted in the form of gravitational waves in such explosions. It is reasonable to assume that the parameters of the system change on the time-scale characteristic for hydrodynamic processes, i.e.

$$\Delta t \sim \left(\frac{R^3}{GM}\right)^{\frac{1}{2}}. \tag{9.81}$$

The maximum amount of the energy emitted can be found by the following argument. The binding energy of the star before the explosion is very small. The final state of the system is a neutron star with a binding energy $\sim 0.1\, Mc^2$. Assume that the whole of this energy is emitted as gravitational waves. Using (9.73), we obtain

$$L^{GW} \sim 10^{-5} L_0 = 3.6 \cdot 10^{54}\ \text{erg/s}, \tag{9.82}$$

and the average wave frequency

$$\langle \nu \rangle = \frac{1}{\Delta t} = \left(\frac{GM}{R^3}\right)^{\frac{1}{2}} \sim 0.03\, \frac{c^3}{GM}. \tag{9.83}$$

Inserting $M = M_\odot$ gives $\langle \nu \rangle \approx 6000$ Hz. If the explosion occurred at the centre of our galaxy, the energy flux on the surface of the Earth would be about $7 \cdot 10^8$ erg/s·cm². This is just a rough estimate of course. It is even more difficult to estimate the energy emitted as gravitational waves during the gravitational collapse of a star. So far, no one has attempted a description of the collapse dynamics with allowance for rotation and the effects due to strong magnetic fields. It is thought that the gravitational energy emitted in a collapse is 10^{-2}–$10^{-3}\, Mc^2$. As in the case of a supernova explosion, the radiation will emerge in a short pulse, of duration $\sim r/c$. Taking $M = 2M_\odot$ we find that an energy of 10^{51}–10^{52} erg is emitted during $\sim 10^{-5}$ s with an average frequency of about 10^4 Hz. Extrapolation of the data obtained for hydrodynamic models of supernova explosion suggests that the first, major pulse is followed by a succession of pulses with a mean frequency of 100 Hz to 10 kHz. Their duration is estimated to be 10^{-3}–1 s. If the collapsing star were at the centre of our galaxy, the energy flux on the surface of the Earth would be 10^7–10^{10} erg/cm²s and the perturbations in the metric would be of order 10^{-19}–10^{-17}.

Supernova explosions and gravitational collapse are probably the most powerful sources of gravitational radiation. The theory of stellar evolution makes it possible to predict how often such processes occur in our galaxy. It is estimated that a supernova explosion occurs at least once in 100 years and that a star collapses gravitationally about equally often. These estimates being rather discouraging, the possibility has been considered of observing such processes in the nearest cluster of galaxies, which lies in the Virgo Constellation at a distance of 19 Mpc. The cluster contains about 2500 galaxies. There, a supernova explosion or a gravitational collapse should occur at least once a month, producing on the Earth an energy flux of 10–10^4 erg/cm²s and metric pertubations of amplitude $\sim 10^{-22}$–10^{-20}.

Neutron stars formed in supernova explosions can rotate with high angular velocities and pulsate. As shown by Thorne (1969), non-radial oscillations are damped very quickly. Model calculations imply that the energy of nonradial oscillations can be as high as $10^{54}\left(\frac{\delta R}{R}\right)$ erg, where $\delta R/R$ is the relative amplitude of the oscillations. Their period is $P\sim 1$–0.1 ms, and hence the characteristic frequency of the wave emitted is $\langle\nu\rangle\sim 1$–10 kHz. Nonradial oscillations are damped very quickly: their decay period is estimated to be $\tau\sim 0.1$–10 s.

Rapid rotation of a star in which the distribution of matter is not symmetric also gives rise to emission of gravitational waves. A rotating neutron star with a non-zero quadrupole moment is a source of monochromatic gravitational waves. The period of rotation of a neutron star can be very short, just a few milliseconds. It is to be expected therefore that the frequency of the emitted wave will be about 1 kHz. If such a source happens to be at a distance of a few hundred parsecs from the Earth, the energy flux it will produce on the Earth will be ~ 1 erg/cm²s and the metric perturbation amplitude $h\sim 10^{-22}$.

As we know, pulsars are rotating magnetic neutron stars. Their strong gravitational fields, rotation and the elastic properties of the crust prevent any substantial departures from axial symmetry. However, if the equatorial cross-section of such a star is an ellipse with a small eccentricity e, the average power of the radiation is given by

$$\left\langle\frac{\mathrm{d}E}{\mathrm{d}t}\right\rangle = -\frac{8}{5}\frac{G}{c^5} I^2 e^4 \left(\frac{2\pi}{P}\right)^6, \tag{9.84}$$

where I is the moment of inertia and P the period of rotation. The elastic properties of the pulsar crust can be estimated from the frequency of the abrupt changes in the period caused by the cracking of the crust; this permits an indirect determination of the eccentricity. For the Crab Nebula pulsar we get $e\sim 10^{-1}$–10^{-3} and $\langle \mathrm{d}E/\mathrm{d}t\rangle \approx 2\cdot 10^{37}$ erg/s, which considering the 1410 pc that separate this pulsar from the Earth, gives the energy flux passing the Earth of about $8\cdot 10^{-8}$ erg/s·cm². Gravitational radiation emitted by other pulsars is much weaker. The power of radiation depends very strongly on the pulsar period.

All nonstationary processes taking place near black holes should typically be accompanied by emission of strong gravitational waves. In such cases the weak-field approximation can no longer be applied and we have to resort to other methods, for example the use of a high-frequency approximation. Only in a few instances have the efficiency of the process and the properties of the emitted gravitational waves been studied in more detail. The largest amount of energy is radiated in a direct collision of two black holes. By the second law of black hole mechanics, the surface area of the resulting black hole cannot be less than the sum of the areas of the colliding holes. For two black holes with equal masses the maximum energy that can be emitted is $(2-\sqrt{2})Mc^2 = 0.58\,Mc^2$.

Most of this energy will be radiated in a short burst of duration $\Delta t\sim r_g/c = 2GM/c^3$. If a body of mass m falls into a black hole of mass $M \gg m$, then, as estimated by Ruffini and Wheeler (1971), the gravitational energy emitted is

$$E = 0.00246\,\frac{m}{M}\,mc^2, \tag{9.85}$$

and the mean frequency of the radiation will be

$$\langle \nu \rangle \sim 0.024 \frac{c^3}{GM} = 4.9 \cdot 10^3 \frac{M_\odot}{M} \text{ Hz}. \tag{9.86}$$

Another case investigated in detail was that of *gravitational synchrotron radiation* (Breuer *et al.*, 1973; Breuer, 1975). The following model was considered: A particle moves on a circular orbit around a black hole. The mass m of the particle is much less than that of the hole. If the velocity of the particle is close to the speed of light, we should expect that, as in the case of a relativistic electron travelling on a circular orbit, the gravitational radiation emitted by the particle will be strongly beamed and concentrated within a small solid angle about the direction of the instantaneous velocity. Chrzanowski and Misner (1974) have shown that gravitational synchrotron radiation can occur only when the velocity of the particle is very high, a particle moving on the innermost stable circular orbit with $v \approx 0.9c$ being a likely example. It is hard to imagine a system in which particles could be accelerated to such high velocities. For this reason, gravitational synchrotron radiation, even though theoretically possible, is of little astrophysical interest.

The last twenty years of radio-astronomical observations have revealed many new powerful sources of electromagnetic radiation, among them quasars and active galactic nuclei (Seyfert galaxies). It is believed that they are also strong sources of gravitational waves. Enormous amounts of energy released in these objects are attributed to processes involving black holes with masses of order 10^6–$10^{10} M_\odot$. If this is true, every quasar and every active galactic nucleus can be expected to emit 1–10 strong pulses of gravitational radiation. According to the estimates of Thorne and Braginsky (1976) a minimum of one such pulse in every 300 years and the maximum of 50 pulses per year should reach the Earth. The average duration of the pulses is supposed to be $\tau \sim 90(M/10^6 M_\odot)$ s, where M is the mass of the black hole. The mean amplitude of the perturbation in the metric due to the pulse is estimated to be $h \sim 2 \cdot 10^{-17}(M/10^6 M_\odot)$.

We have presented only a few possible astrophysical sources of gravitational radiation, beginning with those whose existence is established beyond any doubt and concluding with the probable ones. More details about the properties of the sources discussed above and information about other possible sources can be found in books on gravitation and in review articles; see for example Press and Thorne (1972), Misner, Thorne and Wheeler (1973), Rees, Ruffini and Wheeler (1974), Thorne (1978), and Smarr (1979). There is certainly plenty of room in this field for new, ingenious models.

9.5. Weber's Antenna and Other Attempts to Detect Gravitational Waves

Disputes over the question of whether gravitational waves can exist at all were still in full swing when in 1959 Joseph Weber began his pioneering work on the construction of a gravitational antenna. The original design of the antenna proposed by Weber was so simple and so perfect that most of the antennae constructed since then have been to a large extent a repetition of that first construction. Weber's idea was to use a crystal or a suspended homogeneous metal bar (Weber, 1961, 1968, 1980). A gravitational wave falling on such a system excites mechanical vibrations in it. The vibrations can be transformed into electric oscillations, which are then strongly amplified and recorded.

Just as in electrodynamics, where a good transmitting antenna is also a good receiving antenna, a system to be chosen for the reception of gravitational waves should be an effective gravitational radiator. Weber proposed to use a large aluminium cylinder. A gravitational wave falling on the cylinder will displace different parts of the cylinder differently, giving rise to mechanical vibrations. Let us place a local inertial reference frame at the centre of mass of the cylinder. Relative displacements of different parts of the cylinder can be calculated from the equation of geodesic deviation (2.78), which we can write in the form

$$\frac{d^2n^\mu}{dt^2}+\frac{D^\mu{}_\beta}{m}\frac{dn^\beta}{dt}+\frac{K^\mu{}_\beta}{m}n^\beta=-c^2R^\mu{}_{0\beta0}\,\gamma^\beta, \tag{9.87}$$

where γ^β is the relative position of two elements in equilibrium, $D^\mu{}_\beta$ represents dissipation, $K^\mu{}_\beta$ elastic forces, and m is the mass of either element. Using equation (9.87), we can readily calculate the energy that can be absorbed by the antenna in a unit of time. If the only damping factor is radiative friction, formula (9.46) permits us to find the energy lost by the antenna in a unit of time. By comparing these two quantities Weber has shown that the maximum possible effective cross-section for absorption of gravitational radiation is

$$\sigma=\frac{15}{16\pi}\lambda^2, \tag{9.88}$$

where λ is the wave-length corresponding to the fundamental vibrational mode of the antenna. In a real antenna, besides radiative friction, there are also other dissipative factors, and so the average effective cross-section given by formula (9.88) should be regarded as an upper limit.

By the principle of equipartition of energy, the energy share of each oscillating degree of freedom of the system is kT. The amplitude of thermal fluctuations exceeds the estimated amplitude of gravitational excitations and must be taken into account. In fact, it is only owing to its random variation that observation of gravitational excitations is, in principle, possible at all. To find the total amplitude of the oscillations of the antenna we must modify equation (9.87) by adding to its right-hand side another term—the Nyquist force, which represents the thermal fluctuations. The total amplitude can be written as a sum $\xi=\xi_G+\xi_T$, where ξ_G is the gravitational excitation and ξ_T the amplitude of thermal fluctuations. If ω_n is the frequency of the n-th proper mode, then

$$|\xi_{nT}|\sim\left(\frac{kT}{m\omega_n^2}\right)^{\frac{1}{2}}. \tag{9.89}$$

Let τ_n be the characteristic time of thermal fluctuations. Changes in the amplitude of thermal fluctuations over any time period $\Delta t\ll\tau_n$ obey the stochastic law, so that

$$|\Delta\xi_{nT}|\sim\left(\frac{\Delta t}{\tau_n}\right)^{\frac{1}{2}}|\xi_{nT}|. \tag{9.90}$$

Gravitational excitations will be detectable when

$$|\xi_{nG}| > |\Delta\xi_{nT}|. \tag{9.91}$$

For this condition to be satisfied the effective-cross section of the antenna should be large, the antenna relaxation time should not be less than τ_n and the temperature should be as low as possible. For a thorough discussion of different types of mechanical detectors of gravitational waves see Misner, Thorne, and Wheeler (1973).

Weber used an aluminium cylinder of diameter 0.6 m, length 1.5 m, and weight $1.5 \cdot 10^3$ kg suspended from a metal frame resting on shock absorbers placed on top of one

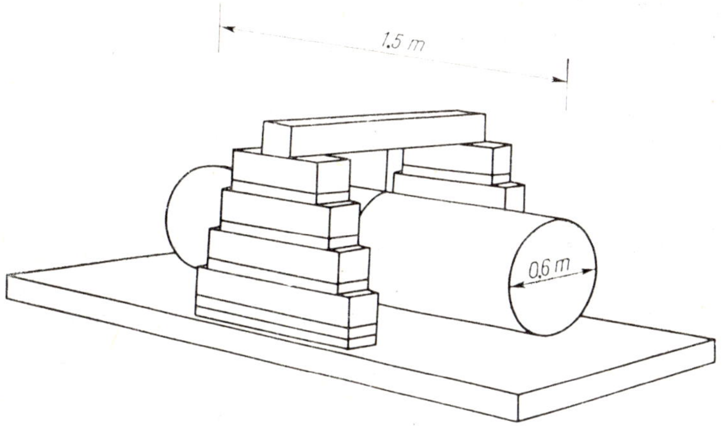

FIG. 9.2. Weber's gravitational antenna. An aluminium cylinder is suspended from a metal frame resting on shock absorbers. The whole is placed on a metal foundation (adapted from Weber, 1961).

another (Fig. 9.2). Inside the cylinder, along its circumference, there are several piezoelectric crystals which transform mechanical vibrations of the antenna into electric oscillations. The whole system is closed in a cylindrical metal vacuum chamber. The electric signals coming from the crystals are amplified and recorded on a magnetic tape.

It was also Weber's idea to use at least two such antennas and to look for coincident excitations. He set up two aluminium cylinders 1000 km apart: one at College Park, Maryland and the other at the Argonne National Laboratory near Chicago. In 1969 Weber reported a coincidence rate between the two detectors exceeding the level of random coincidence. The number of coincident pulses was apparently greater when the antennas were directed towards the centre of the Galaxy.

Similar experiments with other antennas conducted by Braginsky (1974) in Moscow, Bonazzola (1974) in Paris, Drever in Glasgow, Douglas in Rochester, Tyson in Murray Hill and by others have not confirmed Weber's results (see also DeWitt–Morette (1974), and Tyson and Giffard (1978)). It remains unknown what factors produced the coincidences observed by Weber. In any event, they are not gravitational waves. The measurements carried out so far indicate that the intensity of gravitational waves near the surface of the Earth is less than 10^5 erg$\cdot$cm$^{-2}\cdot$Hz^{-1}.

A few years ago William Fairbank undertook the ambitious project of constructing a gravitational antenna whose sensitivity would permit detection of gravitational waves of intensity 0.1 $erg \cdot cm^{-2} \cdot Hz^{-1}$. Resorting to the latest achievements of low-temperature physics, Fairbank (1972) intends to cool down an aluminium antenna of weight $5 \cdot 10^3$ kg to a temperature of $3 \cdot 10^{-3}$ K. With the aid of this antenna it will be possible to record supernova explosions not only in the galaxies of the Local Group but also in the Virgo cluster of galaxies. Other methods of detection of gravitational waves proposed recently are discussed by Braginsky and Rudenko (1978), Douglas and Braginsky (1979), Thorne *et al.* (1979), and Weiss (1979).

9.6. Interaction Between Electromagnetic and Gravitational Waves. Laboratory Sources and Detectors of Gravitational Waves

A detailed study of the properties of gravitational waves will only become possible when we are able to set up a source and a detector of gravitational waves in a laboratory. The problem is extremely difficult but some hopes of success seem to be dawning. They are related to the possibility of conversion of gravitational waves to electromagnetic waves. De Sabbata and his collaborators (Boccaletti, 1970), and subsequently Braginsky, Zeldovich and others (Grishchuk and Polnarev, 1980) have investigated the possibility of using this effect for generation and detection of gravitational waves.

An analysis of the coupled system of Einstein and Maxwell equations

$$G_{\alpha\beta}=\frac{8\pi G}{c^4}T_{\alpha\beta}, \tag{9.92}$$

$$F_{\alpha\ ;\beta}^{\ \beta}=0, \qquad F_{[\alpha\beta;\gamma]}=0,$$

where $T_{\alpha\beta}$ is the electromagnetic energy-momentum tensor, shows that a monochromatic electromagnetic wave propagating in a region with a constant electromagnetic field generates a gravitational wave of identical frequency. Conversely, if a monochromatic gravitational wave falls on a region with a constant electromagnetic field, it produces an electromagnetic wave of identical frequency. Monochromaticity of the waves is not essential; the effect occurs also when wave packets are considered. Let l be the characteristic dimension of the region endowed with a constant electromagnetic field, and A the amplitude of an electromagnetic wave propagating along the x-axis. Then, the energy-density flux of the generated gravitational wave (also propagating along the x-axis) is given, to order of magnitude, by

$$T^{01}=\frac{G}{4\pi c^4}l^2A^2[(H_y+E_z)^2+(E_y+H_z)^2], \tag{9.93}$$

where H_y, H_z, E_y and E_z are the components of the constant field. We can see from this formula that only those components of the constant field which are transverse to the direction of the electromagnetic wave are important in generation of the gravitational wave. When a gravitational wave travelling along the x-axis hits a region with a constant electromagnetic field, it generates an electromagnetic wave with energy density

$$T^{00} \sim \frac{L^2 a^2}{\lambda^2} [(E_y - H_z)^2 + (H_y + E_z)^2], \tag{9.94}$$

where L is the scale of the constant field region, λ is the wavelength and a the amplitude of the incident gravitational wave. As in the previous case, only transverse components of the constant electromagnetic field are important.

We now give some estimates (Grishchuk, 1977). Suppose we construct a cubic resonator with 10 m sides and maintain in it a constant electromagnetic field of intensity $E \sim H \sim 10^5$ gauss. An incident electromagnetic wave of length $\lambda = 10$ m will generate in this system gravitational waves of amplitude $h \sim \frac{G}{c^4} E^2 \lambda^2 \sim 10^{-36}$ and power $L \sim \frac{c^5}{G} h^2 \sim 10^{-13}$ erg/s.

On the other hand, if we want to detect these waves by means of an electromagnetic detector, the signal should be at least of the same order as the noise level of the detector. As follows from Grishchuk's estimates, the amplitude of the incident wave should satisfy

$$h \geqslant \frac{(\hbar c)^{\frac{1}{2}}}{QH\lambda_g^2}, \tag{9.95}$$

where $\hbar$ is the Planck constant, Q is the detector quality factor, H the characteristic field strength in the detector and λ_g the length of the incident gravitational wave. Inserting $\lambda_g = 10$ m, $H = 5 \cdot 10^5$ gauss and $Q = 10^{13}$, we obtain $h \gtrsim 10^{-31}$, while the source produces waves of amplitude $h \sim 10^{-36}$. Thus the amplitude of the gravitational wave emitted by the source is five orders of magnitude too small to be observed by an electromagnetic detector.

The difference, although it gives no ground for undue optimism, is not really large, and for the first time the realization of a gravitational analogue of Hertz's experiment seems to be a possibility (Picasso, 1980). As follows from formula (9.95), we would have to increase Q or the size of the detector, or produce and maintain a stronger electromagnetic field. Each of these requirements presents great technological difficulties and we are likely to wait another decade or two before they are surmounted.

9.7. Binary Pulsar PSR 1913+16: an Indirect Confirmation of the Existence of Gravitational Waves

As was already mentioned in Chapter 6, the binary pulsar PSR1913+16 provides a unique opportunity of testing the predictions of general relativity outside the solar system; what is even more important, with sufficiently long and careful timing measurements it should be possible to check predictions beyond the post-Newtonian approximation. Let us recall that this binary is most probably composed of two neutron stars with masses $M_p = (1.39 \pm 0.15)\ M_\odot$, $M_c = (1.44 \pm 0.15)\ M_\odot$ moving on a fairly eccentric orbit ($e = 0.617155 \pm 0.000007$) quite close to each other; the projected semi-major axis is $a \sin i \approx 7 \cdot 10^{10}$ cm $\approx R_\odot$.

Gravitational waves produced by such a system will carry away energy and angular momentum, causing a secular change in the orbital period. Consider a binary system consisting of two stars with masses M_1 and M_2 moving around the common center of mass with an orbital period P_b. Their relative orbit is characterized by eccentricity e, semi-major axis a, inclination angle i and angle ω between the ascending node and periastron. The amount of energy lost by the system due to emission of gravitational waves is given by equation (9.80). The change in the angular momentum is (Peters and Mathews, 1963)

$$\frac{\mathrm{d}J}{\mathrm{d}t}=-\frac{32}{5}\frac{G^{7/2}M_1^2M_2^2(M_1+M_2)^{\frac{1}{2}}}{c^5a^{\frac{7}{2}}}(1-e^2)^{-2}(1+\tfrac{7}{2}e^2). \tag{9.96}$$

The energy E and the orbital angular momentum J of the binary in the Newtonian approximation are

$$E=-\frac{1}{2}\frac{GM_1M_2}{a}, \quad J^2=\frac{GM_1^2M_2^2}{(M_1+M_2)}a(1-e^2). \tag{9.97}$$

Using Kepler's third law $P_b=2\pi a^{3/2}(G\,M)^{-1/2}$, one can easily derive equations for the rates of change of the orbital period and eccentricity; they are (Wagoner, 1975)

$$\frac{1}{P_b}\frac{\mathrm{d}P_b}{\mathrm{d}t}=-\frac{3}{2E}\frac{\mathrm{d}E}{\mathrm{d}t}=-\frac{96}{5}\frac{G^3M_1M_2(M_1+M_2)}{c^5a^4}(1-e^2)^{-7/2}(1+\tfrac{73}{24}e^2+\tfrac{37}{96}e^4), \tag{9.98}$$

$$\frac{P_b}{e}\frac{\mathrm{d}e}{\mathrm{d}P_b}=\tfrac{19}{18}(1-e^2)(1+\tfrac{121}{304}e^2)(1+\tfrac{73}{24}e^2+\tfrac{37}{96}e^4)^{-1}. \tag{9.99}$$

Substituting here the values characterizing the pulsar PSR 1913+16 we obtain $-\left\langle\frac{\mathrm{d}E}{\mathrm{d}t}\right\rangle$ $=0.63\cdot10^{33}$ erg/s$=0.16L_\odot$, $\frac{\mathrm{d}P_b}{\mathrm{d}t}=-2.4\cdot10^{-12}$ ss^{-1} and $\frac{P_b}{e}\frac{\mathrm{d}e}{\mathrm{d}P_b}=0.34$. The most interesting of these results is the value of the rate of change of the orbital period.

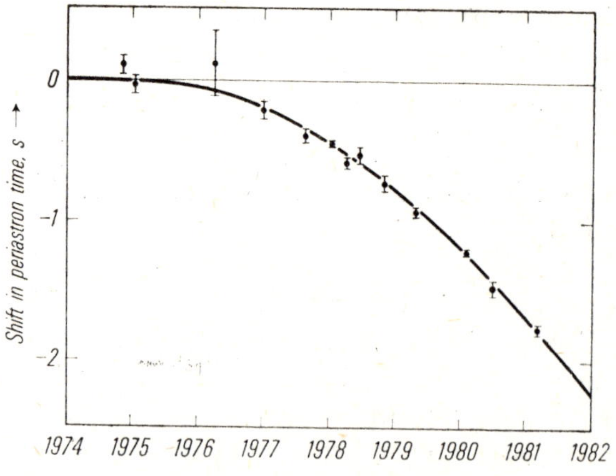

FIG. 9.3. The dots represent the observed orbital phase shifts with respect to the positions predicted on the assumption that P_b is constant (the horizontal line). The curve is the general-relativity prediction, with $M_p=M_0=1.41\,M$ (adapted from Weisberg *et al.*, 1981).

After five years of timing observations of PSR 1913+16, Taylor, Fowler and McCulloch (1979) announced that the rate of change of the orbital period of this system derived from the timing data is $\dot{P}_b = (-3.2 \pm 0.6)\ 10^{-12}\ \text{ss}^{-1}$. If the observed $\dot{P}_b$ is entirely due to gravitational radiation damping, then the observed value of $\dot{P}_b$ agrees with the value derived from general relativity to within a factor of 1.3 ± 0.3.

There are other effects which could change the orbital period, for example tidal interactions and/or mass transfer. If the pulsar really consists of two neutron stars, then there should not be any mass transfer and tidal interactions could not produce the observed change in the orbital period. Crane, Nelson and Tyson (1979) claim to have observed a very faint star just at the position of PSR 1913+16. If confirmed this could mean that the companion star of the pulsar is not a neutron star but rather a hydrogen star, and the change in the orbital period could then be completely due to tidal interactions (Smarr and Blandford, 1976). This question can only be settled by further observations.

In closing, let us note that with the increasing time span of the timing data for PSR 1913+16 the uncertainties in the value of the orbital parameters and masses of the stars decrease, and by now the evidence for the existence of gravitational radiation might be stronger than in 1981.

Literature

BOCCALETTI, D. *et al.* (1970) *Nuovo Cimento* **70B**, 129.

BONAZZOLA, S. *et al.* (1974) *Meudon Gravitational Radiation Experiment*, in *Ondes et Radiations Gravitationelles*, CNRS Coll. 220, Edition CNRS, Paris.

BONDI, H. (1964) *Radiation from an Isolated System*, in *Proceedings on Theory of Gravitation, Conference in Warszawa and Jablonna 1962*, Gauthier-Villars, Paris.

BRAGINSKY, V. B. *et al.* (1974) *Sov. Phys. JETP*, **38**, 865.

BRAGINSKY, V. B. *et al.* (1974) *Sov. Phys. JETP* **39**, 387.

BRAGINSKY, V. B. and MANUKIN, A. B. (1977) *Measurements of Weak Forces in Physics Experiments*, ed.: Drever, R. W. P., University of Chicago Press, Chicago.

BRAGINSKY, V. B. and RUDENKO, V. N., (1978) *Physics Reports* **46**, 165.

BREUER, R. A. (1975) *Gravitational Perturbation Theory and Synchrotron Radiation*, Springer-Verlag, Berlin.

BREUER, R. A. *et al.* (1973) *Phys. Rev.* **D8**, 4309.

BRILL, D. and HARTLE, J. (1964) *Phys. Rev.* **B135**, 271.

CAVES, C. M., THORNE, K. S., DREVER, R. W. P., SANDBERG, V. D. and ZIMMERMAN, M. (1980) *Rev. Mod. Phys.* **52**, 341.

CHRZANOWSKI, P. L. and MISNER, C. W. (1974) *Phys. Rev.* **D10**, 170.

CRANE, P., NELSON, J. E. and TYSON, J. A. (1979) *Nature* **280**, 367.

DEWITT-MORETTE, C. (ed.) (1974) *Gravitational Radiation and Gravitational Collapse*, D. Reidel Publ. Comp., Dordrecht.

DOUGLASS, D. H. and BRAGINSKY, V. B. (1979) *Gravitational Radiation Experiments*, in *General Relativity, An Einstein Centenary Survey*, eds.: Hawking, S. W. and Israel, W., Cambridge University Press, Cambridge.

DYSON, F. (1969) *Astrophys. J.* **156**, 529.

EINSTEIN, A. (1913) *Phys. Z.* **14**, 1249.

FAIRBANK, W. *et al.* (1972) *Search for Gravitational Radiation*, in Proc. Intern. School of Phys. "E. Fermi", Academic Press., N. Y.

GRISHCHUK, L. P. (1977) *Sov. Phys. Uspekhi* **20**, 319.

GRISHCHUK, L. P. and POLNAREV, A. G. (1980) *Gravitational Waves and Their Interaction with Matter and Fields*, in *General Relativity and Gravitation*, ed.: Held, A., Plenum Publ. Corp., New York.

ISAACSON, R. (1968) *Phys. Rev.* **166**, 1263.

MISNER, C. W., THORNE, K. S. and WHEELER, J. A. (1973) *Gravitation*, Freeman, San Francisco.

PETERS, P. C. and MATHEWS, J. (1963) *Phys. Rev.* **131**, 435.

PICASSO, E. (1980) *Experiments on Gravitational Waves with Electromagnetic Detectors*, Proceedings of the 9th GRG Conference, Jena, GDR.

PIRANI, F. (1957) *Phys. Rev.* **105**, 1089.

PRESS, W. H. and THORNE, K. S. (1972) *Ann. Rev. Astron. Astrophys.* **10**, 335.

REES, M., RUFFINI, R. and WHEELER, J. A. (1974) *B'ack Holes, Gravitational Waves and Cosmology*, Gordon and Breach, N. Y.

RUFFINI, R. and WHEELER, J. A. (1971) *Relativistic Cosmology and Space Platforms*, in *The Significance of Space Research for Fundamental Physics*, eds.: Moore, A. F., Hardy, V. ESRO, Paris.

SACHS, R. K. (1962) *Proc. Roy. Soc. London*, **A270**, 103.

SACHS, R. K. (1964) *Gravitational Waves*, in *Relativity, Groups and Topology*, eds.: DeWitt, C. and DeWitt, B., Gordon and Breach, New York.

SMARR, L. L. (ed.) (1979) *Sources of Gravitational Radiation*, Cambridge University Press, Cambridge.

SMARR, L. L. and BLANDFORD, R. (1976) *Astrophys. J.* **207**, 574.

TAYLOR, J. H., FOWLER, L. A. and MCCULLOCH, P. M. (1979) *Nature* **277**, 437.

THORNE, K. S. (1969) *Astrophys, J.* **158**, 1.

THORNE, K. S. (1977) *The Generation of Gravitational Waves: a Review of Computational Techniques*, in *Topics in Theoretical and Experimental Gravitational Physics*, eds.: De Sabbata, V., Weber J., Plenum Press, New York.

THORNE, K. S. (1978) *General Relativistic Astrophysics*, in *Theoretical Principles in Astrophysics and Relativity*, eds.: Lebovitz, N. R., Reid, E. H. and Vandervoort, P. O., The University of Chicago Press, Chicago.

THORNE, K. S. (1980) *Rev. Mod. Phys.* **52**, 285.

THORNE, K. S. and BRAGINSKY, V. B. (1976) *Astrophys. J.* **204**, L1.

THORNE, K. S., CAVES, C. M., SANDBERG, V. D. and ZIMMERMANN, M. (1979) *The Quantum Limit for Gravitational-Wave Detectors and Methods of Circumventing It*, in *Sources of Gravitational Radiation*, ed.: Smarr, L. L., Cambridge University Press, Cambridge.

TRAUTMAN, A. (1958) *Bull. Acad. Polon. Sci.* III, **6**, 403.

TYSON, J. A., GIFFARD, R. P. (1978) *Ann. Rev. Astron. Astrophys.* **16**, 521.

WAGONER, R. V. (1975) *Astrophys., J.*, **196**, L63.

WEBER, J. (1961) *General Relativity and Gravitational Waves*, Wiley–Interscience, N. Y.

WEBER, J. (1964) *Gravitational waves*, in *Gravitation and Relativity*, ed. Chiu, H. Y., Hoffmann, Benjamin, New York.

WEBER, J. (1968) *Physics Today*, April, p. 34.

WEBER, J. (1969) *Phys. Rev. Lett.* **22**, 1320

WEBER, J. (1980) *Search for Gravitational Radiation*, in *General Relativity and Gravitation*, ed.: Held, A., Plenum Publ. Corp., New York.

WEISS, R., (1979) *Gravitational Radiation—The Status of the Experiments and Prospects for the Future*, in *Sources of Gravitational Radiation*, ed.: Smarr, L. L., Cambridge University Press, Cambridge.

WEISBERG, J. M., TAYLOR, J. H. and FOWLER, L. A. (1981) *Scientific American* **245**, October, 74.

WHEELER, J. A. (1962) *Geometrodynamics*, Academic Press, New York.

CHAPTER 10

COSMOLOGY

10.1. Homogeneous and Isotropic Matter Distributions. The Friedman Solution

Before we discuss one of the most important solutions of Einstein's equations, which describes a homogeneous and isotropic distribution of matter filling all of space, let us consider a simple Newtonian model (Milne and McCrea, 1934; Bondi, 1961). Consider two arbitrary particles in a homogeneous and isotropic cloud of dust filling the whole space. The law of universal gravitation implies that only that part of the cloud mass is dynamically involved in the motion of the particles which is contained in the sphere with centre at one of the particles and radius equal to their separation. If $R(t)$ is this separation at time t, the energy conservation law implies that

$$\tfrac{1}{2}\dot{R}^2 - \frac{GM}{R} = -\tfrac{1}{2}kc^2; \tag{10.1}$$

$-\frac{1}{2}kc^2$ stands for the energy per unit of mass. We assume of course that matter is neither created nor destroyed and therefore we can supplement the above equation with the mass conservation law

$$\tfrac{4}{3}\pi\rho R^3 = M = \text{const}. \tag{10.2}$$

Let the relative position of the points at an initial moment be $\boldsymbol{r}_0$. Their relative position at any later time t is then

$$\boldsymbol{r}(t) = \frac{R(t)}{R(t_0)}\,\boldsymbol{r}_0, \tag{10.3}$$

and the velocity

$$\boldsymbol{v}(t) = \frac{\dot{R}(t)}{R(t_0)}\,\boldsymbol{r}_0 = H(t)\,\boldsymbol{r}(t), \tag{10.4}$$

where

$$H(t) = \frac{\dot{R}(t)}{R(t)}. \tag{10.5}$$

From equation (10.1) we can determine the time-dependence of the distance R. This is particularly simple when $k=0$, i.e. when the total energy is equal to zero. We then have

$$R(t)=\left(\frac{9GM}{2}\right)^{1/3} t^{2/3}, \tag{10.6}$$

the time scale having been chosen so that the particle separation at the initial moment is zero. Hence the initial density is infinite. Beginning from this singular state, the system expands and the distances between the particles grow like $t^{2/3}$, while the density of the dust decreases as t^{-2}. Note that near the singularity R changes as $t^{2/3}$ and density as t^{-2} irrespective of whether $k=0$ or $k\neq 0$.

When $k\neq 0$, equation (10.1) is no longer so easily integrable. By a suitable choice of the scale of R we can reduce k to ± 1. For the present, we shall content ourselves with a qualitative discussion of the behaviour of the function $R(t)$. If k is positive, i.e. if the energy per unit mass is negative, the system first expands from the singular state to a maximum R

$$R_{\max}=\frac{2GM}{c^2k}, \tag{10.7}$$

and then contracts, until it reaches the singular state again; the time of one such complete cycle is finite. When the energy per unit mass is positive, the picture is different. In this case, the system will go on expanding forever, and, for large t, R will change in proportion to t.

In 1922, a Soviet mathematician, Alexander Friedman, gave a solution of Einstein's equations which describes a homogeneous and isotropic distribution of matter. The general line element of such a space can be written in the form

$$ds^2=c^2dt^2-R^2(t)\left\{\frac{dr^2}{1-kr^2}+r^2(d\theta^2+\sin^2\theta\, d\varphi^2)\right\}, \tag{10.8}$$

where k can take the values $+1, 0, -1$; $R(t)$ is a function of time only and can be determined from the field equations.

The constant k has a simple geometrical interpretation. If we calculate the curvature of a three-surface $t=t_0=\text{const}$, we obtain

$$^{(3)}R=\frac{6k}{R^2(t_0)}, \tag{10.9}$$

hence, when $k=0$, such three-surfaces are flat, for $k=1$ they are three-spheres, and for $k=-1$ three-dimensional pseudospheres.

If we use the hydrodynamic energy-momentum tensor to describe the distribution of matter, the symmetry conditions imply that density and pressure can depend on time only. The field equations then reduce to

$$\dot{R}^2-\frac{8\pi}{3}G\rho R^2=-kc^2, \tag{10.10}$$

$$2\frac{\ddot{R}}{R}+\frac{\dot{R}^2}{R^2}+\frac{kc^2}{R^2}=-\frac{8\pi G}{c^2}p. \tag{10.11}$$

In addition, we have to specify the equation of state. In the case of a "dust", where $p=0$, equations (10.10) and (10.11) take exactly the same form as in the Newtonian case, and only the interpretation of the constant k is different. In a comoving frame, the equations of motion of matter take the form

$$(\rho R^3)^{\cdot}+(R^3)^{\cdot}\,\frac{p}{c^2}=0 \tag{10.12}$$

and

$$\dot{u}^\alpha=u^\alpha{}_{;\beta}u^\beta=0; \tag{10.13}$$

thus the particles of matter move on geodesics.

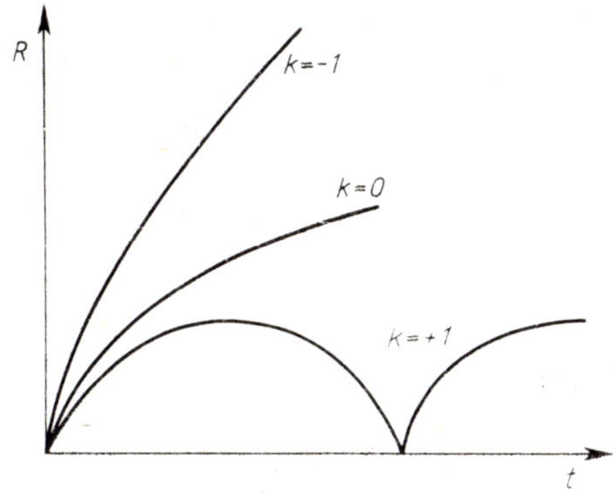

FIG. 10.1. Time-dependence of R for different k. If $k=1$, we have a closed model of the universe; if $k=0$, we have an open model with flat spacelike cross-sections; if $k=-1$, the universe is open.

It follows from (10.12) and (2.101) that the system evolves adiabatically. The field equations can easily be solved when the pressure is zero or negligibly small. In this case, (10.12) yields a simple relationship between density and R:

$$\rho(t)=\rho(t_0)\,\frac{R^3(t_0)}{R^3(t)}. \tag{10.14}$$

The solution of the field equations can now be represented parametrically. It is easy to verify that

$$\begin{aligned} &R=\alpha(1-\cos\omega), \quad t=\alpha/c(\omega-\sin\omega), \quad k=+1,\\ &R=\left(\frac{9GM}{2}\right)^{1/3} t^{2/3}, \quad k=0. \\ &R=\alpha(\cosh\omega-1), \quad t=\alpha/c(\sinh\omega-\omega), \quad k=-1, \end{aligned} \tag{10.15}$$

where $\alpha=\dfrac{4\pi G}{3c^2}\,\rho(t_0)\,R^3(t_0)$. The corresponding curves are shown in Fig. 10.1.

When the curvature of the three-surfaces $t=\text{const}$ is positive, i.e. when they are three-spheres, their volume is finite and equal to $V=2\pi^2R^3$. This model is often referred to as the closed model of the universe. In open, spatially infinite models, the curvature of three-dimensional spacelike sections is either negative or zero.

The line element of a four-dimensional isotropic and homogeneous space can be written in a somewhat different form by introducing a new variable χ such that $d\chi = dr/\sqrt{1-kr^2}$; then, we get

$$ds^2 = c^2 dt^2 - R^2(t)\{d\chi^2 + f^2(\chi)(d\theta^2 + \sin^2\theta \, d\varphi^2)\}, \tag{10.16}$$

where

$$f(\chi) = \begin{cases} \sin\chi, & k=+1, \\ \chi, & k=0, \\ \sinh\chi, & k=-1. \end{cases} \tag{10.17}$$

The fundamental property of the Friedman solution is its non-stationarity. The metric and the curvature of the space depend on time. Distances between particles of matter also change. Friedman's principal assumption that matter is distributed uniformly, with no point or direction distinguished, is an idealization, and only a confrontation of the predictions of his model with observational data can confirm or refute this hypothesis. So far, although the Friedman model is already more than fifty years old, not a single observational fact† contradictory with its predictions has been reported.

10.2. The Red Shift. Horizons

Before we analyse the homogeneous and isotropic model of the universe described by the line element (10.8) in more detail, let us find out how the function $R(t)$ is related to observational data. To this end, consider the simplest type of observation, namely that of light propagation. Suppose that an observer placed at the origin of coordinates observes a distant galaxy located at a point with coordinates r_1, θ_1 and φ_1. At a time $t=t_1$ a light signal is sent from the galaxy, to reach the observer at a later time $t=t_0$. Along the light ray we have

$$ds^2 = 0 = c^2 dt^2 - R^2(t)\frac{dr^2}{1-kr^2}, \tag{10.18}$$

because θ_1 and φ_1 remain constant. Integrating this equation and remembering that the light signal moves towards decreasing r, we obtain

$$c\int_{t_1}^{t_0} \frac{dt}{R(t)} = \int_0^{r_1} \frac{dr}{\sqrt{1-kr^2}}, \tag{10.19}$$

where

$$\int_0^{r_1} \frac{dr}{\sqrt{1-kr^2}} = F(r_1) = \begin{cases} \arcsin r_1, & k=+1 \\ r_1, & k=0, \\ \operatorname{arc\,sinh} r_1, & k=-1. \end{cases} \tag{10.20}$$

† Observations should concern sufficiently large regions. We discuss this in more detail on p. 293.

Suppose that a second signal is sent from the galaxy after a short time δt_1 and reaches the observer at a time $t_0+\delta t_0$; therefore

$$c\int_{t_1+\delta t_1}^{t_0+\delta t_0}\frac{dt}{R(t)}=F(r_1). \tag{10.21}$$

Subtracting (10.19) from (10.21) and assuming δt_0 and δt_1 to be small, we get

$$\frac{\delta t_0}{R(t_0)}=\frac{\delta t_1}{R(t_1)}; \tag{10.22}$$

in deriving this relation we assumed that the function $R(t)$ changes very slowly with time. Relating the time interval δt with the frequency of the emitted light, we get

$$\frac{\delta t_1}{\delta t_0}=\frac{\nu_0}{\nu_1}=\frac{R(t_1)}{R(t_0)}. \tag{10.23}$$

The galaxy emits ligth of frequency ν_1. It is assumed that the mechanism of emission is the same as under laboratory conditions on the Earth, i.e. the emission is produced by transition of an electron from an energy state E_n to an energy state E_m:

$$h\nu_1=E_n-E_m. \tag{10.24}$$

However, the frequency reported by the observer will be ν_0. If $R(t_0)>R(t_1)$, i.e. if the universe is expanding, then $\nu_0<\nu_1$ and the frequency undergoes a red shift, whereas if $R(t_0)<R(t_1)$ and the universe is contracting, $\nu_0>\nu_1$ and the frequency is shifted towards the blue end of the spectrum.

What causes this change in frequency? If the Friedman metric admitted a timelike Killing vector, i.e. if there existed a coordinate system in which the metric was independent of time, the answer would be simple. A photon travelling between the galaxy and the observer would have performed some work and therefore would have lost part of its energy (gravitational red shift). The Friedman metric is not stationary, however, and there is no coordinate system in which the metric is independent of time. On the other hand, we know that a red shift can result from the relative motion of the observer and the light source. In reality, the gravitational red shift plays a minute role and the Doppler effect dominates.

The red shift is conveniently expressed in terms of a red-shift parameter z, defined as

$$z=\frac{\lambda_0-\lambda_1}{\lambda_1}, \tag{10.25}$$

i.e. as the fractional change in wavelength; using (10.23) we can write it in the equivalent form

$$z=\frac{R(t_0)}{R(t_1)}-1. \tag{10.26}$$

Now suppose that the observer and the light source are close to each other, so that

$t_1 = t_0 + \Delta t$, where Δt is small. Then

$$z = \frac{R(t_1+\Delta t)}{R(t_1)} - 1 = \frac{\dot{R}(t_1)}{R(t_1)}\Delta t, \tag{10.27}$$

but since $\Delta t = r/c$,

$$z = \frac{\dot{R}(t_1)}{R(t_1)}\frac{r}{c} = \frac{v}{c}. \tag{10.28}$$

Thus for small distances the relative red shift of the spectral lines is proportional to the relative velocity of the observer and the light source. Further corrections depend on the large-scale distribution of matter.

An observer seeking information about events occurring in the universe may at a given point of time perceive only those events which are within his past cone. The boundary of that cone is formed by the light rays satisfying equation (10.19).

If the integral $\int_{t_1}^{t_0} \frac{dt}{R(t)}$ diverges as $t_1 \to 0$ (or as $t_1 \to -\infty$ for singularity-free space-times), it is possible, in principle, to receive signals sent sufficiently early from any particle in the universe. If it is convergent, the observer's visual field will be limited by the particle horizon (Rindler, 1956). At a fixed time t_0 he will be able to receive signals only from particles which are within the radius $r_H(t_0)$ from him, where

$$\int_0^{r_H(t_0)} \frac{dr}{\sqrt{1-kr^2}} = c\int_0^{t_0} \frac{dt}{R(t)}. \tag{10.29}$$

The proper distance of this horizon is

$$d_H(t_0) = R(t_0)\int_0^{r_H(t_0)} \frac{dr}{\sqrt{1-kr^2}} = cR(t_0)\int_0^{t_0} \frac{dt}{R(t)}. \tag{10.30}$$

Thus, in some cosmological models, no matter how long we waited, we would not be able to see all the events. All the events are only observable when the integral $\int_{t_1}^{t_0} \frac{dt}{R(t)}$ diverges as $t_0 \to \infty$ or $t_0 \to t_{max}$, where t_{max} is such that $R(t_{max}) = 0$. If, instead, the integral is convergent for $t_0 \to \infty$ or $t_0 \to t_{max}$, only those events can be seen for which

$$\int_0^{r_1} \frac{dr}{\sqrt{1-kr^2}} \leqslant c\int_{t_1}^{t_{max}} \frac{dt}{R(t)}. \tag{10.31}$$

The boundary of this region is called the event horizon. For open cosmological models, $R(t)$ increases at least as fast as $t^{2/3}$. In this case there is no event horizon, although of course a particle horizon exists. In closed models of the universe ($k>0$) there always exists a t_{max} such that $R(t_{max}) = 0$, the integral is convergent with respect to the upper limit and there is an event horizon.

10.3. Hubble's Constant

In what way can observations be used to decide which of the basic cosmological models discussed above is actually realized? To answer this question, let us turn back to the general equations describing the time-variation of the function $R(t)$. They are

$$2\frac{\ddot{R}}{R}+\frac{\dot{R}^2}{R^2}+\frac{8\pi G}{c^2}p=-\frac{kc^2}{R^2}, \tag{10.32}$$

$$\frac{\dot{R}^2}{R^2}-\frac{8\pi G}{3}\rho=-\frac{kc^2}{R^2}; \tag{10.33}$$

it is assumed that the distribution of matter is given by the hydrodynamic energy-momentum tensor, and so p is the isotropic hydrodynamic pressure and ρc^2 the energy density.

A number of interesting conclusions can be derived from these equations. First, let us introduce the *Hubble constant*

$$H(t)=\frac{\dot{R}(t)}{R(t)} \tag{10.34}$$

and the *deceleration parameter*

$$q=-\frac{\ddot{R}R}{\dot{R}^2}=-\frac{\ddot{R}}{RH^2}. \tag{10.35}$$

Subtracting (10.33) from (10.32) gives the equation

$$\frac{\ddot{R}}{R}+4\pi G\left(\frac{\rho}{3}+\frac{p}{c^2}\right)=0, \tag{10.36}$$

which can be reduced to the form

$$\rho+\frac{3p}{c^2}=\frac{3H^2q}{4\pi G}. \tag{10.37}$$

Now inserting (10.34) into (10.33), we get

$$\frac{kc^2}{R^2}=\frac{8\pi G\rho}{3}-H^2 \tag{10.38}$$

or, if we use (10.37),

$$\frac{kc^2}{R^2}=\frac{4\pi G}{3q}\left[\rho(2q-1)-\frac{3p}{c^2}\right]. \tag{10.39}$$

This last relation expresses one of the fundamental facts of the general relativity theory, namely that the geometry of the space, represented here by the curvature k/R^2, is determined by the distribution of matter in the universe.

The above relations apply at an arbitrary moment of time. We shall be interested in the present values of the parameters concerned, and shall denote them by adding the subscript 0, for example H_0 is the value of the Hubble constant at the present moment.

Let us now estimate the values of the energy density $c^2\rho_0$ and pressure p_0 (Sandage, 1961). The pressure p_0 is the sum of the radiation pressure $\frac{1}{3}aT_0^4$ and the pressure produced by the chaotic motion of galaxies. The latter is equal to 2/3 of the kinetic energy $\frac{1}{2}\rho v^2$ of this motion, where v is the linear velocity of the motion and, as follows from observations, does not exceed 300 km/s. The energy density is the sum of the mass energy density $\rho_{0m}c^2$ and the radiation energy density aT_0^4. Equation (10.37) can thus be written

$$\rho_{0m}+\frac{2aT_0^4}{c^2}+\frac{\rho_{0m}v^2}{c^2}=\frac{3H_0^2q_0}{4\pi G}. \tag{10.40}$$

At the present stage of the evolution of the universe, the density of the observed (luminous) matter is about 10^{-31} g/cm³. The present temperature of the radiation that fills isotropically the intergalactic space is 3 K. Therefore the radiation term in equation (10.40) is of order 10^{-33} g/cm³ and can be neglected when compared with the matter density. The term representing the contribution of the chaotic galactic motion is smaller still, so its omission will also be fully justified. At the present time the matter filling the universe can be treated as dust; pressure and the influence of radiation can be neglected. Consequently, equations (10.40) and (10.39) simplify to

$$\rho_0=\frac{3H_0^2q_0}{4\pi G} \tag{10.41}$$

and

$$\frac{kc^2}{R_0^2}=H_0^2(2q_0-1). \tag{10.42}$$

Thus, if we manage to determine the Hubble constant H_0 and the deceleration parameter q_0 from observation, we shall be able to calculate the mean density of matter and the curvature of space.

Formula (10.42) implies that if $q_0=\frac{1}{2}$ then $k=0$ and the universe is open with flat Euclidean spacelike cross-sections. The mean density of matter corresponding to this situation,

$$\rho_{cr}=\frac{3H_0^2}{8\pi G} \tag{10.43}$$

is called the critical density. Instead of the present value of the density, ρ_0, we shall often use the dimensionless parameter $\Omega=\rho_0/\rho_{cr}$.

If $q_0>\frac{1}{2}$, i.e. when $\rho_0>\rho_{cr}$ ($\Omega>1$), the universe is closed, whereas if $q_0\leqslant\frac{1}{2}$ ($\rho_0\leqslant\rho_{cr}$, $\Omega\leqslant 1$) the universe is open.

By measuring shifts of the spectral lines of distant galaxies we can determine z. We recall that

$$1+z=\frac{\lambda_{ob}}{\lambda_{em}}=\frac{R(t_{ob})}{R(t_{em})}. \tag{10.44}$$

Expanding $R(t_{em})=R(t_{ob}-\Delta t)$ as a series, we obtain for small Δt

$$cz=dH_0 \tag{10.45}$$

where d is the distance of a given galaxy. The formula is valid only for small values of Δt, and therefore for small z. If we take higher-order corrections into account, then

$$d=\frac{c}{H_0}\left(z+\tfrac{1}{2}(1-q_0)z^2+\dots\right). \tag{10.46}$$

In order to determine the Hubble constant, a single galaxy whose red shift z and distance d are known is all we need. The red shift of the galaxy should be sufficiently large ($z \geqslant 0.01$) for the velocity due to cosmological expansion to be greater than the local velocities. The problem is not simple, however, the main difficulty being the determination of astronomical distances (Sandage, 1961; van den Bergh, 1975; de Vaucouleurs 1979a).

The method which is most commonly used for determining extragalactic distances is the following. In a cluster of galaxies one particular galaxy is chosen according to some criterion, for example the fourth brightest galaxy, and it is assumed that the corresponding galaxies in other clusters have the same absolute luminosity L. The apparent luminosity l is given by the formula

$$l=\frac{L}{4\pi d^2(1+z)^2}; \tag{10.47}$$

here $\frac{1}{4\pi d^2}$ represents the change in luminosity due to the distance of the source. Because of the red shift, the energy of a photon measured by an observer will be $(1+z)$ times less than the energy at the instant of emission. The second factor $(1+z)$ is present because the photon flux at the point of observation is measured relative to the local time standard and not the time standard at the emission point. Using formula (10.46), we can write l in the form

$$l=\frac{LH_0^2}{4\pi c^2 z^2}\left[1+(q_0-1)z+\dots\right]. \tag{10.48}$$

Observational astronomy makes use of the bolometric stellar magnitude m rather than the apparent luminosity l. These two quantities are related by

$$m=-2.5\log l+\text{const}; \tag{10.49}$$

hence

$$m-M=25-5\log H_0+5\log cz+1.086(1-q_0)z+\dots, \tag{10.50}$$

where H_0 is measured in $\text{km}\cdot\text{s}^{-1}\cdot\text{Mpc}^{-1}$ and M is the absolute magnitude. The observed relationship between z and m is shown in Fig. 10.2.

The choice of an appropriate "standard candle" in the sky and the necessity of allowing for various corrections cause astronomers a great deal of trouble (Sandage, 1970, 1975). By means of the largest telescope in operation, the Mt. Palomar telescope, whose diameter is 5 m, it is possible to observe objects with $m=24$. Recently an even more powerful telescope was put into service at Zelenchukskaya in the Caucasus; it permits observation of still more distant objects. To travel the distances that separate us from those furthermost objects light needs at least 10^9 years. Clearly, during such periods of time the objects evolve. We should be very cautious, therefore, if we want to extrapolate the properties of nearby galaxies and assume that they are typical. New corrections introduced now and again continually change the Hubble constant, the updated values getting smaller and smaller. In

1975 Sandage gave the Hubble constant as

$$H_0 = 50 \pm 5 \text{ km} \cdot \text{s}^{-1} \text{Mpc}^{-1}\dagger \tag{10.51}$$

and the deceleration parameter

$$q_0 = 0.28 \pm 0.08 .$$

These values do not permit us yet to give a definitive answer even to so general a question as

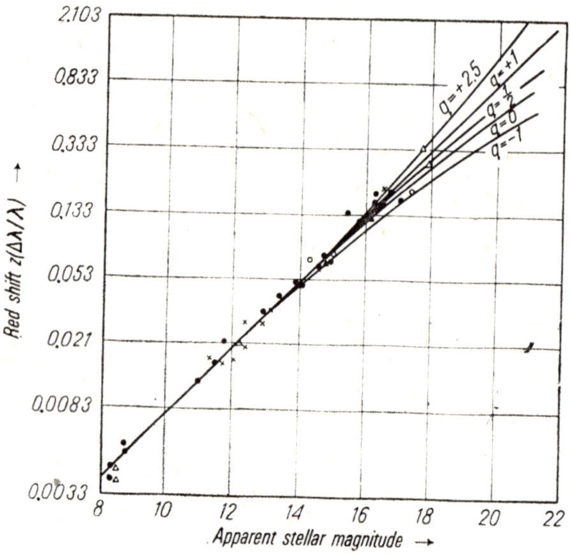

FIG. 10.2. Red shift versus apparent stellar magnitude. The apparent magnitude is modified to allow for the influence of the interstellar dust and the evolution of the object. The obervational data available at present do no permit us to determine the deceleration parameter accurately enough to decide whether the universe fits the open or the closed model.

whether the universe is spatially open or closed. The observational data seem to favour the open model, but there is no certainty (Gott *et al.*, 1974; Gunn and Tinsley, 1975; Gunn, 1977, 1978). The value of q_0 may change substantially when the evolutional effects are taken into account (Tinsley, 1979).

If we take q_0 to be $\frac{1}{2}$, then formula (10.41) gives the mean density of matter in the universe to be $6.1 \cdot 10^{-30}$ g/cm³, a value at least three times that of the mean density of luminous matter determined observationally.

Nevertheless, the present estimates of q_0 make it possible to eliminate certain cosmological models previously regarded as probable. In the late 1940's a great deal of research was devoted to the steady state model of the universe (see Bondi 1961). The mean density of matter in this model was assumed to be constant in time. Matter was supposed to be created

† Recently de Vaucouleurs (1979a, b), excluding effects induced by the Local Supercluster, obtained $H_0 = 100 \pm 10$ km s^{-1}Mpc^{-1}.

continuously so that, in spite of universal expansion, ρ could remain unchanged. In the steady state model $R(t)=e^{t/T}$, where $T=\text{const}$. The Hubble constant is then $H_0=1/T$ and $q_0=-1$. This predicted value of the deceleration parameter is now seen to disagree with the observational data. The existence of the cosmic microwave radiation background furnishes an independent weighty argument against the steady state model.

10.4. Galaxy and Radio Counts

In a Euclidean space, if the absolute luminosity of a galaxy is L, its apparent luminosity l is

$$l=\frac{L}{4\pi r^2}. \tag{10.52}$$

Let N be the number of galaxies contained in a sphere of radius r; if their distribution is uniform, $N\simeq r^3$. Hence $N\simeq l^{-\frac{3}{2}}$, or

$$\log N=-\tfrac{3}{2}\log l+\text{const}. \tag{10.53}$$

Combining this with relation (10.49) between luminosity and stellar magnitude, we get

$$\log N(m)=0.6m+\text{const}. \tag{10.54}$$

To obtain an analogous relation for isotropic cosmological models, note that if n_G is the present galaxy number density and $N(r)$ the number of galaxies at distances between r and $r+dr$, then

$$N(r)\,dr=4\pi n_G\left(\frac{\sin kr}{k}\right)^2 dr, \tag{10.55}$$

here $k=-1, 0$ or $+1$. Let $I(\nu)$ denote the power of a source at frequency ν, so that the power emitted by the source in the frequency range $(\nu, \nu+d\nu)$ is $I(\nu)d\nu$. The observed flux density at a frequency ν_0 is

$$S(\nu_0)=\frac{I(\nu_e)}{4\pi\left(\dfrac{\sin kr}{k}\right)^2(1+z)}, \tag{10.56}$$

where $\nu_e=\nu_0(1+z)$. Using (10.55) and (10.56), and assuming that the number of sources per unit of volume does not vary, we find, after Longair (1971), that the number of sources with flux density greater than S is

$$N(S)=\int_I\int_0^{r(I)} n_G(I)\left(\frac{\sin kr}{k}\right)^2 dr\,dI, \tag{10.57}$$

where $r(I)$ is given implicitly by

$$I[\nu(1+z)]=S(\nu)(1+z)\left(\frac{\sin kr}{k}\right)^2, \tag{10.58}$$

and $n_G(I(\nu))$ is the present number density of sources whose power at frequency ν is $I(\nu)$.

For the power law spectrum $I(\nu) \sim \nu^{-\alpha}$, we have

$$I(\nu) = S(\nu)\left(\frac{\sin kr}{k}\right)^2 (1+z)^{1+\alpha}. \tag{10.59}$$

The first to number-count galaxies was Hubble. By considering faint galaxies of apparent magnitude $m=19$ he showed that formula (10.54) is in good agreement with the observational data. The galaxies concerned are comparatively close to us, however, and this result

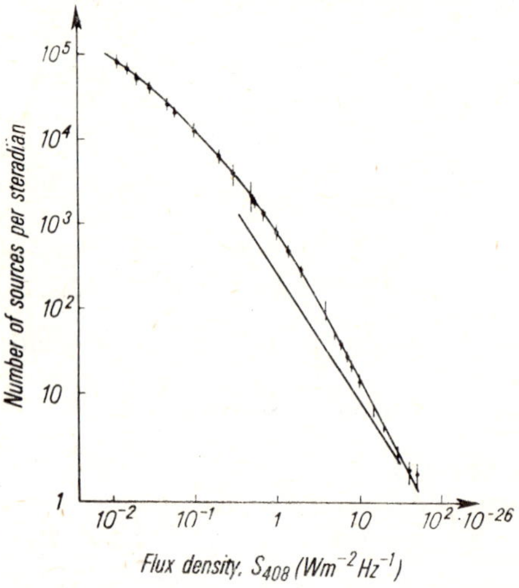

FIG. 10.3. Counts of radio sources indicate a distinct deviation from the flat space model. The lower line represents the predicted relation for the Euclidean model (adapted from Pooley and Ryle, 1968).

cannot be regarded as a good cosmological test. Hubble's observations implied that the nearby galaxies are distributed as in a flat Euclidean space.

Radio sources became objects of cosmological interest when it was found that their distribution away from the galactic plane is isotropic and, consequently, that they are distant extragalactic objects. Only a small fraction of the detected radio sources have been identified with optical objects and only for a few of them has it been possible to determine z. The absolute luminosity of these objects varies considerably, and although they are the furthest observed objects in the sky, they have been of no use in determining the Hubble constant and the deceleration parameter q_0. However, they are very well suited for the purpose of confronting formula (10.57) with observational data.

Radio source counts have now been performed by many different groups, perhaps the most systematic and longstanding investigations being those conducted by a Cambridge group of radioastronomers led by Ryle (Ryle, 1968; Jauncey, 1977; Longair, 1978). The frequency at which they carry their observations is 408 MHz. Radiotelescopes measure the power received per unit antenna area and per unit frequency interval, at a fixed frequency. We shall denote this quantity by S, sometimes making it clear at what frequency the measurement was performed (e.g. S_{408}).

On the assumption that $N(S) \approx S^{-\beta}$, two important facts regarding the exponent β have been established in the Cambridge surveys (Fig. 10.3):

(a) For large S the radio-source counts give $\beta = 1.8$, which indicates an excess of sources as compared with predictions based on Euclidean geometry.

(b) At luminosities below $S_{408} = 0.1 \cdot 10^{-26}$ Wm^{-2}Hz^{-1} the relation changes and $\beta = 0.8$.

In order to compare the observational data with the predictions of cosmological models, it is convenient to use $N(S)/N_0(S)$ instead of $N(S)$, where $N_0(S)$ is calculated for a Euclidean geometry. From formula (10.55) we find that the number of sources with a given I is

$$N(S, I) = \int_0^{r(I)} \rho(I) \left(\frac{\sin kr}{k} \right)^2 \mathrm{d}r = \frac{\rho(I)}{4k^3} (2kr - \sin 2kr). \tag{10.60}$$

where $r(I)$ is implicitly given by (10.59). In the Euclidean case

$$N_0(S, I) = \tfrac{1}{2} \rho(I) (I/S)^{3/2}. \tag{10.61}$$

Using (10.59), we obtain

$$N/N_0 = \frac{3}{4} \frac{2kr - \sin 2kr}{\sin^3 kr (1+z)^{\frac{3}{2}(1+\alpha)}}. \tag{10.62}$$

For every realistic choice of α the value of N/N_0 is smaller than unity. In all models in which the source number per unit of volume is constant, $N(S)/N(S_0)$ should decrease monotonically with decreasing apparent luminosity. Observational data contradict this conclusion

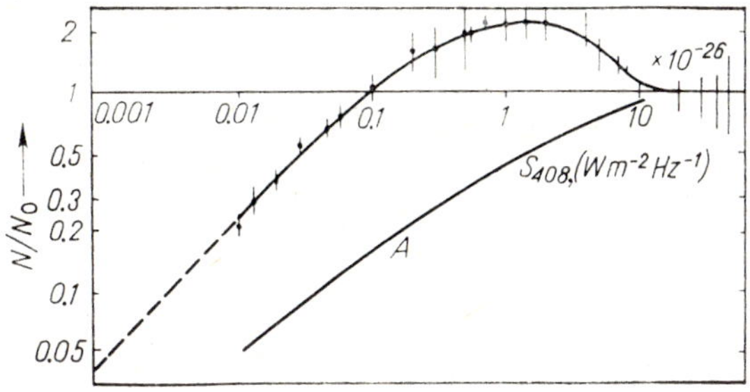

FIG. 10.4. The discrepancy between the observed distribution of radio-sources and the predictions of a typical cosmological model (curve A) is seen more clearly if the data are plotted in the form N/N_0, where N_0 corresponds to a Euclidean universe. The differences indicate that the properties of radio sources change on a cosmological time scale (adapted from Pooley and Ryle, 1968).

(see Fig. 10.4) and can be taken as evidence against "source conserving" models. In particular, the steady state model must be rejected as being contrary to observation.

The source counts seem to imply that the properties of radio sources have changed over a cosmological time scale. No one has succeeded as yet in working out a model of such a slowly evolving radio source.

10.5. Microwave Background Radiation

In 1965, Penzias and Wilson, using a large horn reflector antenna operating at a wavelength of 7.4 cm, discovered isotropic radiation of intensity corresponding to black-body radiation at temperature 3.5±1.0 K (Wilson, 1979). Since then, all ground-based measurements in the wavelength range from 75 cm to 8 mm confirmed the existence of 3 K background radiation. All of these data refer to the Rayleigh–Jeans part of the spectrum, which is well approximated by the relation $I(\nu)\sim\nu^2$. However, two measurements carried out at a wavelength of 3.3 mm depart from $I(\nu)\sim\nu^2$ and provide strong evidence that the actual distribution of the background is consistent with a black-body spectrum. The results of the measurements are shown in Fig. 10.5.

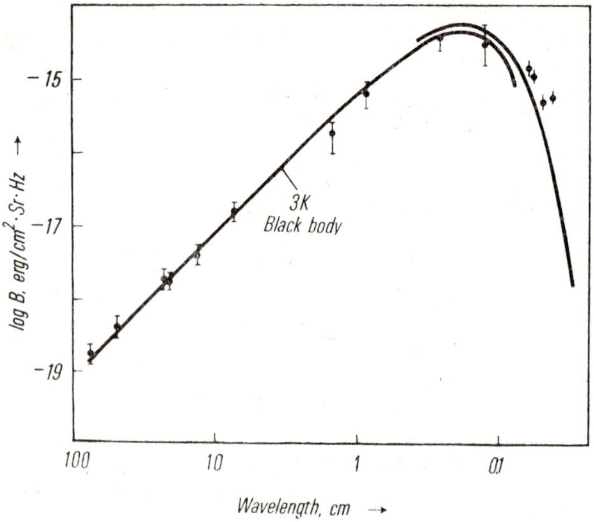

FIG. 10.5. The observed spectrum of relic radiation. Measurements at the short-wave part of the spectrum confirm the Planck character of the distribution (adapted from Wilson, 1979).

The existence of black-body background radiation was predicted in 1948 by Gamow, and in the same year Alpher and Herman estimated the present radiation temperature to be 5 K; a remarkable prediction indeed. Hoping to explain the observed abundance of elements in the universe, Gamow considered an expanding cosmological model and assumed that matter was initially very hot and very dense. There was a period in the evolution of the universe when matter and radiation were in very good thermal contact. Radiation has since undergone thermalization and its spectrum now corresponds to black-body radiation and is isotropic and unpolarized. We call it relic radiation.

Observational data imply that the universe expands. In what way does this expansion affect the properties of relic radiation? Our earlier discussion of light propagation (Sec. 10.2) showed that the photon wavelength measured by an observer comoving with matter increases proportionally to $R(t)$. Qualitatively, this result is quite clear if we remember that a photon moving away from the point of emission bypasses observers moving with increasing escape velocities, and from the point of view of those observers the photon wavelength will become increasingly red-shifted. More precisely, we can write

$$\frac{\lambda_0}{\lambda_e}=\frac{R(t_0)}{R(t_e)}, \tag{10.63}$$

where λ_e is the photon wavelength at the moment of emission. Now suppose that at a certain moment t_e the universe was filled with blackbody radiation

$$I(\nu)=\frac{2h\nu^3}{c^2\left(\exp\frac{h\nu}{kT}-1\right)}. \tag{10.64}$$

If we choose the scale $R(t)$ so that $R(t_e)=1$, the frequency ν_0 equals $\frac{\nu_e}{R(t_0)}$. When there is no radiation–matter interaction, the number of photons is conserved, i.e.

$$\frac{I(\nu_0)\,d\nu_0\,R^3(t_0)}{h\nu_0}=\frac{I(\nu_e)\,d\nu_e}{h\nu_e}. \tag{10.65}$$

Therefore

$$I(\nu_0)=\frac{\nu_0\,d\nu_e}{R^3(t_0)\,d\nu_0}I(\nu_e)=\frac{2h\nu_0^3}{c^2\left(\exp\frac{h\nu_0 R(t_0)}{kT_e}-1\right)}, \tag{10.66}$$

which corresponds to black-body radiation with temperature

$$T_0=\frac{T_e}{R(t_0)}. \tag{10.67}$$

Thus, if radiation does not interact with matter, the Planck distribution is maintained during the expansion of the universe (Dicke *et al.*, 1965; Peebles, 1971; Weinberg, 1972). The radiation spectrum is shifted towards longer and longer wavelengths in proportion to $R(t)$. In the earlier epochs the relic radiation temperature was higher, for example when R was 10^{-3} of the present value, i.e. when $z=1000$, the radiation temperature was 3000 K. That is very important, for then there was an abundance of photons with energy $h\nu>13.6$ eV, able to photoionize hydrogen atoms. Plasma was the main component of matter, and owing to Compton collisions matter and radiation were in thermodynamic equilibrium.

The problem now is to find the function $T(R)$. As long as the radiation pressure $\frac{1}{3}aT^4$ is much higher than the gas pressure nkT (n is the baryon number density), the energy transfer between matter and radiation is negligibly small. We can then use the adiabatic expansion law

$$d(aT^4R^3)+\tfrac{1}{3}aT^4d(R^3)=0, \tag{10.68}$$

which implies that TR=const. Relation (10.68) also applies to epochs when the radiation energy was dominant.

An observational confirmation that relic radiation is Planck-distributed became a matter of fundamental importance. Radiometer measurements at a wavelength of 3.3 mm showed departures from the Rayleigh–Jeans law $I(\nu)\sim\nu^2$ in favour of black-body spectrum, but this alone was not yet a convincing argument. Other independent evidence for the existence of a characteristic Planck fall-off in the spectrum has been furnished by observations of absorption lines of interstellar molecules.

It was known in the early 1940's that a considerable proportion of cyanogen (CN) molecules present in interstellar clouds are in the first rotational state. For a long time this observational fact puzzled astronomers. Interstellar atoms do have sufficient energy for these states to be excited in collision with ions, electrons, or atoms, but, at the density appropriate to the interstellar medium, collisions are so rare that the particles have enough time to return to the ground state, emitting photons at a wavelength of 0.26 cm. It was estimated that the collisions could account merely for one hundredth of the observed excitations; most molecules should therefore occur in the ground rotational state if there were no other excitation mechanism (McKellar, 1941).

The discovery of relic radiation suggested that the excitations in the spectrum of the star ζ Ophiuchi are caused by relic radiation photons at wavelength of 0.26 cm. It was found later that similar absorption lines also occur in other interstellar clouds. The degree of excitation is very similar in all these cases. Since the clouds in question are separated from one another by hundreds of light years, the excitation mechanism is unlikely to be local. Various

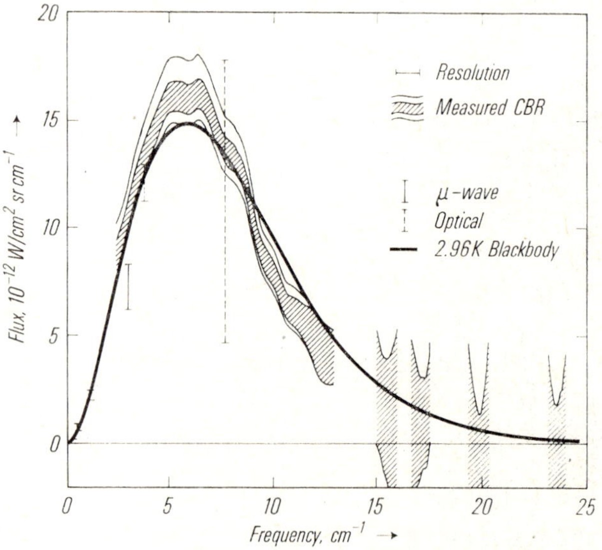

FIG. 10.6. Measured spectrum of the cosmic background radiation. The shaded region includes contributions to the error which are uncorrelated across the spectrum. Selected microwave and optical measurements are also shown. The 2.96 K blackbody spectrum is plotted for comparison. Gaps in the data correspond to frequencies at which the atmospheric contribution is very large (adapted from Woody and Richards, 1979.)

excitation mechanisms were proposed, for example fluorescence, but none of them except relic radiation could explain the observed degree of excitation.

Measurements of the relative intensity of the absorption lines indicate that the CN molecules are in a heat bath of temperature $T=2.83\pm0.15$ K. Studies of other absorption lines in the optical spectrum of CN, CH and CH^+ molecules present in the interstellar space gave no precise radiation temperature but allowed upper limits to be set on the temperature at several short wavelengths (Fig. 10.5). The limits confirm the Planck bending of the spectrum (at $\lambda = 1.32$ mm, $T_0 < 4.74$ K; at $\lambda = 0.56$ mm, $T_0 < 5.43$ K; and at $\lambda = 0.36$ mm, $T_0 < 8.8$ K) (Field, 1969; Thaddeus, 1972).

Measurements in the millimeter band were also conducted with the use of balloons and rockets. For a time, the data acquired in this way were inconsistent with the results of observation of interstellar gas molecules. Discrepancies were large. However, the latest measurements carried out by Woody and Richards (1979) by means of a balloon-borne spectrophotometer at an altitude of 41 km have confirmed the Planck pattern of relic radiation at wavelengths 0.25–5 mm (Fig. 10.6). For the first time information has been obtained about the short-wave region of the spectrum, $\lambda < 0.5$ mm. The measurements of Woody and Richards give the present relic radiation temperature as $2.96^{+0.04}_{-0.06}$ K and locate the maximum intensity at $\lambda_{max} = 1.66$ mm. Earlier measurements gave somewhat lower temperatures (Danese and De Zotti, 1978). Also, small departures have been noticed from the Planck distribution: in the maximum intensity region, $\lambda \approx 1.66$, the observed intensity is about 10% higher and at $\lambda \approx 0.9$ mm about 20% lower than for a Planck distribution. These results are very important as a source of information about processes that occurred during recombination and subsequent periods.

The observed relic radiation is almost exactly isotropic and there are no significant departures from isotropy on any size-scale. The upper limit to anisotropy of the radiation temperature does not exceed $\Delta T/T < 10^{-3}$ on any scale (Boynton and Partridge, 1973; Partridge, 1979). Recent measurements carried out by Pariski sharpened this restriction considerably. For fluctuations on scales 5′ to 3° Pariski (1973, 1978) obtained $\Delta T/K < 8\cdot 10^{-5} - 1.3\cdot 10^{-5}$; there are, however, some reservations regarding this result (Partridge, 1979).

The existence of relic radiation and its isotropy allow us to distinguish a unique reference frame in which relic radiation is isotropic. Suppose that the velocity of an Earth-bound observer relative to this fundamental frame is v. If the observer measures the background temperature at an angle θ to his velocity, he will find that

$$T(\theta) = T_0 \frac{\left(1 - \dfrac{v^2}{c^2}\right)^{\frac{1}{2}}}{1 - \dfrac{v}{c}\cos\theta} \simeq T_0\left(1 + \frac{v}{c}\cos\theta\right), \tag{10.69}$$

provided $v/c \ll 1$. It follows that the temperature should change with periods of 24 and 12 hours. The 24-hr variation is a consequence of the 24-hr-motion of the Earth, whereas the 12-hr variation may reflect the anisotropy of expansion (Peebles 1971). Measurements of the anisotropy are summarized in Table 10.1. The 24-hr motion of the Earth is a vector sum of the Earth's motion with respect to the Sun, the Sun's motion with respect to the centre of the Galaxy and the Galaxy's motion in the local group of galaxies. Observations show that the projection of the resultant velocity on the plane of observations is about 400 km/s. According to the latest data given by Smoot, Gorenstein and Muller (1977), the Earth moves at a velocity of 390 ± 60 km/s towards the Leo constellation and the Galaxy at about

600 km/s towards the Hydra constellation. The peculiar velocity of our Galaxy is unexpectedly high.

The absence of a detectable anisotropy with a 12-hr period is an evidence that at the time when recombination took place the universe was expanding isotropically. Relic radiation furnishes here the most convincing argument for large-scale isotropy of the universe.

TABLE 10.1. *Measurements of anisotropies in relic radiation*

Place Year	Authors	Declination	λ cm	Amplitude 10^{-3} K	Right ascension of maximum
			24-hr anisotropy		
Princeton 1967	Partridge Wilkinson	−8	3.2	2.2±1.8	17^h
Yuma 1968	Dismukes Wilkinson	0	3.2	2.2±2.1	2^h
	Partridge	42	3.2	1.5±2.7	8^h
White Mt. 1972	Conklin	32	3.8	2.3±0.9	11^h
Princeton 1971	Boughn Fram Partridge	0	0.86	7.5±11.6	6^h
Texas 1971	Henry	−30	2.9	3.2±0.8	$10–11^h$
Los Alamos 1971	Beery Wilkinson Partridge	0	3.2	0.7±1.2	16^h
			12-hr anisotropy		
Princeton 1967	Partridge Wilkinson	−8	3.2	2.7±1.9	7^h, 19^h
Yuma 1968	Dismukes Wilkinson	0	3.2	2.1±2.0	5^h, 17^h
	Partridge	42	3.2	4.0±2.4	8^h, 20^h
White Mt. 1972	Conklin	32	3.8	1.35±0.8	6^h, 18^h
Princeton	Boughn Fram Partridge	0	0.86	5.5±6.6	0^h, 12^h
Los Alamos 1968	Beery Wilkinson Partridge	0	3.2	1.9±1.2	9^h, 21^h

It is generally believed that galaxies and groups of galaxies were formed as a result of growth of density perturbations, whose amplitude in the recombination period was very small. Such perturbations should have had an effect on the isotropy of relic radiation, giving rise to local anisotropies. The observed limits on local fluctuations of the relic radiation temperature provide information about the amplitude of density perturbations in the recombination period (Zeldovich, 1979; Wilson and Silk, 1981; Sunyaev and Zeldovich, 1980).

The high degree of isotropy of relic radiation and the Planck character of its spectrum are the most convincing arguments in support of the so-called hot big bang model of the universe (Peebles, 1971; Weinberg, 1972; Zeldovich and Novikov, 1983). According to this model, at the initial stages of the evolution of the universe matter was very dense and very hot. The thermal radiation dominated the very early evolution of the universe and, as the universe expanded adiabatically, cooled down to a temperature of 3 K; this is why we call it relic radiation.

An analysis of the physical processes occurring in such a hot universe (Danese and De-Zotti 1977, Zeldovich 1979), which we present in detail in the following sections, leads to the conclusion that the observed distribution of relic radiation could not have emerged later than at $z \sim 10^5$, $t \sim 3 \cdot 10^9$ s ($T \sim 3 \cdot 10^5$ K). In a cold model of the universe, a release of a sufficiently large amount of energy at $z > 10^5$ would also have given rise to thermal radiation. If this release had occurred in later epochs, the produced photons would not have had enough time to thermalize (Zeldovich and Sunyaev, 1969; Zeldovich *et al.*, 1972).

Finally, one may ask whether relic radiation is of any astrophysical significance at the present time. If the cosmic rays are of extragalactic origin, then, for example, very energetic cosmic ray protons can lose part of their energy in collisions with the relic radiation photons. For an observer moving with a cosmic proton the background photons carry considerable energy, which may exceed the 135 MeV threshold for pion production. This will happen when the energy of the proton is greater than $E_{cr} = 10^{20}$ eV. The effective cross-section for pion production reaches a maximum at 300 MeV, which corresponds to a proton energy of $3 \cdot 10^{20}$ eV. Thus the energy spectrum of the cosmic ray protons should have a sharp cutoff at energies above 10^{20} eV.

Similarly, inverse Compton scattering of high-energy cosmic ray electrons on the background photons may produce X-rays. This radiation should be isotropic and contribute to the X-ray component of the radiation background.

10.6. The Hot Big Bang Model of the Universe

The discovery of the microwave background radiation furnished an additional argument for the thesis that at the early stages of the evolution the matter filling the universe was very hot. In fact, the view that the initial state was very dense and very hot dates to Hubble's discovery of universal expansion. In 1948, Gamow tried to explain the origin of the light elements, starting from the assumption that in early times the temperature of matter was very high (Alpher and Herman, 1949). This view, however, did not gain general acceptance; it was rejected by the followers of the steady state theory, who denied the evolution of the universe in general. The observational data available at present rule out the steady state model and, what is more, clearly speak in favour of the hot big bang model.

One of the arguments for a hot and dense origin of the universe is the observed number of photons per particle (baryon) (Novikov and Zeldovich, 1967, 1973; Zeldovich and Novikov, 1983; Harrison, 1973). The entropy density s of radiation is given by the formula

$$s = \frac{4}{3} a T^3, \tag{10.70}$$

where $a = \frac{4\sigma}{c}$ and σ is the Stefan–Boltzmann constant. As a result of adiabatic expansion the temperature T changes as R^{-1}; therefore $s \sim R^{-3}$, but the particle number density n_b changes likewise, which implies that the ratio s/n_b is constant. For convenience, instead of using entropy we shall introduce a dimensionless quantity $\Sigma = s/k$, where k is the Boltzmann constant; thus $\Sigma/n_b = \text{const}$. At the present stage of the evolution of the universe

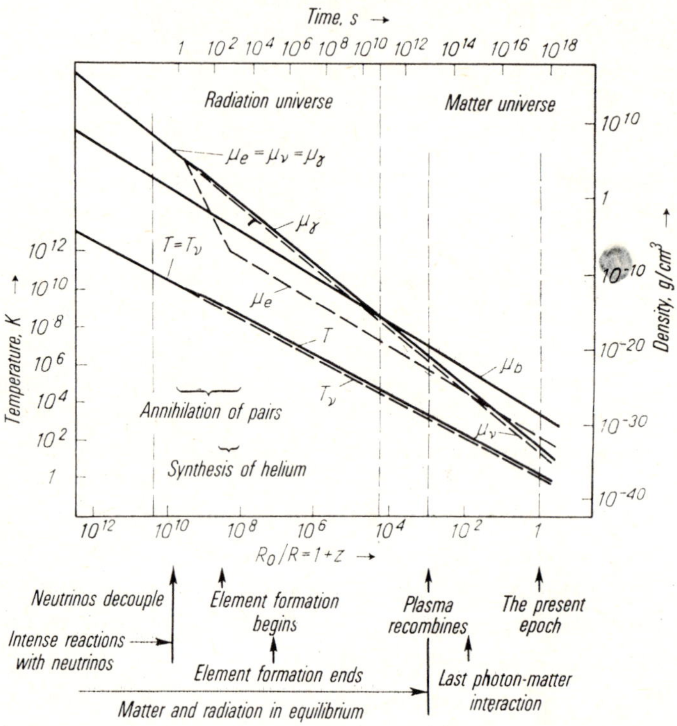

FIG. 10.7. Thermal history of the universe in the big bang model (adapted from Kundt, 1968).

$T_0 = 3$ K and $n_b \approx 10^{-6}$ cm^{-3}, so $\Sigma/n_b \sim 10^9$. Note that the photon number density n_γ is proportional to the third power of the temperature and therefore the ratio n_γ/n_b should also be constant. Since $n_\gamma = 20T^3$, we obtain for the present epoch $n_\gamma/n_b \approx 10^9$. The number of photons in the universe is thus much larger than the number of baryons.

If the present temperature of the relic radiation is 3 K, then from formula (10.67), which can be rewritten as

$$T_0 = \frac{T}{1+z}, \tag{10.71}$$

we conclude that in earlier epochs the temperature was very high. In the previous section we noted that this formula also applies to epochs in which matter and radiation were in thermal contact.

The general picture of the evolution is described below (Zeldovich, 1967, 1979; Kundt, 1971; Weinberg, 1972; Harrison, 1973; Zeldovich and Novikov, 1983).

Very little is known about the initial state of the universe. For the present we shall content ourselves with stating that at that time matter was very dense and very hot, and the dynamics of expansion of the universe was determined by the energy density of radiation. When the temperature dropped to 10^{13} K as a result of the expansion, a whole group of hadrons annihilated and matter consisted mainly of leptons, antileptons and photons, and also of nucleons and a decreasing number of pions. The lepton era began.

Muon pairs annihilated when the temperature dropped below $\frac{m_\mu c^2}{k} \approx 8 \cdot 10^{11}$ K. Soon after that the density decreased so much that muon and electron neutrinos practically ceased to interact with matter. Electrons, positrons and photons continued in thermal equilibrium until the temperature dropped to $\frac{m_e c^2}{k} \approx 4 \cdot 10^9$ K. Then electron pairs annihilated and the universe entered the radiation era.

The radiation era lasted until the temperature dropped to 3000 K and hydrogen recombination, which began at $T \approx 4000$ K, was completed. Throughout most of the radiation era, the radiation energy density was greater than matter density. As follows from formula (10.71), the former decreases as $(1+z)^{-4}$ and the latter as $(1+z)^{-3}$. Depending on the initial conditions, by the end of the radiation era matter density might exceed radiation density. From that moment the dynamics of the expansion of the universe was governed by the density of matter. The radiation era did not necessarily end when matter assumed the dominant role: radiation could continue to be the main source of pressure in the universe. At the beginning of the radiation era helium and other light elements were synthesized. During the radiation era matter was in a state of completely ionized plasma, and owing to Compton scattering was in continuous thermal contact with radiation, so that the plasma and the radiation had the same temperature.

After the recombination follows the era of galaxy formation. Radiation no longer has any effect on the dynamics of expansion, and its contribution to pressure is practically nil. Galactic clusters, galaxies and stars emerge. In stellar interiors the light elements combine to form heavy ones. Galactic clusters and galaxies undergo slow evolution. If the universe is open, the expansion will continue for ever and the average density of matter will decrease to zero (Dyson, 1979). In the closed model, the universe, after reaching some maximum dimension, will begin to contract gradually and the whole cycle will be repeated approximately in the reversed order.

10.7. The Problem of Initial Conditions. The Hadron Era

The initial state of the universe cannot be described by means of classical general-relativistic equations. Consider three quantities that can be formed from the universal constants G, $\hbar$ and c: one having the dimension of length, $l_p = (G\hbar c^{-3})^{\frac{1}{2}} = 1.7 \cdot 10^{-33}$ cm, one having the dimension of time $t_p = (G\hbar/c^5)^{\frac{1}{2}} = 5 \cdot 3 \cdot 10^{-44}$ s, and one having the dimension of mass density†, $\rho_p = c^5/\hbar G^2 \approx 5 \cdot 10^{93}$ g/cm^3. When the curvature of space is less than l_p or the density is greater than ρ_p, or both, quantum effects become essential, and we simply do not know how to describe the properties of space-time.

† Quantity having the dimension of mass can also be formed; it is called Planck mass $m_p = (\hbar c/G)^{1/2} = 22 \cdot 10^{-5}$ g.

The only way out of this situation is to assume some initial conditions near the singular state. There are two extreme approaches to the problem. In the first, it is supposed that the initial state of the universe was completely chaotic and that evolution has smoothed out the original inhomogeneities, producing in effect the distribution of matter as it is observed now (Misner, 1967a, b, 1969; Barrow, 1977). According to the second theory, the initial distribution of matter was homogeneous, and the inhomogeneities observed at present are the result of growth of small perturbations whose origin can be traced back to the initial state (Zeldovich and Novikov, 1983).

On the basis of the existing observational data it is difficult to decide which of these two approaches is right. The high degree of isotropy of relic radiation and observed abundance of primordial light elements merely indicates that even if the initial state was completely chaotic, inhomogeneities must have decayed comparatively fast. In the sequel we shall assume that the initial state of the universe was homogeneous, the main reason for adopting this approach being its simplicity.

We begin our considerations from the moment when particles come on the scene. Quantum mechanics sets restrictions on the size of particles. To a particle of mass m corresponds the Compton wavelength h/mc. To be able to speak of particles, one must wait until the horizon becomes greater than a hadron diameter, i.e. until $h/m_p c = ct_h$ or

$$t_h = \frac{\hbar}{m_p c^2} \approx 10^{-23}\ \mathrm{s}. \tag{10.72}$$

Note that t_h is a typical time of strong interactions.

No one has yet succeeded in giving a consistent theory of processes that occurred in the hadron era. Two attempts deserve to be mentioned. Assuming that the particle mass spectrum is exponential, Hagedorn (1965) derived an equation of state which implies that there exists an absolute maximum to temperature: $T_0 \approx 1.8 \cdot 10^{12}$ K. Carlitz, Frautschi and Nahm (1973) modified Hagedorn's equation and used it to describe the processes of the hadron era. In their theory most particles move with non-relativistic velocities and with kinetic energies of order kT. The time scales of decay and diffusion are comparable to the expansion time-scale, and therefore the system never achieves thermodynamic equilibrium. The matter is composed mainly of large-mass resonances. The hadron era is longer than in Hagedorn's model and the inhomogeneities in the distribution of matter generated in that time are likely to survive the radiation era and to give rise to galaxies and clusters of galaxies.

Both in deriving the equation of state and in their subsequent considerations the authors made several speculative assumptions, and although the idea of relating the observed distribution of matter to the processes that occurred at the initial stages of evolution of the universe is attractive, the models cannot be taken seriously in all their details.

Owing to the latest achievements of elementary particle physics we are now able to say a good deal more about the hadron era (see for example Greenberg, 1978; Weinberg, 1979; Turner and Schramm, 1979; Wagoner, 1979; Steigman, 1979; Hu, 1980; Dolgov and Zeldovich, 1980, Ellis *et al.*, 1980., Kolb and Wolfram, 1980). It has turned out that hadrons consist of still more elementary components: quarks. The successes of the unified theory of the weak and electromagnetic interactions (the Weinberg–Salam model) have stimulated wide interest in the unification program of strong, weak and electromagnetic interactions. Both in the Weinberg–Salam model and in Grand Unified Theories it is postulated that there exist particles much heavier than those which the most powerful accelerators of the present day can produce. This means that a direct experimental verification of these theories will not be possible in the immediate future. Nevertheless, it is certainly worthwhile to investigate how they can contribute to our understanding of the processes that occurred in very early times. An idea of Weinberg (1979), subsequently developed by Turner and Schramm (1979), deserves particular attention. Using the fact that in Grand Unification Theories the baryon number is not conserved, Weinberg points out that departures from thermal equilibrium at early stages of the evolution could have led to the present baryon number density and the observed photon to baryon number density ratio. The idea is strongly appealing, for, if accepted, it provides a clue to one of the fundamental problems of cosmology, namely that of the origin of the excess of matter over antimatter (excess of baryons over antibaryons).

10.8. The Lepton Era

Beginning at the lepton era we can use conventional laws of physics to describe physical processes. For example, we can determine the neutron-proton abundance ratio. At temperatures above $2 \cdot 10^{10}$ K, neutrons and protons can transform into one another, via the reactions

$$\begin{gathered} e^{+}+n \leftrightarrow p+\tilde{\nu}_{e}, \\ \nu_{e}+n \leftrightarrow p+e^{-}. \end{gathered} \tag{10.73}$$

Since the characteristic time for these reactions is short compared with the expansion time-scale, a state of thermal equilibrium is established between neutrons and protons, with the neutron–proton ratio

$$\frac{n_n}{n_p} = e^{-\frac{\Delta mc^2}{kT}}, \tag{10.74}$$

where $\Delta m = m_n - m_p$. By the end of the lepton era reactions (10.73) slow down considerably and they cannot sustain thermal equilibrium between neutrons, protons and leptons (Hayashi, 1950; Peebles, 1966; Weinberg, 1972). The relative number density of neutrons to protons becomes frozen at the decoupling value and it is slowly changing due to decay of neutrons. In the first approximation this process can be neglected since the lifetime of neutrons is much larger than the characteristic time of expansion. At the time of decoupling neutrons constitute about 16% and protons about 84% of nucleons.

The main components of matter in the lepton era are photons, electrons and positrons, muons, electron and muon neutrinos, and, as has been shown recently, also heavier leptons and their associated neutrinos. All these particles are in thermal equilibrium, which is maintained owing to the fact that the reactions

$$\begin{aligned}&\gamma+\gamma\leftrightarrow\mu^{+}+\mu^{-}\leftrightarrow\nu_{\mu}+\tilde{\nu}_{\mu}\leftrightarrow\gamma+\gamma,\\&\gamma+\gamma\leftrightarrow e^{+}+e^{-}\leftrightarrow\nu_{e}+\tilde{\nu}_{e}\leftrightarrow\gamma+\gamma\\&\gamma+\gamma\leftrightarrow\tau^{+}+\tau^{-}\leftrightarrow\nu_{\tau}+\tilde{\nu}_{\tau}\leftrightarrow\gamma+\gamma\end{aligned}\tag{10.75}$$

and reactions between leptons, such as

$$\begin{aligned}&e^{+}+\mu^{-}\leftrightarrow\tilde{\nu}_{e}+\nu_{\mu},\quad e^{-}+\mu^{+}\leftrightarrow\nu_{e}+\tilde{\nu}_{\mu},\quad e^{+}+\tau^{-}\leftrightarrow\tilde{\nu}_{e}+\nu_{\tau},\\&e^{+}+\nu_{e}\leftrightarrow\mu^{+}+\nu_{\mu},\quad e^{-}+\tilde{\nu}_{e}\leftrightarrow\mu^{-}+\tilde{\nu}_{\mu},\quad e^{+}+\nu_{e}\leftrightarrow\tau^{+}+\nu_{\tau}\end{aligned}\tag{10.76}$$

proceed much faster than the expansion.

At a temperature of 10^{12} K the muon pairs annihilate (the τ-pairs annihilate earlier) and as the universe continues to expand and the temperature drops to $5\cdot10^{9}$ K, annihilation of electron pairs takes place.

The characteristic times of electron-neutrino and muon-neutrino reactions are

$$\tau_{e\nu}^{-1}=n_{e}c\sigma_{e\nu},\qquad \tau_{\mu\nu}^{-1}=n_{\mu}c\sigma_{\mu\nu},\tag{10.77}$$

where $\sigma_{e\nu}$ and $\sigma_{\mu\nu}$ denote the corresponding effective cross-sections. Both $\tau_{e\nu}$ and $\tau_{\mu\nu}$ increase quickly as the universe expands, and when they become greater than the characteristic time of expansion $\tau=H^{-1}$, the reactions cannot longer sustain thermal equilibrium and neutrinos decouple. Due to expansion they are adiabatically cooled and therefore after decoupling both kinds of neutrinos have nearly the same temperature.

Annihilation of electron pairs is accompanied by production of photons. This supply of photons keeps the temperature of the photon gas above that of the neutrino gas.

To find the relation between the neutrino gas and the photon gas temperatures, note that before the annihilation of pairs the two temperatures are equal. When the annihilation takes place, the temperature of the neutrino gas remains unchanged whereas the temperature of the photon gas increases. To avoid the complications of an exact analysis, let us use a simplified model, in which the annihilation is assumed to occur instantaneously, so that the photon gas temperature undergoes a sudden jump. Before the annihilation, the major part of the total entropy is contributed by the photons, electrons, and positrons; furthermore, the electrons and positrons are relativistic. A relativistic electron gas in thermal equilibrium with radiation can be described by the equation of state

$$\varepsilon=\frac{7}{4}aT^{4},\qquad p=\frac{1}{3}\varepsilon,\tag{10.78}$$

where ε is the energy density. The entropy density of the electron gas and the radiation is then given by

$$s=\frac{4}{3T}(\varepsilon_{\gamma}+\varepsilon_{e})=\frac{11}{3}aT^{3}.\tag{10.79}$$

After the annihilation of pairs, photons and neutrinos are almost the only particles left, and the entropy density of photons becomes

$$s=\frac{4}{3}aT^3. \tag{10.80}$$

Before annihilation, the neutrino and the radiation temperatures changed in the same way. The annihilation increases the photon temperature, but the entropy remains the same, and so

$$\frac{11}{3}aT_\nu^3=\frac{4}{3}aT^3. \tag{10.81}$$

Hence

$$T_\nu=\left(\frac{4}{11}\right)^{1/3}T=0.7\,T, \tag{10.82}$$

which implies that at the present era, when $T=3$ K, the relic neutrino background should have a temperature of 2 K. The present methods of detection of neutrinos are not yet sensitive enough to provide a direct confirmation of the existence of a relic neutrino background.

10.9. The Radiation Era. Formation of Light Elements

Annihilation of the electron pairs marks the end of the lepton era. There begins a long era during which matter consisting mainly of protons, neutrons and electrons, which have to be present to maintain charge neutrality, remains in contact with radiation. This coupling between matter and radiation is due to the effectiveness of the Compton scattering. The dynamics of expansion is still determined by the radiation density.

From the energy conservation law

$$\frac{\mathrm{d}}{\mathrm{d}t}(\varepsilon R^3)+p\frac{\mathrm{d}}{\mathrm{d}t}R^3=0 \tag{10.83}$$

and from the equation of state $p=\frac{1}{3}\varepsilon$ we obtain in this case

$$\frac{\mathrm{d}}{\mathrm{d}t}(\varepsilon R^4)=0, \tag{10.84}$$

and since $\varepsilon=aT^4$, we get $T\sim R^{-1}$. To find the time-dependence of the temperature, we have to solve the system of Friedman's equations (10.10) and (10.11). It can easily be verified that when the dynamics of expansion is determined by the density of radiation, $R\sim t^{1/2}$ for small t, irrespective of the curvature of space. Then $T\sim t^{-1/2}$ and the expansion time-scale $\tau=H^{-1}\sim t\sim T^{-2}$.

The radiation era corresponds to a temperature interval 13.6 eV $<kT<m_ec^2=0.5$ MeV. At the beginning, the expansion time-scale is shorter than the lifetime of free neutrons ($t_n=935$ s), and the number of neutrons is maintained at the equilibrium level appropriate to the moment of decoupling of electron neutrinos. When the temperature falls to $2.5\cdot10^9$ K,

the reactions which have kept the neutron–proton ratio in equilibrium become so slow that the equilibrium is upset and the reaction

$$n \to p + e^- + \tilde{\nu} \tag{10.85}$$

begins to emerge as a dominant one. However, every free neutron is quickly captured by a proton to form deuterium. At temperatures higher than 10^9 K the deuterium is rapidly photodisintegrated. The characteristic time of rapid interactions is much shorter than the lifetime of free neutrons, and so the number of neutrons remains near the equilibrium level, $n_n = 0.19\, n_p$. At temperatures below 10^9 K, however, when violent photodisintegration of deuterium stops, the amount of deuterium is sufficient for further reactions leading to the formation of helium to occur. One such possible chain of reactions is

$$D(D, n)^3\mathrm{He}(n, p)\,T\,(p, \gamma)^4\mathrm{He}. \tag{10.86}$$

When the continuing expansion reduces the temperature to below 10^8 K, the kinetic energies of the particles become much lower than the treshold energies of the reactions and the composition of matter is in effect "frozen in".

Assuming now very roughly that every two neutrons produce a helium nucleus, we find that the expected primordial helium abundance $X(^4\mathrm{He})$ is

$$X(^4\mathrm{He}) = \frac{2n_n}{n_n + n_p} = 0.32. \tag{10.87}$$

Thus by very general quantitative estimates we have come to the conclusion that pregalactic matter should contain about 32% helium.

A very detailed analysis of nucleosynthesis in the early universe was carried out by Wagoner, Fowler and Hoyle (1967). They considered 144 reactions between the lightest elements, up to neon. These calculations have been repeated by Wagoner (1973) for the light elements up to oxygen, with the use of the latest data on effective cross-sections. The analysis starts from the point where the temperature is 10^{11} K, at which the relative abundance of the elements is determined by thermal equilibrium.

The reactions taken into account are presented in Fig. 10.8. These very complex calculations have led to the following conclusions: In all isotropic cosmological models, in which $0.01 \leqslant \Omega \leqslant 10$, the helium abundance is 20–30% by mass. The elements heavier than helium are produced in very small quantities, for example for $\Omega = 0.1$, $X(^2\mathrm{H}) = 1.3 \cdot 10^{-5}$, $X(^3\mathrm{He}) = 1.2 \cdot 10^{-5}$, $X(^6\mathrm{Li}) < 10^{-12}$, $X(^7\mathrm{Li}) = 3.5 \cdot 10^{-9}$, $X(^{11}\mathrm{B}) < 10^{-12}$, $X(A > 12) < 10^{-12}$, while $X(^4\mathrm{He}) = 0.246$. The amount of helium produced varies weakly with Ω, the reason being that the dynamics of expansion and the rates of the processes involving weak interactions are determined by the distribution of radiation energy of which the baryon share is insignificant. Thus, pregalactic matter in the big-bang model consisted mainly of hydrogen and helium, with a helium content of 20 to 30%. Heavier elements are formed in nuclear processes in the interiors of stars.

The present helium abundance in young stars and in the interstellar gas around the Sun is well known. There are basically three different methods by which it can be measured.

The theory of young stellar structures provides a functional relation between the mass, radius, luminosity and chemical composition, in particular the helium content, of a star

(Aller, 1958, 1961). By combining this relation with the observational data it has been found that $X(^4\text{He})=0.25\pm0.05$.

Studying the shape of helium absorption lines in the spectra of the atmospheres of young hot stars is another method of determining $X(^4\text{He})$. The earliest attempts to measure

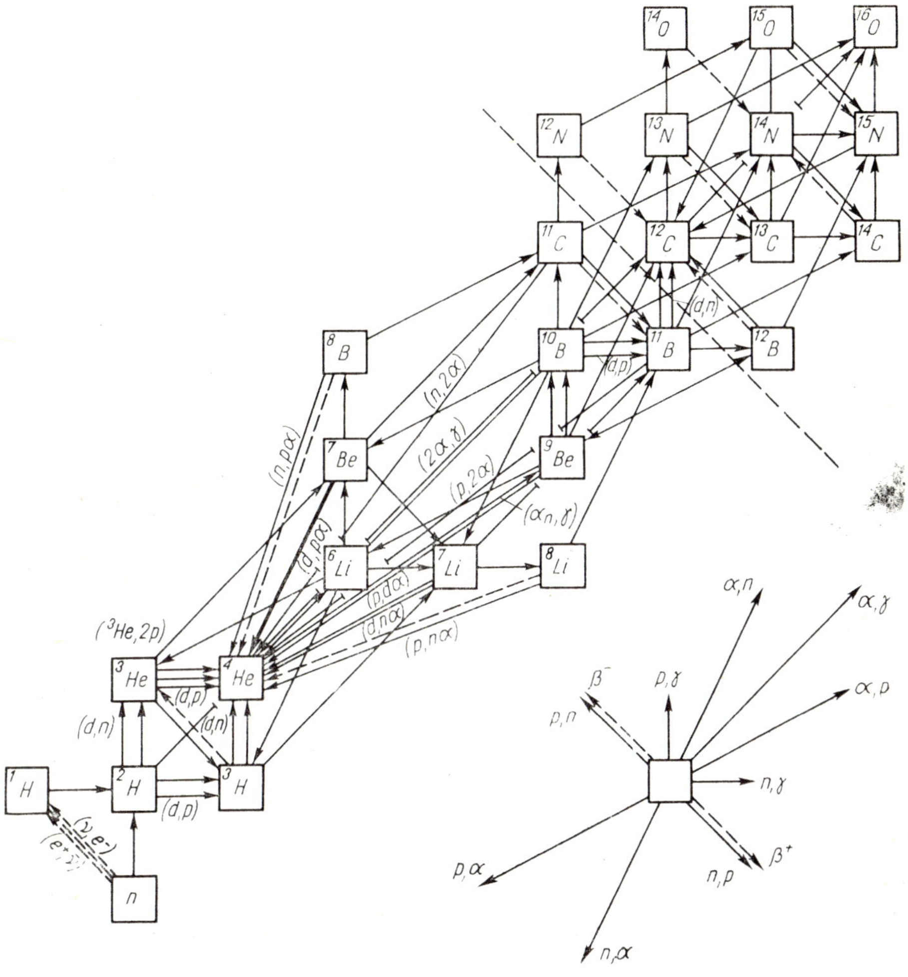

FIG. 10.8. Nuclear reactions in the Wagoner–Fowler–Hoyle analysis of the synthesis of the light elements in the hot big bang model of the universe (adapted from Wagoner, 1973).

the helium content by using this method were based on a very approximate theory of spectral line formation and the results were uncertain. A considerable theoretical advance made in the recent years permitted new, more precise estimations, which yielded $X(^4\text{He})=0.30 \pm0.04$. However, there still remains the problem of how to interpret these results. It is believed that they reflect the composition of the matter from which a given star was formed. The excess in the helium content could have arisen from the nuclear processes occurring in the star. It is more difficult to explain the cases where the helium content is below 25%.

The most drastic example is the star 3 Cen A. At its surface $X(^4He)=0.04$, and the helium is found mainly in the form ^3He. There are also other stars in the neighbourhood of the Sun whose chemical composition is anomalous. Their existence is a warning that caution

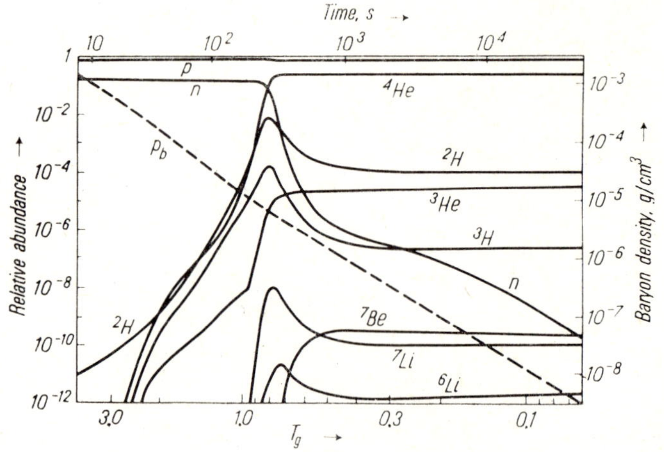

FIG. 10.9. Evolution of the relative abundance of some elements. After 10^3 s the helium to hydrogen ratio becomes practically constant (adapted from Wagoner, 1973).

must be used when drawing conclusions about the initial chemical composition of a star from the chemical composition of its atmosphere.

The helium abundance can also be deduced from the relative intensity of emission lines of the ionized gas surrounding young hot stars. The ratio of the He^+ ion number density to the proton number density can be found directly from the intensity ratio of the recombination lines He I and H I. Pairs of these lines can be observed in the optical and the radio parts of the spectrum. Both the observational data and the theoretical predictions agree that in regions of ionized hydrogen surrounding young stars double-ionized helium is scarce and either most of the helium is in the atomic form or almost all atoms are single-ionized. Cases where atomic helium and single-ionized helium appear in comparable quantities are very rare. On averaging the values obtained for a few regions in which optical frequency recombination lines are observed we find $n(He^+)/n(H^+)=0.090\pm0.010$. A similar averaging in radio frequencies leads to $n(He^+)/n(H^+)=0.084\pm0.003$. If the neutral helium content in these regions is negligible, this gives $X(He)=0.26\pm0.01$. It is safer to regard this value as a lower limit, because some small admixture of neutral helium, difficult to estimate, is to be expected here.

All these methods taken together set the relative abundance of helium in the neighbourhood of the Sun within the limits $0.26<X(He)<0.32$ (Danziger, 1970; Searle and Sargent, 1972; Reeves *et al.*, 1973; Trimble, 1975).

The helium content in a galaxy at early stages of its evolution can be estimated from the chemical composition of stars and ionized gas clouds. The existing stellar models allow us to predict evolutionary changes in the luminosity and radius of a given star. For a system of stars with different masses which were formed at the same time the theory predicts the distribution of luminosities and radii in relation to the initial helium content. Comparison of these predictions with observational data makes it possible to determine the age of the system and the initial helium content. All calculations of this kind lead to the conclusion that the initial helium content in the oldest and least metal-contaminated stars is about 30%.

In a seeming conflict with these results are some data obtained by spectroscopic methods. Several old stars have been found to have a much lower helium content. The accuracy of the measurements leaves no doubt that such stars exist; it is not certain, however, whether the composition of their atmospheres is the same as that of the material from which they were formed. All these stars are peculiar as far as their chemical composition is concerned. Their heavy-element content has most certainly changed since early times and there is no reason why their low helium content should be related to the original helium abundance in pregalactic matter. Observations imply that the oldest stars in our Galaxy were formed from matter containing as much helium as the young stars in the neighbourhood of the Sun (Danziger, 1970).

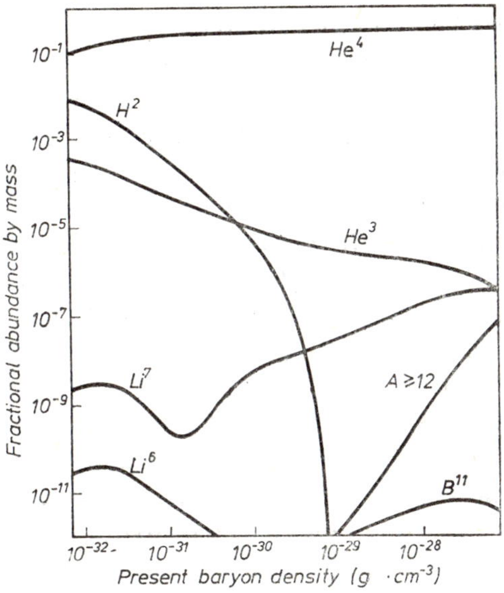

FIG. 10.10. Fractional abundance of leight elements and all elements of mass number 12 and larger as a function of present baryon density (adapted from Schramm and Wagoner, 1974).

The homogeneity of the helium distribution outside our Galaxy can be studied with the use of the emission lines of ionized gas surrounding young stars in spiral galaxies. The method has been used to determine the helium content for a few galaxies of the Local Group and a few external galaxies. In each case the helium content basically agrees with that observed in the neighbourhood of the Sun or in our Galaxy. The measurements give $X(\mathrm{He})=0.28 \pm 0.06$ (Peimbert, 1975).

Searle and Sargent (1972b) have found two galaxies in which the mass of the interstellar gas is ten times the mass of the stars themselves; the galaxies are thus very young. Measurements of their helium content have given $X(\mathrm{He})=0.29$.

The existing data on the helium content cover only a small portion of the universe. No one has yet succeeded in determining the helium content of galaxies which are more than 50 Mpc away from us. Bearing in mind these limitations, we can nevertheless say that the universal helium abundance predicted by the theory of nucleosynthesis in the hot model of the universe is in good agreement with the observational data (Schramm and Wagoner, 1977; Wagoner, 1979).

According to the theory of nucleosynthesis the cosmological production of deuterium strongly depends on the value of Ω (see Fig. 10.10). The evidence pertaining to the deuterium abundance can therefore provide very important information about the value of Ω and thus help to answer the crucial question of whether the universe is open or closed. Some caution is required, however, in interpreting the results of observations, because some of the primordial deuterium could have been consumed in thermonuclear processes in the stars, and new deuterium could have been produced in various processes that occur in galaxies. To measure the deuterium abundance in interstellar gas clouds seems to the be most reasonable proposition, because the deuterium content there should not be drastically different from its primordial abundance. The direct observations of the Lyman spectral lines of atomic deuterium carried out from the Copernicus satellite have so far provided the most reliable information about the abundance of deuterium. According to those observations, the interstellar gas around the Sun contains $X(\mathrm{D})=X(^2\mathrm{H})=2\cdot10^{-5}$ of deuterium by mass. Taking this to be the primordial abundance of deuterium, we obtain the mean density of baryons $\rho_0\leqslant6\cdot10^{-31}$ g/cm^3 (Schramm and Wagoner, 1974, 1977). If the cosmological constant is zero, which we always assume, then $\Omega=1.0$ and $q_0=0.2$. It follows that the universe is open and its age is $1.6\cdot10^{10}$ years (Penzias, 1978). It should be remembered, however, that this result is based on the assumed interpretation of the observed abundance of deuterium and on the vanishing of the cosmological constant. The conclusions which can be drawn from observational data when the cosmological constant is non-zero have been discussed by Gunn and Tinsley (1975), and by Tinsley (1977).

The abundances of light elements in pregalactic matter depend in an essential way on the expansion rate of the universe at the time the elements were formed. Even comparatively small departures from the Friedman model can have a drastic effect on the predicted helium abundance, changing its value to the extremes: 0 or 100% (Partridge, 1979; Barrow, 1976). The very good agreement between the predicted and the observed helium abundance is evidence that during the element production period the universe was almost exactly isotropic (Barrow and Matzner, 1977).

10.10. The Recombination Period

When the temperature drops to 10^8 K the Coulomb barriers prevent further nucleosynthesis. There follows a long and quiet period durig which the hydrogen and helium plasma is in thermal equilibrium with radiation. Radiation energy density falls with time as R^{-4}, while matter density changes as R^{-3}. Although entropy per baryon is very high, matter becomes dynamically important before the recombination process begins. To a good approximation we can assume that this change in the dynamical balance occurs instantaneously. Radiation energy density loses its dynamic significance very quickly, even though the main contribution to pressure will still continue to come from radiation for a time. Owing to

Compton scattering, matter and radiation are kept at the same temperature. The amount of energy transmitted to matter in this process is negligibly small and does not affect the radiation spectrum.

Zeldovich and Sunyaev (1969) (Zeldovich, 1979) asked how the radiation spectrum would change if some of the gas were reheated, as a result of dissipative processes, to a temperature T_g exceeding the radiation temperature T_r. Compton collisions would increase the energy of photons. The frequency of photons would rise by $\Delta\nu \approx \nu v^2/c^2$. The photon energy would show a noticeable change, $\Delta\nu/\nu \gtrsim 1$, if

$$n_e \sigma r_H \left(\frac{mc^2}{kT_g}\right) \gtrsim 1, \tag{10.88}$$

where σ_T is the Thomson cross-section for scattering, r_H is the dimension of the horizon and m the electron mass. If the gas were reheated sufficiently early ($z \gtrsim 10^5$), then, for example, thermal bremsstrahlung could produce such quantities of low-energy photons that the Planck form of the radiation spectrum would be preserved. Distortions would occur if large amounts of energy were released at $z < 10^5$.

When the radiation temperature falls to $T_r = 4000$ K at $z = 1500$, hydrogen recombines forming atoms. The main complication is now that the plasma density is very low and therefore the relations which describe thermodynamic equilibrium cannot be used.

Initially, the main products of the recombination are excited hydrogen atoms. The ionization energy for a hydrogen atom is 13.6 eV, and all the excited states are distributed within an energy span of 3.4 eV. During the recombination period most of the thermal photons have energies within this range. When the temperature is of the order of 4000 K and the recombination rate is maximum, there are about 10^5 thermal photons per nucleon with energies higher than 3.4 eV, but very few with energies higher than the ionization energy. Direct recombination to the ground state is negligibly small. The recombination rate depends on how fast the photon frequencies are red-shifted relative to the resonant frequency, making radiative transitions to the ground state possible, and on the rate of two-photon transitions from the 2s state to the ground state. The large number of resonance and excited levels complicates the problem considerably.

A detailed analysis of the whole process of recombination has been carried out by Peebles (1968) and by Sunyaev and Zeldovich (1970). They obtained the following formula for the degree of ionization as a function of z:

$$X(p) = \frac{n_p}{n_p + n_H} = \frac{A}{\Omega^{\frac{1}{2}} z} e^{-B/z}, \tag{10.89}$$

where $A = 6 \cdot 10^6$ and $B = 1.458 \cdot 10^4$. The formula implies that when $z = 700$, i.e. when $T_r \approx 2000$ K, the recombination is almost completed and $X(p) \approx 10^{-5}$. The plasma is now completely recombined but matter and radiation are still in close thermal contact. Though only one atom in 10^5 is ionized, the remaining electrons continue to be strongly coupled with the thermal background radiation and can effectively pass their energy to neutral particles through collisions. As a result, the gas temperature falls much more slowly than might be expected, and only at about $z = 100$ matter and radiation decouple and undergo independent adiabatic cooling. The gas and the radiation temperatures fall as $(1+z)^2$ and $(1+z)$, respectively. It might be expected, therefore, that the present temperature of the intergalactic gas should be much lower than the relic radiation temperature. However, the observational data show that this is not so. As a result of further evolutionary processes the intergalactic gas gets heated, often to very high temperatures. Limited space does not permit us to look at this problem more closely. It is worthwhile, however, to consider the effect that energy released at later stages of evolution can have on the relic radiation spectrum (Danese and De Zotti, 1977). Let us introduce, after Zeldovich and Sunyaev (1969), a new parameter y, describing the relative increase in photon energy due to Compton collisions with hot intergalactic gas during the period between a given z and now. It is defined as

$$y=\int_{0}^{\tau(z_0)} \frac{kT_g(z)}{mc^2} d\tau, \tag{10.90}$$

where τ is the photon mean free path and $T_g(z)$ is the gas temperature. Zeldovich and Sunyaev show that for small y the effective temperature of the longwave part of the spectrum is decreased from the Planck temperature T_0 to

$$T_{R-J}=T_0 e^{-2y}$$

and the energy density is increased to $\varepsilon=aT_0^4 e^{4y}$; since we in fact measure T_{R-J} and not T_0, we can write

$$\varepsilon=aT_{R-J}^4 e^{12y} \tag{10.92}$$

Thus, an injection of energy leads to an increase in radiation energy density in the millimeter region of the spectrum, leaving its longwave part essentially unchanged (De Zotti, 1979). Observational data set the upper limit $y \leqslant 0.02$ (Danese and De Zotti 1978; De Zotti, 1979). More precise information about the amount of energy released in the reheating of intergalactic gas will hopefully be provided by further measurements of radiation energy density in the millimeter region of the spectrum.

The observed departures of the relic radiation spectrum from the Planck distribution are of different type than the theoretical imprints of pre- or post-recombination injections of energy (cf. Fig. 10.6) (Woody and Richards 1979). An excess (of about 10%) of photons is found at energies near the maximum of the Planck curve, while in the shortwave region their number falls below the Planck value (by about 20% at $\lambda \sim 0.1$ cm). Rowan-Robinson, Negroponte and Silk (1979) suggest that the observed departures could arise if, for $z \geqslant 100$, a small part of the matter was in the form of dust (mainly silicate). The dust could appear if there existed at sufficiently early times, (i.e. before the formation of galaxies), massive stars with rapid thermonuclear processes going on in their interiors (Rees, 1978). It is an interesting but controversial idea.

10.11. Perturbations of an Isotropic Newtonian Cosmological Model. Jeans Instability

The observed isotropy of the microwave backround radiation leads to the conclusion that the distribution of matter during the recombination period was very uniform. The characteristic condensations of matter which exist now – stars, galaxies and clusters of galaxies – must have appeared later. Compare the density of matter in small regions with the mean density of visible matter in the universe: for the Solar system $\delta\rho/\langle\rho\rangle \approx 10^{19}$, for a typical galaxy $\delta\rho/\langle\rho\rangle \approx 10^6$, and for a cluster of galaxies $\delta\rho/\langle\rho\rangle \approx 10^3$. These figures show clearly how non-uniform the present Universe is on a small scale. The distribution of matter in small regions is not isotropic either: stars are grouped into galaxies and these form clusters. Only an averaged matter distribution over regions containing many clusters of galaxies is isotropic. An analysis of the distribution of radio sources (Webster, 1977) shows that their number, and hence (probably) the matter density, vary by no more than 3% among arbitrary regions with linear dimensions of 10^9 pc$=10^3$ Mpc$=1$ Gpc or more. It is worth realizing that the present universe can hold only a few dozen of these regions.

It is generally believed that the observed small-scale distribution of matter developed slowly from density perturbations whose amplitude at the time of recombination was too small to introduce noticeable distortions into the microwave background.

To discuss this hypothesis in greater detail, we begin with the cosmological model based on Newton's theory of gravitation first considered by Milne and McCrea (1934). The model is hydrodynamic, so the equations of motion are

$$\frac{\partial \boldsymbol{u}}{\partial t} + (\boldsymbol{u}\cdot\nabla)\boldsymbol{u} = -\nabla\phi - \frac{1}{\rho}\nabla p, \tag{10.93}$$

$$\frac{\partial \rho}{\partial t} + \nabla(\rho\boldsymbol{u}) = 0, \tag{10.94}$$

$$\Delta\phi = 4\pi G\rho. \tag{10.95}$$

To pass to the comoving coordinate system

$$\boldsymbol{r} = \frac{R(t)}{R_0}\boldsymbol{r}_0, \qquad \boldsymbol{u} = H\cdot\boldsymbol{r} \tag{10.96}$$

we transform our equations according to

$$\frac{\partial}{\partial t} + \boldsymbol{u}\cdot\nabla \rightarrow \frac{\mathrm{d}}{\mathrm{d}t}, \qquad \nabla \rightarrow \frac{R_0}{R}\nabla_0. \tag{10.97}$$

The result is

$$\frac{R}{R_0}\frac{\mathrm{d}\boldsymbol{u}}{\mathrm{d}t} = -\nabla_0\phi - \rho^{-1}\nabla_0 p, \tag{10.98}$$

$$\frac{R}{R_0}\frac{\partial\rho}{\partial t} = -\nabla_0(\rho\boldsymbol{u}), \tag{10.99}$$

$$\left(\frac{R_0}{R}\right)^2 \Delta_0\phi = 4\pi G\rho. \tag{10.100}$$

In an isotropic model $\nabla_0 p = 0$ and the equations can be written

$$3\ddot{R} + 4\pi G\rho R = 0 \tag{10.101}$$

and

$$\dot{R}^2 - \tfrac{8}{3}\pi G\rho R^2 = -kc^2, \tag{10.102}$$

where kc^2 is an integration constant. Introducing the constant $\alpha = \frac{8\pi G}{3}\frac{\rho(t_0)}{c^2}R^3(t_0)$, we can write the general solution in the form

$$R=\alpha\omega^2, \quad t=\frac{2}{3}\frac{\alpha}{c}\omega^3 \quad \text{for} \quad k=0, \tag{10.103}$$

$$R=\alpha(1-\cos\omega), \quad t=\frac{\alpha}{c}(\omega-\sin\omega) \quad \text{for } k=1, \tag{10.104}$$

$$R=\alpha(\cosh\omega-1), \quad t=\frac{\alpha}{c}(\sinh\omega-\omega) \quad \text{for } k=-1. \tag{10.105}$$

We now perturb the background solution, introducing

$$\boldsymbol{u}\to\boldsymbol{u}+\delta\boldsymbol{u}, \quad \rho\to\rho+\delta\rho, \quad p\to p+\delta p, \quad \phi\to\phi+\delta\phi. \tag{10.106}$$

Linearising we obtain the system of equations

$$\frac{\partial}{\partial t}\delta\boldsymbol{u}'+\boldsymbol{u}\cdot\nabla\delta\boldsymbol{u}'+\delta\boldsymbol{u}'\cdot\nabla\boldsymbol{u}+\nabla\delta\phi+\rho^{-1}\nabla\delta p=0, \tag{10.107}$$

$$\left(\frac{\partial}{\partial t}+\boldsymbol{u}\cdot\nabla+\nabla\cdot\boldsymbol{u}\right)\delta\rho+\rho\nabla\delta\boldsymbol{u}=0, \tag{10.108}$$

$$\Delta\delta\phi=4\pi G\delta\rho. \tag{10.109}$$

In comoving coordinates they read

$$\frac{\mathrm{d}}{\mathrm{d}t}\left(\frac{R^2}{R_0^2}\delta\boldsymbol{u}\right)+\nabla_0\,\delta\phi+\rho^{-1}\nabla_0\,\delta p=0, \tag{10.110}$$

$$\frac{\mathrm{d}}{\mathrm{d}t}\left(\frac{\delta\rho}{\rho}\right)+\nabla_0\,\delta\boldsymbol{u}=0, \tag{10.111}$$

$$\frac{R_0^2}{R^2}\nabla_0^2\,\delta\phi=4\pi G\delta\rho, \tag{10.112}$$

where δu is the perturbation in the particle velocity and $\delta u'=\frac{R}{R_0}\delta u$.

Taking the curl of both sides of (10.110) gives

$$\frac{\mathrm{d}}{\mathrm{d}t}\left(\frac{R^2}{R_0^2}\nabla\times\delta\boldsymbol{u}\right)=0, \tag{10.113}$$

i.e. the vorticity conservation law. In the following we shall be interested in potential perturbations only; we assume therefore that

$$\delta\boldsymbol{u}=\nabla\chi. \tag{10.114}$$

Taking the divergence of equation (10.110) and using (10.111) and (10.112), we get

$$\left\{\left(\frac{\mathrm{d}}{\mathrm{d}t}+2\frac{\dot{R}}{R}\right)\frac{\mathrm{d}}{\mathrm{d}t}-4\pi G\rho\right\}\frac{\delta\rho}{\rho}=\frac{R_0^2}{R^2}\frac{p}{\rho}\Delta\frac{\delta p}{p}. \tag{10.115}$$

This is the fundamental equation for small perturbations in the homogeneous Newtonian cosmological model (Harrison, 1967). To describe a perturbation, we have to find the four functions δp, $\delta\rho$, $\delta\phi$ and χ, for which we have three equations. One more equation is thus needed; usually, we add the equation of state, defining the relation between p and ρ.

We shall consider two kinds of perturbation: adiabatic and isothermal. Let $L(\rho, p)$ be the energy lost by a volume element of matter per unit time. If $\mathrm{d}Q$ is the energy per unit mass gained by this volume, we can write

$$\rho\frac{\mathrm{d}Q}{\mathrm{d}t}=-L, \tag{10.116}$$

and since by the first law of thermodynamics $\mathrm{d}Q=\mathrm{d}U+p\mathrm{d}V$, we have

$$\frac{\rho^{\Gamma}}{\Gamma-1}\frac{\mathrm{d}}{\mathrm{d}t}\left(\frac{p}{\rho^{\Gamma}}\right)+L=0. \tag{10.117}$$

The equation for small perturbations becomes

$$\frac{1}{\Gamma-1}\frac{\mathrm{d}}{\mathrm{d}t}\left(\frac{\delta p}{p}-\Gamma\frac{\delta\rho}{\rho}\right)+\delta\frac{L}{p}=0. \tag{10.118}$$

In the case of adiabatic fluctuations this gives

$$\frac{\delta p}{\delta\rho}=\Gamma\frac{p}{\rho}=v_s^2, \tag{10.119}$$

where v_s is the speed of sound. Assuming that $\dfrac{\delta\rho}{\rho}=\xi(t)\dfrac{\mathrm{e}^{-ikr}}{r}$, we finally obtain the system of equations

$$\ddot{\xi}+2\frac{\dot{R}}{R}\dot{\xi}+\left[v_s^2k^2\frac{R_0^2}{R^2}-4\pi G\rho\right]\xi=0, \tag{10.120}$$

$$\dot{\xi}=k^2\chi, \tag{10.121}$$

$$4\pi G\rho\xi=-\frac{R_0^2}{R^2}k^2\delta\phi. \tag{10.122}$$

Equation (10.120) leads to some interesting conclusions. Suppose that at time $t=t_0$, $\xi>0$ and $\dot{\xi}>0$, i.e. the perturbation grows. This growth will continue until ξ reaches a maximum, i.e. until the moment when $\dot{\xi}=0$ and $\ddot{\xi}<0$. In an expanding universe $\dot{R}/R>0$, and we see from (10.120) that the condition for a maximum cannot be fulfilled if

$$v_s^2k^2\frac{R_0^2}{R^2}-4\pi G\rho<0. \tag{10.123}$$

If inequality (10.123) holds, perturbations which grew initially, will grow unboundedly. Setting $\lambda=\frac{2\pi}{k}\frac{R}{R_0}$, we obtain for the scale of such ever-growing perturbations the condition

$$\lambda^2>\frac{\pi v_s^2}{G\rho}. \tag{10.124}$$

The critical value $\lambda_J=\sqrt{\frac{\pi v_s^2}{G\rho}}$ is called the Jeans wavelength. Before we examine the time-variation of perturbations with characteristic size greater than the Jeans wavelength, let us consider isothermal perturbations. To this end, assume that the thermal conductivity κ of matter is non-zero. The energy loss per unit volume and unit time due to conduction is

$$L=-\nabla(\kappa\nabla T)=-\frac{R_0^2}{R^2}\nabla_0(\kappa\nabla_0 T), \tag{10.125}$$

so

$$\delta L=-\frac{R_0^2}{R^2}\kappa\Delta\delta T; \tag{10.126}$$

thermal conductivity has been taken as constant and unaffected by perturbations. For spherical perturbations we get

$$\delta L=\left(\frac{R_0}{R}\right)^2\kappa k^2\delta T. \tag{10.127}$$

Assuming that the fluid satisfies the perfect gas equation $p=\frac{R}{\mu}\rho T$, where μ is the mean molecular weight, we have

$$\frac{\delta T}{T}=\frac{\delta p}{p}-\frac{\delta\rho}{\rho}. \tag{10.128}$$

If in the unperturbed case the evolution proceeds adiabatically, we can write equation (10.118) in the form

$$\frac{\mathrm{d}}{\mathrm{d}t}\left(\frac{\delta p}{p}-\Gamma\frac{\delta\rho}{\rho}\right)+\alpha\left(\frac{\delta p}{p}-\frac{\delta\rho}{\rho}\right)=0, \tag{10.129}$$

where $\alpha=(\Gamma-1)\frac{T}{p}\frac{R_0^2}{R^2}$. Combining this equation with (10.115) gives

$$\left\{\left(\frac{\mathrm{d}}{\mathrm{d}t}+\alpha\right)\frac{\rho R^2}{pR_0^2}\left[\left(\frac{\mathrm{d}}{\mathrm{d}t}+2\frac{\dot{R}}{R}\right)\frac{\mathrm{d}}{\mathrm{d}t}-4\pi G\rho\right]+k^2\left(\Gamma\frac{\mathrm{d}}{\mathrm{d}t}+\alpha\right)\right\}\xi=0. \tag{10.130}$$

Thus, allowing for heat conduction leads to a third-order equation for ξ. From the physical point of view, heat conduction becomes important when the time-scale on which thermal energy is carried over distances comparable to k^{-1} is much shorter than the time-scale for the growth of density perturbations. Hence, heat conduction can indeed produce small-scale pressure perturbations δp, but then the density contrast $\delta\rho/\rho$ can be regarded as constant. In this case, equation (10.129) yields

$$\frac{\delta p}{p}-\left(\frac{\delta \rho}{\rho}\right)_0=\left(\frac{\delta p}{p}-\frac{\delta \rho}{\rho}\right)_0 \mathrm{e}^{-\alpha(t-t_0)}, \qquad (10.131)$$

hence if $t-t_0 \gg \alpha^{-1}$, then $\dfrac{\delta p}{p}=\dfrac{\delta \rho}{\rho}$ and (10.128) implies that $\delta T=0$, i.e. the perturbations are isothermal.

We have thus come to the conclusion that small-scale perturbations are isothermal. On the other hand, since the role of heat conduction decreases as the scale of perturbations grows, large-scale perturbations are adiabatic.

Now to find how the density perturbations depend on time, let us discuss the simplest case where particles move at parabolic velocities, so that $R \sim t^{2/3}$. Using the explicit solution (10.103) and considering only growing perturbations, i.e. perturbations with sufficiently small k, we obtain the equation

$$\omega^2 \xi'' + 2\omega\xi' - 6\xi = 0; \qquad (10.132)$$

a prime indicates differentiation with respect to ω. It has two linearly independent solutions, ω^{-3} and ω^2, so finally

$$\xi \sim a_1 t^{2/3} + a_2 t^{-1}. \qquad (10.133)$$

This result is surprising. One would have expected the perturbation amplitude to grow exponentially with time rather than according to a power law (Bonnor, 1957). Note that effectively $\delta\rho$ decreases with time as $t^{-4/3}$; that the density contrast $\delta\rho/\rho$ grows in spite of that is due to the density falling with time as t^{-2}.

In cases where $k = \pm 1$, only the exponents of powers change but the character of the relation remains the same. The general conclusion is that density perturbations in a homogeneous universe grow comparatively slowly. This fact is one of the main difficulties which we encounter in trying to explain the observed distribution of matter on the scale of galaxies and clusters of galaxies, starting from an isotropic and homogeneous model of the universe. We go further into this problem in the next section, where we consider a genaral-relativistic model.

10.12. Perturbations of the Friedman Model

A full relativistic analysis of perturbations in a homogeneous and isotropic Friedman model was given in a classic paper by Lifshitz in 1946 (Lifshitz and Khalatnikov, 1963; Hawking, 1966; Field and Shepley, 1968; Bardeen, 1980). The calculations are much more complicated than in a model based on Newton's theory of gravitation, and so we shall restrict ourselves to a general discussion of the method and its results.

In the Friedman model the coordinate system is chosen in such a way that it is synchronous, i.e. the trajectories of material particles are orthogonal to the surfaces t=const, and comoving, so that the particle velocity four-vector has only one non-zero component – the one along the time-axis. In the perturbed model, where $g_{\alpha\beta} \to g_{\alpha\beta} + h_{\alpha\beta}$ with $h_{\alpha\beta}$ small, we can also choose a synchronous coordinate system. In such a system, $h_{00} = 0 = h_{0\alpha}$. This time, however, it is not, in general, a comoving system, and δu^α – the three-velocity perturbation – ma y be non-zero.

A spacelike cross-section of the Friedman space-time is a surface of constant curvature: for $k=1$ it is a three-sphere, for $k=-1$ it is a three-dimensional pseudo-sphere, and for $k=0$ a three-dimensional Euclidean space. Owing to this property, perturbations of the Friedman model can be resolved into harmonic functions, which are simply eigenfunctions of the Laplace operator Δ in the constant-curvature spaces just mentioned. We shall be interested only in scalar, vector and tensor harmonic functions.

Using the scalar harmonic functions Q defined by

$$\Delta Q = -(n^2 \mp 1) Q, \tag{10.134}$$

where the sign "–" is taken when Q is spherical ($k=-1$) and "+" when it is pseudo-spherical ($k=1$ or 0), we construct three-dimensional tensors

$$Q^a_b = \tfrac{1}{3}\delta^a_b Q, \quad P^a_b = \frac{1}{n^2 \mp 1} Q_{,b}{}^{;a} + Q^a_b. \tag{10.135}$$

By definition $Q^a_a = Q$ and $P^a_a = 0$. We also introduce the vectorial quantity

$$P_a = \frac{1}{n^2 \mp 1} Q_{,a}, \tag{10.136}$$

chosen so that $P^a{}_{;a} = -Q$. Scalar harmonic functions represent perturbations in the gravitational field, density and particle velocity.

The harmonic vector-functions S_a, defined by

$$\Delta S_a = -(n^2 \mp 2) S_a, \quad S^a{}_{;a} = 0 \tag{10.137}$$

give rise to the tensors

$$S^a_b = S_b{}^{;a} + S^a{}_{;b}, \tag{10.138}$$

but cannot be used to construct any scalar quantities. They represent those metric and particle velocity perturbations which leave matter density distribution unaffected.

Tensorial harmonic functions G^a_b are defined by the conditions

$$\Delta G^a{}_b = -(n^2 \mp 3) G^a{}_b, \quad G^a{}_a = 0, \quad G^a{}_{b;a} = 0. \tag{10.139}$$

No scalar or vector quantities can be constructed from G^a_b. They represent those perturbations in the gravitational field which have no effect on the motion of particles or the distribution of matter. Such can only be gravitational waves.

The density perturbations induce the metric perturbations

$$h^a{}_b = \lambda(\omega) P^a{}_b + \mu(\omega) Q^a{}_b, \tag{10.140}$$

where the parametr ω is related to time in the same way as in the Newtonian case. Normal mode analysis of the perturbations allows us to separate out the spatial depedence. The remaining equations contain only functions of time. In order to solve them, we have to specify the equation of state.

At early stages of the evolution, a good approximation to the equation of state is furnished by the relativistic relation $p = \frac{1}{3}\rho c^2$. It is valid for small times, so we can assume that $\omega \ll 1$. As in the Newtonian case, let us consider large-scale and small-scale perturbations. Large-scale perturbations are characterized by small values of n and, conversely, small-scale perturbations have large n.

We first take the case where n is not too large, so that $n\omega \ll 1$. Perturbations in the metric, density and particle velocity in the open model then have the form

$$h^a{}_b = 3c_1\omega^{-1}P^a{}_b + c_2(Q^a{}_b + P^a{}_b), \tag{10.141}$$

$$\frac{\delta\rho}{\rho} = \frac{n^2+4}{9}(c_1\omega + c_2\omega^2), \tag{10.142}$$

$$\delta u^a = \frac{n^2+4}{12}\left(3c_1 + c_2\frac{n^2+1}{9}\omega^3\right)P^a, \tag{10.143}$$

where c_1 and c_2 are constants chosen so that the perturbations are small at the initial moment. Note that the velocity perturbation δu^a is potential, because $\delta u_a \sim Q_{,a}$. For small ω, $t \sim \omega^2$, and therefore the density perturbations grow no faster than linearly with time. The perturbations remain small throughout the period of applicability of the approximation, i.e. up to $\omega \sim 1/n$.

If n is large, so that $n\omega \gg 1$, which corresponds to small-scale perturbations, the metric corrections, the density contrast and the particle velocity perturbation are given by

$$h^a{}_b = \alpha n^{-2}\omega^{-2}(P^a{}_b - 2Q^a{}_b)\exp(in\omega/\sqrt{3}), \tag{10.144}$$

$$\frac{\delta\rho}{\rho} = -\frac{\alpha}{9}Q\exp(in\omega/\sqrt{3}), \tag{10.145}$$

$$\delta u^a = \frac{\alpha i n}{12\sqrt{3}}P^a\exp(in\omega/\sqrt{3}), \tag{10.146}$$

where α is a complex number, $|\alpha| \ll 1$. As in the Newtonian case, we have obtained an oscillating solution. Metric perturbations are of the sound-wave type, with decaying amplitude. The density perturbations oscillate with a constant amplitude.

At late stages of the evolution of the universe radiation pressure and gas pressure are negligibly small and it is a good approximation to regard matter as a non-interacting dust with equation of state $p=0$. The coordinate system can then be chosen so to still be comoving in the perturbed state. The corrections to the metric and the density perturbations in an epoch when the radius of curvature is small compared with the present value are

$$h^a{}_b = \frac{8c_2}{3\omega^3}[(n^2+1)P^a{}_b - (n^2+4)Q^a{}_b], \tag{10.147}$$

$$\frac{\delta\rho}{\rho} = \frac{4(n^2+4)}{3\omega^3}c_2 Q; \tag{10.148}$$

here $t \sim \frac{1}{6}\omega^3$. At later times, when $\omega \gg 1$, the corresponding solution is

$$h^a{}_b = \frac{4c_2}{3}[(n^2+1)P^a{}_b - (n^2+4)Q^a{}_b]e^{-\omega}, \tag{10.149}$$

$$\frac{\delta\rho}{\rho} = \frac{n^2+4}{3}c_2 Q e^{-\omega}, \tag{10.150}$$

and time t is related to the parameter ω by $t \sim \frac{1}{2}e^{\omega}$. The above solutions show that perturbations of this type, both at early stages of evolution and later, decay in inverse proportion to time.

The second, linearly independent solution grows with time as $t^{2/3}$, just as in the Newtonian case. For small times, the cases where n is small and where it is large must be considered separately. For small n and $\omega \ll 1$ we obtain

$$h^a{}_b = \frac{c_1}{2}(P^a{}_b + Q^a{}_b), \qquad \frac{\delta\rho}{\rho} = \frac{n^2+4}{60} c_1 \omega^2 Q, \tag{10.151}$$

and for large n ($1/n \ll \omega \ll 1$)

$$h^a{}_b = \frac{c_1 n^2}{30}\omega^2(P^a{}_b - Q^a{}_b), \qquad \frac{\delta\rho}{\rho} = \frac{c_1 n^2}{60}\omega^2 Q. \tag{10.152}$$

In either case the density perturbations grow like $t^{2/3}$, but in the first case $\delta\rho/\rho$ remains small because n and c_1 are small, whereas in the second case, with n large and near the limit of the applicability of the approximation ($\omega \sim 1$), $\delta\rho/\rho$ can also be large. We then have a true instability. It is, however, of little significance because large n correspond to small-scale perturbations, which, as we shall see later, are strongly damped during the radiation era.

At late stages of the evolution, when $\omega \gg 1$, the metric and the density perturbations are given by the formulae

$$h^a{}_b = 2c_1[-(n^2+1)Q^a{}_b + (n^2+4)P^a{}_b]\,\omega e^{-\omega}, \tag{10.153}$$

$$\frac{\delta\rho}{\rho} = \frac{n^2+4}{6} c_1 Q. \tag{10.154}$$

The density perturbations tend asymptotically to a constant value and do not lead to any instability.

Throughout the foregoing considerations we have used the parameter ω instead of time. It would be interesting to find its present value. For the open model we have the relation

$$H^2 - 8\pi G\rho/3 = c^2/R^2, \tag{10.155}$$

but since

$$H = \frac{\dot{R}}{R} = \frac{c \sinh\omega}{\alpha(\cosh\omega - 1)^2},$$

we find

$$\cosh\frac{\omega}{2} = H\sqrt{\frac{3}{8\pi G\rho}}. \tag{10.156}$$

Inserting the present value of $H = 50\ \text{km}\cdot\text{s}^{-1}\cdot\text{Mpc}^{-1}$, and putting $\rho \sim 10^{-30}\ \text{g/cm}^3$, we obtain $\omega \approx 3$, a value between the asymptotic regions $\omega \ll 1$ and $\omega \gg 1$.

To complete the picture, let us consider perturbations which do not produce any changes in the distribution of matter. These include rotational perturbations and purely gravitational perturbations.

Rotational perturbations, which we describe using vector harmonic functions, produce small changes in the metric,

$$h^a{}_b=\begin{cases}\dfrac{-8d}{\omega^3}S^a{}_b, & \omega\ll 1,\\ -4de^{-\omega}S^a{}_b, & \omega\gg 1,\end{cases}\tag{10.157}$$

and changes in the particle velocity,

$$\delta u^a=\frac{d}{12(\cosh\omega-1)}S^a,\tag{10.158}$$

where d is a constant and pressure has been neglected. Both the metric and the velocity perturbations decrease with time, the latter in inverse proportion to R. This follows also from the vorticity conservation law.

Perturbations in the gravitational field which cause no changes in the distribution or motion of matter can be recognized as gravitational waves. At late stages of the evolution, when pressure can be neglected, purely gravitational perturbations are given by

$$h^a{}_b=\frac{in\alpha}{\sinh\omega/2}e^{in\omega}G^a{}_b,\tag{10.159}$$

which corresponds to a gravitational wave propagating with the speed of light and with an amplitude decreasing as R^{-1}.

In all the cases discussed above, perturbations were assumed to be adiabatic and dissipative processes were neglected. Allowing, for example, for heat conduction does not introduce any essential changes and the results are similar to those obtained in the Newtonian model. One class of perturbations which we have not considered here are perturbations of entropy (isothermal perturbations), i.e. perturbations which affect the distribution of matter but preserve the isotropy of radiation.

10.13. Influence of Radiation on the Evolution of Perturbations in the Expanding Universe

Analysis of small perturbations in the expanding universe leads to the conclusion that adiabatic density perturbations on scales greater than the Jeans wavelength grow with time while those with smaller scales oscillate with constant amplitude. The critical Jeans wavelength is

$$\lambda_J=\left(\frac{\pi v_s^2}{G\rho}\right)^{1/2}.\tag{10.160}$$

The mass contained in a sphere of radius λ_J is called the *Jeans mass*:

$$M_J=\tfrac{4}{3}\pi\rho_m\lambda_J^3=\tfrac{4}{3}\pi\rho_m\left(\frac{\pi v_s^2}{G\rho}\right)^{3/2},\tag{10.161}$$

where ρ_m is the density of matter. When radiation is dynamically dominant, we have

$v_s = \frac{1}{\sqrt{3}}c$, $c^2\rho = aT^4$ and $M_J \sim (1+z)^{-3}$, so during the radiation era the Jeans mass increases. When the recombination of hydrogen takes place, the speed of sound drops drastically, radiation becomes dynamically insignificant and the properties of the model are determined by the density of matter. The speed of sound is then $v_s^2 = \frac{5}{3}\frac{kT}{m}$ and for the Jeans mass we get the expression

$$M_J = 4\left(\frac{\pi}{3}\right)^{5/2}\left(\frac{5kT}{Gm}\right)^{3/2}\rho_m^{-1/2}, \qquad (10.162)$$

where T is the gas temperature and m the mass of a hydrogen atom. For a monatomic perfect gas expanding adiabatically $T \sim R^{-2}$, and since the matter density is proportional to R^{-3}, the Jeans mass decreases as $R^{-3/2}$.

Let us estimate the value of the Jeans mass before and after recombination. We assume that at present $T_0 = 3$ K and $\rho_m = 10^{-30}$ g/cm^3, and that recombination took place at $z \approx 1300$ when the temperature was 4000 K. Inserting these values into (10.161) gives the Jeans mass $M_J \sim 6 \cdot 10^{19}\, M_\odot$ before recombination and $M_J \sim 4.8 \cdot 10^6\, M_\odot$ just after it. Thus after recombination the Jeans mass drops precipitously and the scale of the growing perturbations is comparable to that of globular clusters (Peebles, 1965). These estimates, showing that typical masses of growing perturbations before recombination are much larger than, and those after recombination much smaller than, any galactic mass, indicate that the Jeans mass is of little relevance to the theory of galaxy formation.

A clue to the cause for the observed galactic mass distribution has however been provided by studies of the dissipative damping of certain types and scales of perturbations. News of the discovery of relic radiation had just broken when Peebles (1965) pointed out that interaction between radiation and matter can have a very strong effect on the development of small-scale perturbations. Before recombination radiation interacts strongly with matter through Compton scattering on free electrons, and radiation pressure dominates over gas pressure. Consider, very roughly, what happens to small oscillating perturbations. Assuming that they are adiabatic, we have the relation $\frac{\delta\rho}{\rho} = 3\frac{\delta T}{T}$ and therefore changes in density are accompanied by changes in temperature. But if the photon mean free path is comparable with the perturbation scale, the system cannot be in thermal equilibrium. For such perturbations δT tends to zero, and so does $\delta\rho$. Thus small-scale perturbations are damped.

The process was analysed in detail by Peebles and Yu (1970), by Silk (1968, 1974) and by Weinberg (1971) (see also Peebles (1980)). During the radiation era, the coupling between matter and radiation, though strong, was not perfect. There was always some viscosity and heat conduction. Small perturbations, which would have behaved like sound waves in a perfect gas had the coupling been perfect, were in fact subject to dissipative processes.

By solving the relativistic transport equation – a relativistic analogue of the Boltzmann equation – we can determine the coefficients of viscosity and heat conduction. The distribution function can be expanded in a series with respect to a small parameter, namely the ratio of the photon mean free path to the characteristic size of the perturbation. In the first approximation the following expresions are obtained for the coefficients of viscosity and heat conduction:

$$\zeta=4aT^4\tau\left(\frac{1}{3}-\frac{1}{c^2}\left(\frac{\partial p}{\partial \rho}\right)_s\right)^2,$$
$$\eta=\tfrac{4}{15}aT^4\tau, \qquad (10.163)$$
$$\kappa=\tfrac{4}{3}aT^3\tau,$$

where τ is the photon mean free time.

Knowing ζ, η and κ, we then introduce dissipative terms into the equations of motion. Resolving the perturbation into plane waves, we obtain the following dispersion relation for longitudinal waves:

$$0=\omega^3-k^2\omega v_s^2+ik^2\omega^2(\zeta+\tfrac{4}{3}\eta)\frac{1}{\rho+p/c^2}+\frac{i\kappa}{(\rho+p/c^2)\left(\frac{\partial \rho}{\partial T}\right)_s}\left[\omega^4\frac{T}{c^2}\left(\frac{\partial \rho}{\partial T}\right)_s\right.$$
$$\left.+k^2\omega^2(\rho+p/c^2)-2k^2\omega^2\frac{T}{c^2}\left(\frac{\partial p}{\partial T}\right)_s-k^4n\left(\frac{\partial p}{\partial n}\right)_T\right] \qquad (10.164)$$

(n is the particle number density). In the linear approximation in ζ, η and κ we find that

$$\omega=|k|v_s-ik^2D, \qquad (10.165)$$

$$D=\frac{1}{2(\rho+p/c^2)}\left\{\zeta+\tfrac{4}{3}\eta+\kappa\left(\frac{\delta\rho}{\delta T}\right)_s^{-1}\left[\rho+p/c^2-2\frac{T}{c^2}\left(\frac{\partial p}{\partial T}\right)_s\right.\right.$$
$$\left.\left.+\left(\frac{v_s}{c}\right)^2T\left(\frac{\partial \rho}{\partial T}\right)_s-\frac{n}{v_s^2}\left(\frac{\partial p}{\partial n}\right)_T\right]\right\} \qquad (10.166)$$

The imaginary part which has now appeared in the dispersion relation shows that the oscillations are damped in amplitude at a rate $\Gamma=Dk^2$. Subsituting for ζ, η and κ using (10.163), and using the fact that $\tau^{-1}=nc\sigma_T$, where σ_T is the Thomson cross-section, we find

$$\Gamma=\frac{k^2aT^4}{6cn\sigma_T[nm_H+4aT^4/3c^2]}\left\{\frac{16}{15}+\frac{n^2m_H^2c^2}{aT^4[nm_H+4aT^4/3c^2]}\right\}. \qquad (10.167)$$

If the present matter density is below a certain critical value, the dynamics of the expansion at the time of recombination were determined by the radiation energy distribution. Before recombination, radiation was dominant and in that period

$$\Gamma=\frac{2}{15}\frac{kc^2}{n\sigma_T}. \qquad (10.168)$$

Hence during the radiation era the amplitude decreased by a factor

$$d=\exp\left(-\int_0^{t_r}\Gamma dt\right)=\exp\left(-\frac{2k_r^2ct_r}{15\sigma_T n_r}\right)=\exp\left(-\left(\frac{M_s}{M}\right)^{2/3}\right), \tag{10.169}$$

where t_r is the recombination time and the subscript r refers a given parameter to that time. The mass M is defined by the standard formula

$$M=\tfrac{4}{3}\pi n m_H\left(\frac{2\pi}{|k|}\right)^3, \tag{10.170}$$

and the critical mass M_S, known as the Silk mass, is given by

$$M_s=\frac{32}{3}\pi^4\left(\frac{4}{45}\frac{m_H c}{\sigma_T}\right)^{3/2}(15.5\pi a T_r^4 G/c^2)^{-3/4}(n_r m_H)^{-1/2}. \tag{10.171}$$

If the mean density of visible matter, i.e. $\rho=nm_H=10^{-30}$ g/cm^3, is taken as the present mass density, we get the Silk mass $M_s=2.1\cdot10^{14}$ M. The amplitude of a perturbation of mass 10^{11} $M_\odot$ would have decreased by a factor of 10^{-72}. We can thus see that only perturbations with masses greater than the Silk mass, that is those corresponding to clusters of galaxies, have a chance of surviving the radiation era.

The latest calculations of Press and Vishniac (1980), who used a modified two fluid (baryon, photon) model for the interaction of radiation and plasma, show that adiabatic perturbations are strongly damped during recombination, and that this effect cannot be neglected. The resulting Silk mass is considerably larger, for example for $\Omega=0.1$, $M_S=2\cdot10^{16}$ $M_\odot$, about 100 times as large as the previous estimates.

10.14. Theories of Galaxy Formation

The problem we are facing now is very difficult and very important. It is to explain the origin of galaxies. This central question of contemporary cosmology has not yet been answered satisfactorily. There are a few competing theories but each of them has some weak points. Basically, in all the theories proposed so far it is supposed that the observed distribution of matter in the universe has slowly developed from small perturbations which existed in the radiation era. None of the existing theories explains the origin of those perturbations (see however Press and Vishniac, 1980; Lukash, 1981; Kompaneets *et al.*, 1982). The natural conjecture that they were ordinary statistical fluctuations must be rejected, because the amplitude of such fluctuations for a system of N particles is $1/\sqrt{N}$, which for a typical galaxy amounts to about 10^{-34}. Such perturbations are too small to give rise to galaxies. If galaxies were formed comparatively late, say at $z\approx10$, it is easy to estimate what should be the value of $\delta\rho/\rho$ just after recombination in order that $\delta\rho/\rho\approx1$ at $z\approx10$. After recombination $\frac{\delta\rho}{\rho}(z)\sim t^{2/3}\sim(1+z)^{-1}$, and therefore $\frac{\delta\rho}{\rho}(z_r)=\frac{\delta\rho}{\rho}(z_0)\frac{1+z_0}{1+z_r}\sim\frac{11}{1300}\sim10^{-2}$. We do not know a mechanism that could have produced perturbations of such amplitude. The most attractive proposition is that the perturbations were generated in quantum processes of particle creation (Parker 1976, Lukash *et al.*, 1976; Zeldovich, 1979) or in non-equilibrium processes in quark protoplasma (Press, 1979; Turner and Schramm, 1979). Unfortunately, these are still only hypotheses, and therefore the theories of galaxy formation simply take it for granted that the amplitude of the perturbations at the time of recombination was large enough to eventually give rise to the galaxies and clusters of galaxies observed now.

A good theory of the formation of galaxies should account for the basic observational facts, such as galactic masses and angular momenta, the distribution of galaxies in space – particularly their clumping into clusters, and the thermal history of the intergalactic gas.

As we already know, small perturbations in the isotropic and homogeneous Friedman model can be divided into adiabatic, isothermal (entropy) and rotational perturbations (we omit gravitational waves). Different models favour different types of perturbations. The literature of the subject has grown extensive in the recent years (Zeldovich and Novikov 1983; Ozernoi, 1978; Rees, 1971, 1975; Peebles, 1980; Gott, 1977; Fall, 1979; Field, 1975; Jones, 1976); here we shall present only two models, namely that of Doroshkevich, Sunyaev and Zeldovich (1974) and the White–Rees model (1978).

The view of Doroshkevich, Sunyaev and Zeldovich is that galaxies, or rather clusters of galaxies, have emerged as a result of the growth of small adiabatic perturbations in the Friedman model. During recombination the density perturbations were too small to cause any noticeable departures from isotropy in the relic radiation. After recombination the perturbations grow slowly but their amplitude remains less than unity until $z \approx 10$. Clouds of matter with masses of the order of the Silk mass (10^{13}–10^{15} $M_\odot$) undergo further contraction. Their evolution can no longer be described by the linear approximation. As the contraction goes on, there emerge nearly plane shock waves and dense discs which subsequently fragment. Following this, quasars and galaxies are formed. The shock waves produce turbulence and vortex motions, which become a source of angular momentum. Part of the cold gas which has escaped compression by the shock waves gets heated by the radiation generated in these waves and in young galaxies and quasars, forming the clouds of hot intergalactic gas which we now observe.

We now pass to a more detailed account of the successive stages of this evolution. According to Zeldovich and his co-workers, the initial conditions for perturbations of mass M should refer to the moment t_M when the mass within the horizon is M, i.e. when $\rho(t_M)\cdot(ct_M)^3 \sim M$. If we assume that for given M the amplitude of corresponding perturbation at the time t_M is 10^{-4}, i.e. $\left(\frac{\delta\rho}{\rho}\right)_{t_M} = 10^{-4}$, we find that at the time of recombination

$$\frac{\delta\rho}{\rho} = 10^{-4} e^{-\left(\frac{M_s}{M}\right)^{2/3}} \tag{10.172}$$

for perturbations smaller than the Silk mass M_S, $\frac{\delta\rho}{\rho} \sim 10^{-4}$ for perturbations with masses greater than M_S but less than the Jeans mass M_J, and

$$\frac{\delta\rho}{\rho} = 10^{-4}\left(\frac{M_J}{M}\right)^{2/3} \tag{10.173}$$

for masses greater than M_J. Recall that the characteristic masses change with time as

$M_J \sim t^{3/2}$, $M_S \sim t^{9/4}$, and therefore large-scale perturbations grow in proportion to time (Fig. 10.11). Their amplitude increases rapidly during recombination, and just after it $\frac{\delta\rho}{\rho} \sim 10^{-2}$. As long as $\delta\rho/\rho$ is less than unity, the growth of the perturbations is adequately described in the linear approximation. The position of a particle can then be given as

$$\boldsymbol{r} = \boldsymbol{r}_0 \frac{R(t)}{R(t_r)} - b(t)\,\boldsymbol{S}(\boldsymbol{r}_0), \tag{10.174}$$

where $\boldsymbol{r}_0$ is the position at recombination time t_r in comoving coordinates, $b(t)$ is a function of time only, representing the amplitude of the perturbation, and $\boldsymbol{S}(\boldsymbol{r}_0)$ is the particle displacement vector. Zeldovich assumes that formula (10.174) is applicable even when non-linear effects become significant.

The velocity of the particles can be found by differentiating the position vector with respect to time:

$$\boldsymbol{u} = \frac{d\boldsymbol{r}}{dt} = H\boldsymbol{r}_0 \frac{R(t_r)}{R(t)} - \dot{b}(t)\,\boldsymbol{S}(\boldsymbol{r}_0), \tag{10.175}$$

and the particle density is

$$\rho = \rho_0(t_r) / \frac{\partial(\boldsymbol{r})}{\partial(\boldsymbol{r}_0)}, \tag{10.176}$$

where $\rho_0(t_r)$ is the unperturbed density at recombination and $\partial(\boldsymbol{r})/\partial(\boldsymbol{r}_0)$ is the Jacobian of the coordinate transformation $\boldsymbol{r} = \boldsymbol{r}(\boldsymbol{r}_0, t)$. The shear tensor is defined in the same way as in continuum mechanics:

$$D_{ik} = \frac{\partial r_i}{\partial r_{0k}} = \frac{R(t)}{R(t_r)_i} \delta_{ik} - b(t)\, d_{ik}, \tag{10.177}$$

where $d_{ik} = \frac{\partial S_i}{\partial r_{0k}} = \frac{\partial S_k}{\partial r_{0i}}$. It is symmetric, because we consider only potential perturbations of the velocity. Let λ_1, λ_2 and λ_3 be eigenvalues of the tensor d_{ik}; then

$$\frac{\partial(\boldsymbol{r})}{\partial(\boldsymbol{r}_0)} = \mathrm{Det}\left\| \frac{R(t)}{R(t_r)} \delta_{ik} - b(t)\, d_{ik} \right\| = \frac{R^3(t)}{R^3(t_r)} (1 - B(t)\lambda_1)(1 - B(t)\lambda_2)(1 - B(t)\lambda_3); \tag{10.178}$$

here $B(t) = b(t) \frac{R(t_r)}{R(t)}$ is an increasing function of time. A gas cloud representing a growing perturbation will contract most rapidly along the principal direction corresponding to the largest eigenvalue. Suppose that this is the direction of the x-axis. Even if the cloud is initially spherical, it will deform to a disc in the plane y, z. The dynamics of contraction depend on the form of the particle displacement vector. Its components are random variables characterized by mean values and dispersion. To continue with a qualitative rather than a full description, we shall assume that $b(t) = \frac{R^2(t)}{R^2(t_r)}$, which is correct only when $\rho = \rho_{cr}$ $(\Omega = 1)$ but remains a good approximation when $\rho_{cr} \geqslant \rho$. At the moment when the Jacobian of the transformation $\boldsymbol{r}=\boldsymbol{r}(\boldsymbol{r}_0, t)$ becomes close to zero, the density in the symmetry plane of the disc increases considerably and a shock wave begins to propagate both in the positive and in the negative direction of the x-axis. Further and further layers of the falling matter are compressed by the shock wave front. Part of their kinetic energy is converted into thermal

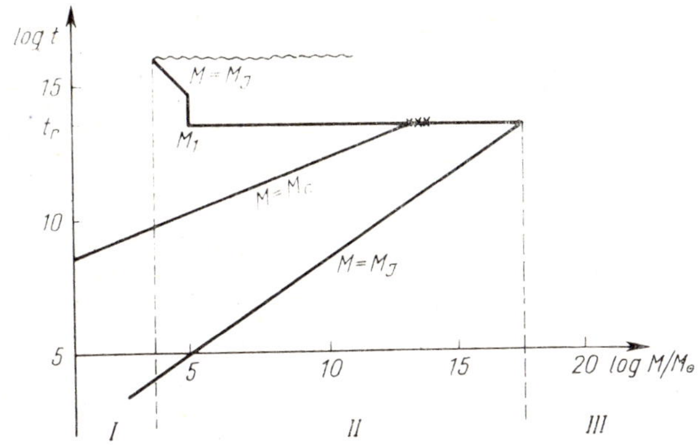

FIG. 10.11. Evolution of M_J and M_S during the radiation era. After recombination, M_J drops drastically (adapted from Zeldovich and Novikov, 1983).

energy, heating the gas to very high temperatures. The dense core of the disc cools rapidly while the layers near the wave front are still hot.

The gas cloud compressed by the shock wave has a mass of about $10^{14}\ M_\odot$, which corresponds to a cluster of galaxies. How does the system break up and how does vorticity come about? Let us begin with the second question. At first it was thought that a theory attributing the formation of galaxies to slowly growing adiabatic perturbations would not be able to account for the observed angular momentum of galaxies. The main difficulty is that, according to the Helmholtz theorem of vorticity conservation, a conservative force field cannot produce vorticity. The Helmholtz theorem does not apply to shock waves. Suppose that the shock front is a plane perpendicular to the x-axis. Before the front the motion is potential and $\nabla \times \boldsymbol{u}=0$. In passing through the shock wave the tangential components of the velocity are unchanged, but the normal component suffers a discontinuity. Approximately, we can assume that the component u_x of the particle velocity is zero after the passage of the shock wave. Therefore in this region $\nabla \times \boldsymbol{u} \neq 0$, provided that the motion is not homogeneous, i.e. the velocity of a particle depends on its position. In the case where the potential motion of matter is homogeneous, vorticity will arise if the shock front is not plane (Doroshkevich, 1972; Binney, 1974). The vorticity generated in this way can explain the observed angular momentum of galaxies.

The disc lying in the y, z plane undergoes further evolution. Its central part cools down to a temperature of order 10^4 K, the hydrogen recombines and the disc breaks up as a result of thermal and gravitational instabilities. Clouds appear with masses 10^5–$10^6\ M_\odot$, corresponding to the Jeans mass, and fragments with masses 10^{10}–$10^{12}\ M_\odot$. Near the centre of the disc, large galaxies with masses of order $10^{12}\ M_\odot$ and small angular momentum are formed. They can be identified with elliptic galaxies. Further away from the centre, where vorticity is stronger, turbulence gives rise to spiral galaxies with masses 10^{10}–$10^{11}\ M_\odot$. Thus the outer parts of a cluster of galaxies should consist mainly of spiral galaxies. Rotation of the whole cluster can be explained by tidal forces.

The thermal history of the intergalactic gas is very important from the astronomical viewpoint. Here the theoretical predictions rest on an even frailer basis. The part of the gas which was compressed by the shock wave cooled down and formed galaxies. The remaining part cooled much more slowly: hence the clouds of ionized gas present in clusters of galaxies.

The theory of galaxy formation of Doroshkevich, Sunyaev and Zeldovich can be regarded as an important step in the right direction. Its essential advantage is simplicity. For a fixed Ω, that is for a defined mean density of matter, the only free parameter is the perturbation amplitude before recombination. Indeed, the spectrum of the initial perturbations favours no particular mass scale. The assumed adiabaticity of perturbations, as supported by the latest results of Weinberg (1979) is also an advantage rather than a disadvantage of the theory. Furthermore, the theory leads to very definite conclusions: first to emerge were systems with masses typical of clusters of galaxies, and only later did galaxies develop in the process of fragmentation. A serious difficulty that the adiabatic theory of galaxy formation may have to face has been noted by Press and Vishniac (1980). According to them, the Silk mass is about $10^{16}\ M_\odot$ (for $\Omega=0.1$) and not $10^{14}\ M_\odot$, as was estimated

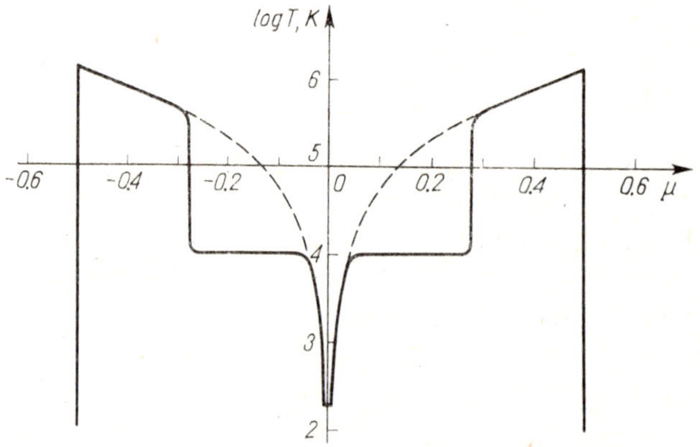

FIG. 10.12. Thermal history of a gas cloud compressed by a shock wave; μ is the ratio of the compressed mass to the total mass of the cloud (adapted from Zeldovich and Novikov, 1983).

previously. In this case the smallest perturbations which could survive the radiation era would have masses much larger than clusters of galaxies.

An entirely different picture of galaxy formation has been conceived by White and Rees (1978). Their theory takes account of two very important pieces of information about galaxies and their distribution which were obtained in the last few years by precise radioastronomical observations. It appears that the mass of a galaxy can be much larger than was supposed before, and that an appreciable part of this mass, up to 90%, may be contained in an extended "halo" composed mainly of non-luminous matter (cold dwarfs, neutron stars or black holes) (Valentijn and van der Laan, 1978; van den Bergh, 1978).

The distribution of galaxies can be described by means of a two-point correlation function defined as

$$\xi(r)=\langle n(\boldsymbol{x})\, n(\boldsymbol{x}+\boldsymbol{r})\rangle/\langle n\rangle^2-1, \tag{10.179}$$

where $n(r)$ is the galaxy density. The function $\xi(r)$ is related to the probability $\delta P(r)$ of finding a galaxy in each of two arbitrary regions with volumes δV_1 and δV_2 separated by a distance r by the equation

$$\delta P(r)=\langle n\rangle^2(1+\xi(r))\,\delta V_1\,\delta V_2. \tag{10.180}$$

A systematic analysis of the distribution of galaxies carried out by Peebles and hi collaborators (Peebles 1980) has shown that for $100h^{-1}\mathrm{kpc}\leqslant r\leqslant 10h^{-1}\mathrm{Mpc}$, where $h=H/100$, $\xi(r)=(r_0/r)^\gamma$, with $r_0\approx 5h^{-1}\mathrm{Mpc}$ and $\gamma=1.77\pm 0.04$. This means that the grouping of galaxies into clusters occurs in even larger regions than had been expected. This result also supports the hypothesis of the hierarchic structure of galaxy distribution, in which galaxies group into clusters, clusters into superclusters and so on.

Inspired by this information, White and Rees assume that the process of formation of the present distribution of matter was essentially governed by isothermal perturbations, which were not subject to damping during the radiation era. After recombination, isothermal perturbations, grew in the same way as adiabatic perturbations, but since they had not been damped, they could develop into systems with masses much smaller than M_s, and even comparable to stellar masses. White and Rees contend that a very large number of stars were formed shortly after recombination. Some of them were small stars, in which only a small fraction of mass was converted to radiation; others were very massive stars, which evolved very rapidly into dark cold dwarfs, neutron stars and black holes. The process of stellar formation was so effective that only about 10% of the matter was left in the form of gas. The practically point-like "granules" of dark matter so formed underwent further condensation, depending on the spectrum of the initial isothermal perturbations. Eventually, there emerged large concentrations of matter – galaxies, clusters of galaxies and so on. The gas which escaped the post-recombination condensation settled slowly in the central regions of galaxies, forming dense clouds which subsequently gave rise to a new generation of stars – the luminous component of galaxies. It is possible that the time-scale of gas and star condensation was very short (rapid relaxation). Dissipative processes could also play an important part. The isothermal theory of galaxy formation sees in these two effects the source of angular momentum and the reason why spiral galaxies were formed. It remains to explain the origin of large elliptic galaxies. In the course of further evolution galaxies could merge, in particular larger galaxies could swallow smaller ones (galactic cannibalism). As shown recently by Fall (1979), large elliptic galaxies could indeed develop from the coalescence of several smaller spiral galaxies.

According to the White–Rees theory, the first-generation stars, amounting to about 90% of the total mass, were formed very early – just after recombination. The thermonuclear processes occurring in those stars must have produced enormous amounts of energy; therefore the remaining gas must have been heated and, although this took place so early, it seems that the gas did not have enough time to cool down to the extent that the new stars might be formed from it. Here lies the main difficulty of the isothermal theory of galaxy formation.

Theories of galaxy formation based on rotational perturbations appear to have the most difficulties. Firstly, they force us to reject the Friedman model near the singularity. Secondly, the observed abundance of light elements sets very stringent limits on the amplitude of rotational perturbations. Finally, rotational perturbations capable of producing galaxies should also have caused much greater fluctuations of the relic radiation temperature than those actually observed.

The theories proposed so far use only linear approximations. Possible modifications may include post-recombination nonlinear interactions, which are negligible to start with but eventually lead to the generation of shortwave perturbations from the initial, purely adiabatic longwave perturbations. As a result, the spectrum of the initial perturbations would change, whereas in the linear theory only the amplitude of perturbations changes while their spectrum is unaffected. Although no concrete model of galaxy formation involving nonlinear interactions has yet been proposed, this line of approach seems very promising and may lead to a reconciliation between the adherents of the adiabatic theory and those of the isothermal theory.

10.15. Structure and Evolution of Galaxies

Systematic astronomical observation of galaxies began about fifty years ago. A great deal of information has accumulated since then (Blaauw and Schmidt, 1965; Sandage *et al.*, 1975; Setti, 1975; Berkhuijsen and Wielebinski, 1978). On the basis of the observed shapes, Hubble introduced a classification scheme, in which he distinguished four classes of galaxies: ellipticals (E), spirals (S), barred spirals (SB) and irregulars (Irr). Hubble's morphological classification is usually represented by the diagram

$$\mathrm{E0}-\mathrm{E7}-\mathrm{S0}\begin{matrix} \nearrow \mathrm{S}a-\mathrm{S}b-\mathrm{Sc}\mathrm{D} \searrow \\ \searrow \mathrm{SB}a-\mathrm{SB}b-\mathrm{SB}c \nearrow \end{matrix}\mathrm{Irr.}$$

Elliptical galaxies are denoted by En, where n is a measure of flatness defined as $n=10(a-b)/b$, with a and b the semi-major and the semi-minor axes of the ellipse. No galaxies with n greater than 7 have been observed so far. Elliptical galaxies contain practically no dust or young blue supergiants of type 0 or B. Only small quantities of gas are present. The isophotes (lines of constant luminosity) of elliptical galaxies form a family of similar ellipses. The distributions of matter and surface luminosity show no irregularities.

The characteristic features of spiral galaxies are central condensations of stars – spherical nuclei, and more or less distinct spiral arms in outer regions. The nuclei of barred spiral galaxies (SB) are not spherical but have the form of elongated, cigar-shaped cross-bars. Depending on the degree of winding of the spiral arms and resolution of the arms into bright stars S and SB galaxies are further classified into a, b and c types; galaxies of the type Sa or SBa have the most convolute arms. SO galaxies constitute an intermediate class between ellipticals and spirals. They are disc-shaped, with no distinguishable spiral arms.

Spiral galaxies contain large amounts of dust. The proportional content of neutral hydrogen increases as we pass from S*a* (SB*a*) to S*c*(SB*c*) galaxies. In contrast with ellipticals, spiral galaxies often contain globular clusters – nearly spherical systems of up to several hundred thousand stars. Young bright stars are another feature of spiral galaxies.

Galaxies that cannot be classified either as elliptical or spiral are called irregular. Their shapes and basic characteristics vary considerably.

Spiral and barred spiral galaxies constitute about 80% of all the galaxies observed; the least numerous are irregular galaxies, at about 5%. As regards size, spiral galaxies are the largest, their diameters reach a few dozen kpc. The most massive galaxies, on the other hand, are elliptical galaxies, whose masses may be as large as $10^{13}\ M_{\odot}$.

The classification outlined above is purely observational and is not intended to correspond to the successive stages of the evolution of galaxies. It seems that the first galaxies to emerge were mainly spiral. Elliptical galaxies could arise, for example, from a fusion of a number of spiral galaxies. Interactions between different galaxies (tidal effects, collisions) can also play an important part in their evolution.

Over the last few years an increasing amount of data has accumulated which indicates that galaxies have very extensive halos – up to 50 kpc in diameter, composed of non-luminous matter, whose mass may be as much as 9 times the mass of the luminous component of a given galaxy. If this proves true, the so-called missing mass problem will be solved. The problem sprang up when several galactic masses were estimated on the basis of the virial theorem and on the assumption that clusters of galaxies are stable (gravitationally bound). The values obtained in that way were found to be from a few times to a few dozen times larger than the estimated mass of the luminous matter. The issue is most striking when we compare the mass-luminosity ratio for a large and bright ellipticall galaxy, typically $M/L \approx 10 \sim 100$, where M and L are expressed in solar units, with the corresponding ratio for galactic clusters, which is several hundred and sometimes even more, if the mass is calculated on the basis of the virial theorem.

Detailed information about the properties of galaxies can be found in many extensive review articles, books and monographs (de Vaucouleurs, 1959; Sandage *et al.*, 1975; Setti, 1975; Hodge, 1966; Tayler, 1978; and others).

The best-studied galaxy is of course our Galaxy, which is one of the largest spiral galaxies of type S*b*. The stars and the interstellar matter contained in our Galaxy form a disc, with a dense, spherical centre and spiral arms. The diameter of the disc is 25 kpc and its thickness about 2 kpc. The Galaxy has a spherical halo with a diameter of about 30 kpc. The total mass is estimated to be $1.4 \cdot 10^{11}\ M_{\odot}$; this should be regarded as a lower limit. The Sun is at a distance of about 10 kpc from the centre. The stars are distributed within about 1 kpc of each side of the plane of symmetry of the disc, whereas the interstellar matter is concentrated nearer the plane and is almost non-existent at distances exceeding 200 pc. The disc is thus strongly flattened. The interstellar matter appears in the form of clouds with diameters of order 10 pc. The density of matter in these clouds is from 3 to 10 times higher than in the neighbouring regions. The Galaxy rotates about the centre with an angular velocity varying approximately in inverse proportion to the distance from the centre, so that the linear velocities are almost the same everywhere. The linear velocity is estimated to be about 250 km·s^{-1}. The orbital period of a star 8 kpc away from the centre is 200 million years, and that of a star at 4 kpc from the centre – 100 million years.

The distribution of stars and gas in our Galaxy is typical of all normal spiral galaxies. Spiral galaxies are flattened and disc-shaped, typically with two, but in some cases more, spiral arms, more or less clearly developed but always present. It is not easy to explain the origin and nature of spiral structure of galaxies. The fundamental question is whether the spiral arms are lasting forms of matter distribution or only density waves. In the first case, the rotation of the arms about the centre of the galaxy would be the actual rotation of the matter they contain; in the second case a density wave would move relative to the background of matter forming the galactic disc.

A spiral structure can be described comparatively easily: it is enough to subject a stationary axisymmetric system to a small perturbation. Let us consider small perturbations of a thin gaseous disc. Choose a cylindrical coordinate system (ρ, z, φ) so that the symmetry plane of the disc is $z=0$ and the coordinate origin coincides with the disc centre. The unperturbed disc is in a stationary state, rotating inhomogeneously about the centre. The angular velocity of the rotation, $\Omega(\rho)$, falls with the increasing radial distance. Suppose that the disc is infinitely thin, so that the velocity component in the z-direction u_z is 0. Let the surface density of matter be $\sigma(\rho, \varphi, t)$.

The continuity equation for matter density can be written in the form

$$\frac{\partial\sigma}{\partial t}+\frac{1}{\rho}\left[\frac{\partial}{\partial\rho}(\sigma\rho u_\rho)+\frac{\partial}{\partial\varphi}(\sigma u_\varphi+\sigma\rho\Omega)\right]=0. \tag{10.181}$$

Now if we put $\sigma=\sigma_0(\rho)+\delta\sigma$, treat $\delta\sigma$, u_ρ and u_φ as infinitesimal quantities and retain only terms linear in $\delta\sigma$, we obtain

$$\frac{\partial}{\partial t}\delta\sigma+\frac{1}{\rho}\left[\frac{\partial}{\partial\rho}(\sigma_0\rho u_\rho)+\frac{\partial}{\partial\varphi}(\sigma_0 u_\varphi+\delta\sigma\rho\Omega)\right]=0. \tag{10.182}$$

Since equation (10.182) is linear and its coefficients depend only on ρ, we can seek a solution of the form

$$\delta\sigma(\rho, \varphi, t)=\mathrm{Re}\{A(\rho)\exp[i(\omega t-m\varphi+\Phi(\rho))]\}, \tag{10.183}$$

where ω is a constant and m is an integer. If $A(\rho)$ is a slowly-varying function and $\Phi(\rho)=\Lambda f(\rho)$, where $f(\rho)$ also varies slowly and Λ is a large parameter, we obtain a tightly wound spiral (Fig. 10.13). To see this, notice that $\delta\sigma$ is approximately constant along

$$\omega t-m\varphi+\Phi(\rho)=C=\mathrm{const} \tag{10.184}$$

and is maximum when C is a multiple of 2π. Equation (10.183) describes a spiral structure which at a given moment of time is given by

$$m(\varphi-\varphi_0)=\Phi(\rho)-\Phi(\rho_0) \tag{10.185}$$

and, as we can see, consists of m arms. The spiral rotates about the centre at a constant angular velocity $\Omega_p=\omega/m$.

The structure of a spiral arm can be described more precisely by giving at each point the inclination angle i between the tangent to the arm and the tangent to the central circle passing through that point.

Using (10.182), we find

$$\varphi(\rho)-\varphi(\rho_0)=\frac{(\rho-\rho_0)}{m}\left.\frac{d\Phi}{d\rho}\right|_{\rho=\rho_0}. \tag{10.186}$$

The derivative of the function $\Phi(\rho)$ is called the radial wave number. The arm inclination

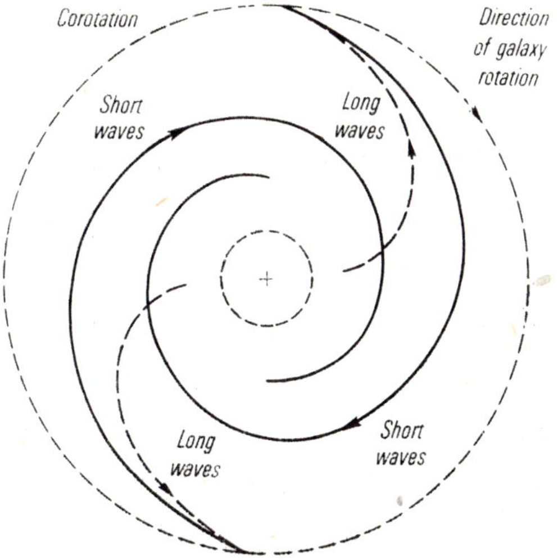

FIG. 10.13. Depending on the form of the function $\Phi(\rho)$ either short waves and a tightly wound spiral or long waves and a more open spiral are obtained.

angle at a point $\rho=\rho_0$ can be found from

$$\tan i=\frac{\rho-\rho_0}{\rho_0(\varphi-\varphi_0)}=\frac{m}{\rho_0\Phi'(\rho_0)}=\frac{m}{\rho_0 k(\rho_0)}. \tag{10.187}$$

If $k(\rho)$ is large, the arm is strongly wound. According as $k<0$ or $k>0$, the arm lags behind or leads the rotation.

Using the continuity equation alone, we have shown that the density distribution in a thin disc can have the form of density waves. It needs verifying whether such a configuration is consistent with the Poisson equation. If the density distribution is assumed to be of the form $\sigma(\rho,\varphi,t)=\sigma(\rho)\cdot\exp i(\omega t-m\varphi+\Phi(\rho))$, the solution of the Poisson equation

$$\Delta\psi=4\pi G\sigma(\rho,\varphi,t)\delta(z) \tag{10.188}$$

is

$$\psi(\rho,\varphi,z,t)=\tilde{\psi}(\rho,z)\exp[i(\omega t-m\varphi+\Phi(\rho))],$$

where

$$\tilde{\psi}(\rho,z)=\frac{2\pi\tilde{\sigma}(\rho)}{k(\rho)|}e^{-|k(\rho)|\cdot|z|}. \tag{10.189}$$

Analyzing propagation of small perturbations, one can derive the dispersion relation, i.e. the relation between the wavelength and frequency. A strict derivation is rather complicated, so here we shall content ourselves with a heuristic argument. For a gas cloud, in which only gravitational forces and pressure play any significant role, the Jeans dispersion relation has the familiar form

$$\omega^2 = k^2 v_s^2 - 4\pi G\rho\,, \tag{10.190}$$

and for a rotating disc, where rotation of matter must also be taken into account, we get

$$\omega^2 = \kappa^2(\rho) + k^2 v_s^2 - 2\pi G\sigma(\rho)\,|k|\,, \tag{10.191}$$

where κ is the epicyclic frequency. In the density wave theory, if the coordinate system is chosen so as to rotate at an angular velocity that is multiple of the angular velocity of the rotation of matter Ω_c, the dispersion relation is

$$(\omega - m\Omega_c(\rho))^2 = \kappa^2(\rho) - 2\pi G\sigma(\rho)\,|k|\,\mathscr{F}_\nu(\chi)\,, \tag{10.192}$$

where $\nu = \dfrac{\omega - m\Omega}{\kappa}$, $\chi = \dfrac{k^2 u^2}{\kappa^2}$ and u is the peculiar velocity of the star. The function $\mathscr{F}_\nu(\chi)$ is called the reduction factor. When $\chi = 0$, $\mathscr{F}_\nu(0) = 1$, and for $\chi \neq 0$ $\mathscr{F}_\nu(\chi) \leqslant 1$. When $\sigma(\rho) \to 0$, the dispersion relation becomes particularly simple:

$$(\omega - m\Omega_c)^2 = \kappa^2\,, \tag{10.193}$$

which implies that in general

$$\Omega_c(\rho) - \frac{\kappa(\rho)}{m} < \Omega_p < \Omega_c(\rho) + \frac{\kappa(\rho)}{m}\,. \tag{10.194}$$

These limiting values of the angular velocity of the spiral are referred to as Lindblad resonances. The matter of the disc feels gravitational field varying periodically with the frequency ω.

Density waves occur only in that region of the disc in which inequality (10.194) holds. The smaller m, the larger this region is. Disregarding the two special cases $m=0$ and $m=1$, we can conclude that the density wave region is maximum in two-arm structures ($m=2$). Observations show that the spiral arms of a typical galaxy extend over the whole galaxy. When $m>2$, with any admissible Ω_p, the region where density waves can occur is very small. The fact that most spiral galaxies possess only two arms may be connected with just this property of density waves.

With m fixed, equation (10.192) becomes a relation between ω and k, so for a given ω, we can find corresponding k. Physically, this means exciting oscillations of frequency ω; propagating, they form a wave. Wavelength of this wave can be determined from the dispersion relation. If the medium is inhomogeneous, the wavelength depends on position. In a system rotating differentially, spiral waves rotate rigidly, that is their angular velocity is independent of the distance from the rotation axis. Owing to this, density waves can explain the existence and lasting character of spiral structure. In a system rotating differentially, spiral waves rotate rigidly, that is their angular velocity is independent of the distance from the rotation axis.

Predictions of the density wave theory can be tested by comparison with observational data. The distribution of neutral hydrogen in the disc of the Galaxy obtained by radio measurements (the so-called *HI* regions (Fig. 10.14)) shows good agreement with the theory if we assume that $\Omega_p = 11\text{–}13\ \text{km}\cdot\text{s}^{-1}\cdot\text{kpc}^{-1}$. Density waves should cause departures from the purely rotational motion of the gas. Indeed, the gas velocity distribution in the Galaxy does show such departures. Particle velocities exhibit additional systematic changes of order 10 $\text{km}\cdot\text{s}^{-1}$, which may well be attributed to density waves.

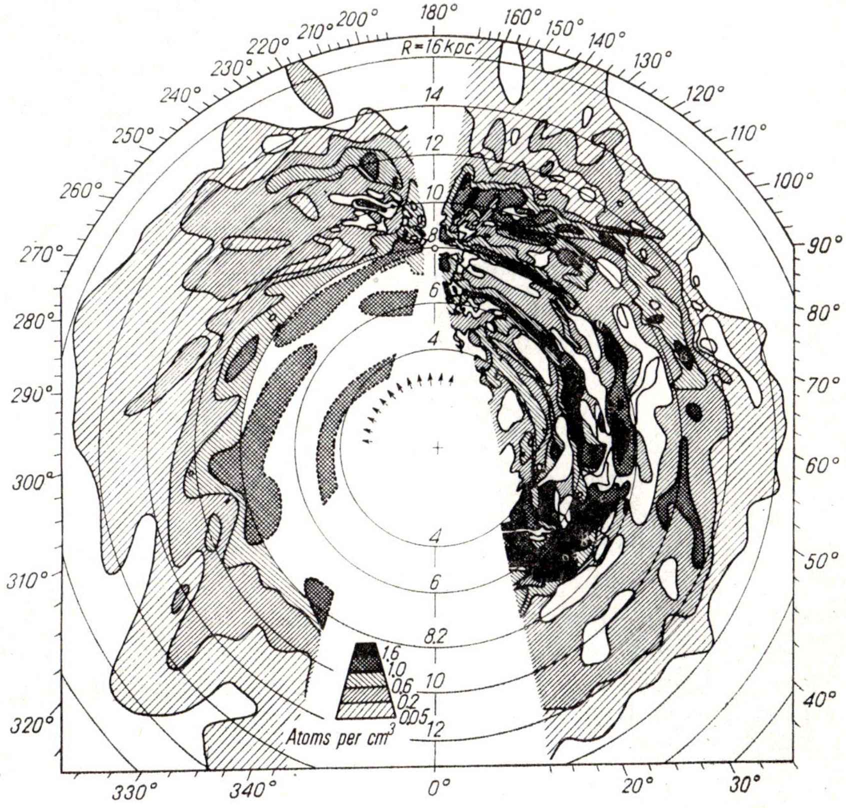

FIG. 10.14. The radio-observed distribution of hydrogen in the plane of our galaxy (adapted from Kerr and Westerhout, 1965).

It is also known that the distribution of young stars coincides with the spiral structure. Stars are probably formed in regions with the highest concentration of gas, and since this occurs in the spiral arms, it is in the arms that young stars should be most numerous. Matter rotates about the centre of the Galaxy, and therefore the stars which are formed in the spiral arms will migrate from them, because $\Omega_c > \Omega_p$. If we take $\Omega_p = 13\ \text{km}\cdot\text{s}^{-1}\cdot\text{kpc}^{-1}$, the relative displacement of the stars and the gaseous arm in the neighbourhood of the Sun during 10 million years will be about 1.2 kpc, with $\Omega(\rho_\odot) = 25\ \text{km}\cdot\text{sec}^{-1}\cdot\text{kpc}^{-1}$. Considering the small inclination angle of the arm, the relative radial displacement will be less by a factor of 0.1. The effect is hardly observable and for all practical purposes we can assume that young stars lie within the spiral arms.

Despite its undeniable successes, the density wave theory encounters serious difficulties. In the first place, there is the unexplained wave generation mechanism. Numerical models, in which the collision-free motion of several hundred particles is used as a flat two-dimensional approximation of a galaxy, show that the particles tend to form a spiral structure very quickly. Their peculiar velocities grow rapidly and become equal to the rotational velocity. The whole system remains in equilibrium owing to pressure and not to rotation. Further evolution makes the spiral structure decay (Hohl and Hockney, 1969).

A more precise analysis of the propagation of density waves – allowing for viscosity – has led to the conclusion that density waves rapidly die out with time. To avoid this difficulty, Toomre (1969) has suggested that wave packets should be used instead of single density waves. The spiral structure would then be formed from wave packets and their evolution would be determined by the group velocity v_g. Even in this case, however, the spiral approaches the Lindblad resonance after no more than a few revolutions and the spiral structure ceases to exist as a global property.

The theory of density waves is developing vigorously and it may be hoped that the difficulties it now encounters will be overcome. Detailed information can be found in original papers and review articles (Shu, 1970; Lin and Shu, 1968, 1972; Lin, 1967, 1975; Marochnik and Suchkov, 1974; Rohlfs, 1977). Several related problems also call for closer study. Of particular interest is the behaviour of gas when it meets a density wave. Shock waves and turbulent motions which can arise in this process will enhance condensation of the gas and thus contribute to the formation of stars. The theory of density waves may also add to our understanding of the structure of interstellar dust clouds and their evolution.

Evolution of galaxies, an issue of great cosmological importance, has been and continues to be the subject of extensive studies. So far there is no comprehensive theory which could be termed a theory of galaxy evolution. Particularly promising in this context are current investigations of changes in the total luminosity of galaxies and galactic clusters on a cosmological time scale. We refer the readers interested in these problems to a collection of articles edited by Tinsley and Larson (1977).

10.16. Other Cosmological Models. The Mixmaster Universe

The Friedman model is obtained as a cosmological solution of Einstein's equations on the assumption that the distribution of matter is homogeneous and isotropic. This very simple assumption leads to a surprisingly good description of the large-scale evolution of the universe. In the present epoch the distribution of matter over large regions is isotropic and, as follows from the relic radiation anisotropy measurements, the anisotropy of matter distribution at the time of recombination was very small. The measurements only give upper bounds anyway. However, the question of whether or not the Friedman model provides an adequate description of the early universe has not been definitely answered by observation. Its predictions do not conflict with observational data, but it is not the only model theoretically possible. It is conceivable for example that the initial distribution of matter was highly anisotropic and that this initial anisotropy decayed rapidly with time.

For simplicity, let us consider models in which the requirement of isotropy has been dropped but which are still homogeneous. A space-time is homogeneous if there is a class of observers who are unable, at any time, to distinguish any point in the space. More precisely, we say that a space-time is spatially homogeneous if it admits at least one three-parameter group of motions, transitive on spacelike three-surfaces. Let us recall that a group of motions is transitive on a space S if for every two points of S there is an element in the group which carries one of the points to the other. Bianchi (1897) classified all possible three-parameter groups transitive on spacelike three-surfaces. There are nine types of such groups; we call them *Bianchi types* (Taub, 1951; Ryan and Shepley, 1975; MacCallum, 1971, 1979a, b).

The coordinate system in a spatially homogeneous space-time can be chosen so that the surfaces of transitivity coincide with the surfaces t=const. Time is measured along timelike geodesics perpendicular to these surfaces. The line element can then be written

$$ds^2 = c^2 dt^2 - \gamma_{ab}(t)\, e^a{}_i(x^k)\, e^b{}_j(x^k)\, dx^i dx^j, \tag{10.195}$$

where $\gamma_{ab}(t)$ is a symmetric, non-singular matrix 3×3, and $e^a_i(x^k)dx^i = \omega^a$ is a triad of one-forms such that

$$d\omega^a = \tfrac{1}{2} C^a_{bc}\, \omega^b \wedge \omega^c\,; \tag{10.196}$$

C^a_{bc} are structural constants defining the Bianchi type. Einstein's equations for a perfect fluid are

$$\tfrac{1}{2}(\ln\gamma)^{\cdot\cdot} + \tfrac{1}{4}\dot{\gamma}_{ab}\gamma^{ab\cdot}\dot{\gamma}_{cd}\gamma^{cd} = \frac{8\pi G}{c^4}\left(T^0_0 - \tfrac{1}{2}T\right), \tag{10.197}$$

$$\tfrac{1}{2}\dot{\gamma}_{ab}\gamma^{ad}\left(C^b_{dc} - \delta^b{}_c C^f_{fd}\right) = \frac{8\pi G}{c^4}\, T^0_c \tag{10.198}$$

$$R^*_{cd} + \tfrac{1}{2}\ddot{\gamma}_{cd} - \tfrac{1}{2}\dot{\gamma}_{ca}\gamma^{ab\cdot}\dot{\gamma}_{bd} + \tfrac{1}{4}\dot{\gamma}_{cd}(\ln\gamma)^{\cdot} = \frac{8\pi G}{c^4}\left(T_{cd} - \delta_{cd}T\right), \tag{10.199}$$

where

$$R^*{}_{cd} = a^f{}_{cg}\, a^g{}_{df} + C^f_{fg}\, a^g_{cd}, \tag{10.200}$$

$$a^f{}_{cg} = \tfrac{1}{2}\left(C^f_{cg} + C^f_{cg} + C^f_{gc}\right), \tag{10.201}$$

and $T = T^\alpha_\alpha$, $\gamma = \mathrm{Det}\,\|\gamma_{ab}\|$. The equations of motion of the fluid, $T^{\alpha\beta}_{;\beta}=0$, can be written as a system of equations

$$\begin{gathered} -(\rho c^2 + p)\,\dot{p}_c p^c = \dot{p} p_c p^c, \\ (\rho c^2 + p)(u_0 \dot{p}_c + p_a p^b C^a_{cb}) + u_0 p_c \dot{p} = 0, \\ (\rho c^2 + p)(\dot{u}_0 + p^c C^b_{bc} + \tfrac{1}{2} u_0 \dot{\gamma}/\gamma) + c^2 \dot{\rho} u_0 = 0; \end{gathered} \tag{10.202}$$

$p_c = e^i_c u_i$ are the spatial components of the velocity.

We shall only discuss the two most interesting cases, namely the types I and IX. A Bianchi I space admits a group isomorphic with the group of translations in a 3-dimensional Euclidean space; the group of motions in a space of type IX is isomorphic with the group of rotations $O^+(3)$.

For the *Bianchi I type* the structural constants are all zero. Coordinates on the three-surfaces of transitivity can be chosen so that $e^a_i(x^k) = \delta^a_i$. A particular case of a metric of type I was given by Kasner (1921):

$$ds^2 = c^2 dt^2 - \left(\frac{t}{\tau}\right)^{2p_1} dx^2 - \left(\frac{t}{\tau}\right)^{2p_2} dy^2 - \left(\frac{t}{\tau}\right)^{2p_3} dz^2; \tag{10.203}$$

the vacuum solution of Einstein's equations is obtained if

$$p_1^2 + p_2^2 + p_3^2 = p_1 + p_2 + p_3 = 1. \tag{10.204}$$

The constants p_1, p_2 and p_3 cannot be all equal, and equality of two of them is possible

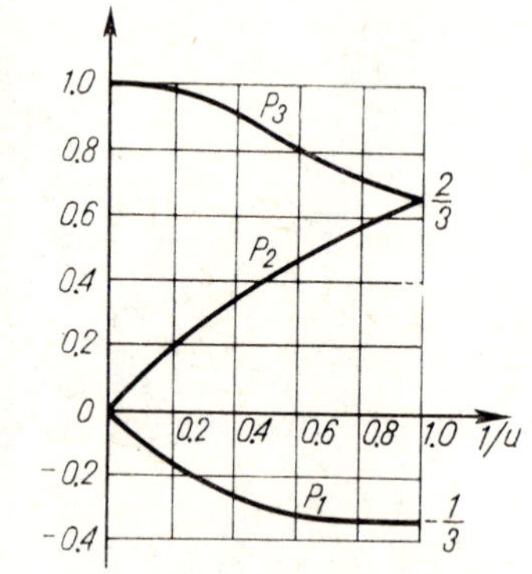

FIG. 10.15. Kasner parameters in relation to u.

only in the combinations (0, 0, 1) and $(-\frac{1}{3}, \frac{2}{3}, \frac{2}{3})$. In all other cases p_1, p_2 and p_3 are different and it can be assumed that $p_1 < p_2 < p_3$. Their respective ranges are

$$-\tfrac{1}{3} \leqslant p_1 \leqslant 0, \quad 0 \leqslant p_2 \leqslant \tfrac{2}{3}, \quad \tfrac{2}{3} \leqslant p_3 \leqslant 1$$

(Fig. 10.15). It is convenient to use a parametric representation

$$p_1(u) = \frac{-u}{1+u+u^2}, \quad p_2(u) = \frac{1+u}{1+u+u^2}, \quad p_3(u) = \frac{u(1+u)}{1+u+u^2}. \tag{10.205}$$

If $u > 1$, $p_1 < p_2 < p_3$; for $u < 1$ the same order is obtained after the transformation

$$p_1\left(\frac{1}{u}\right) \to p_1(u), \quad p_2\left(\frac{1}{u}\right) \to p_3(u), \quad p_3\left(\frac{1}{u}\right) \to p_2(u). \tag{10.206}$$

Kasner's model describes an empty space-time, with no matter, but of course we can consider motion of test particles in it. It then turns out that a test dust cloud would contract along the x-axis and expand at different rates along the axes y and z. The volume of the cloud would grow with time as t. In this simple case the anisotropy of the expansion does not decay with time because p_1, p_2 and p_3 are constant.

To consider a more general case, we shall introduce a different parametrization of the metric, namely

$$ds^2 = c^2 dt^2 - R_0^2 e^{-2\Omega} e_{ij}^{2\beta} \omega^i \omega^j, \tag{10.207}$$

where β is a symmetric, traceless matrix 3×3, and Ω is a scalar. Both β and Ω depend on time. R_0 is a constant whose value depends on the choice of units. If Ω is a monotonic function of time, we can introduce a new time coordinate $\tau=\Omega(t)$. In the case of Bianchi I spaces, without loss of generality, the matrix β can be assumed diagonal, and then

$$ds^2=c^2dt^2-e^{-2\Omega}[e^{2(\beta_++\sqrt{3}\beta_-)}dx^2+e^{2(\beta_+-\sqrt{3}\beta_-)}dy^2+e^{-4\beta_+}dz^2]. \quad (10.208)$$

To describe the distribution of matter we use the hydrodynamic energy-momentum tensor, with equation of state $p=(\gamma-1)\rho c^2$. From the equations of motion $T^{\alpha\beta}_{;\beta}=0$ it follows that

$$\rho=\rho_0\, e^{3\gamma\Omega}. \quad (10.209)$$

Einstein's equations for the metric (10.206) can be reduced to Hamilton's equations, with the Hamiltonian given by

$$H^2=p_+^2+p_-^2+\rho_0\, e^{-3\gamma\Omega}; \quad (10.210)$$

here Ω plays the role of time and p_+ and p_- are the momenta conjugate to β_+ and β_-, respectively. The Hamiltonian depends on time explicity but is independent of the generalized coordinates β_+ and β_-. Consequently, p_+ and p_- are constant and the history of the universe on the plane β_+, β_- is represented by a point moving in a straight line. Its velocity is

$$\left[\left(\frac{d\beta_+}{d\Omega}\right)^2+\left(\frac{d\beta_-}{d\Omega}\right)^2\right]^{\frac{1}{2}}=(p_+^2+p_-^2)^{\frac{1}{2}}H^{-1}. \quad (10\ 211)$$

As $\Omega\to\infty$, $H^2\to p_+^2+p_-^2$, and therefore the velocity tends to unity. The limit $\Omega\to\infty$ corresponds to a singularity. As we can see from (10.209), the density of matter then grows to infinity and the volume element $\omega^1\wedge\omega^2\wedge\omega^3=e^{-3\Omega}dx^1\wedge dx^2\wedge dx^3$ approaches zero. The time-dependence of Ω can be found from the equation

$$\frac{d\Omega}{c\,dt}=He^{3\Omega}. \quad (10.212)$$

Near the singularity H=const and we get $\Omega\approx-\frac{1}{3}\ln t$.

To show that the anisotropy of expansion in Bianchi I spaces decays with time, note that it is given by $n_i\sigma^{ij}n_j$, where n_i is the unit vector in the direction of observation and

$$\sigma_{ij}=\frac{1}{2}\left[e_{si}^{-\beta}\frac{d}{dt}\,e_{sj}^{\beta}+e_{sj}^{-\beta}\frac{d}{dt}\,e_{si}^{\beta}\right] \quad (10.2\ 3)$$

is the shear tensor. For metric (10.208), the scalar $\sigma^2=\sigma_{ij}\sigma^{ij}$ is found to be

$$\sigma^2=6\left[\left(\frac{d\beta_+}{d\Omega}\right)^2+\left(\frac{d\beta_-}{d\Omega}\right)^2\right]\left(\frac{d\Omega}{cdt}\right)^2=6(p_+^2+p_-^2)e^{2\Omega}. \quad (10.214)$$

Thus as Ω decreases, i.e. as time increases, the anisotropy declines, but it vanishes only asymptotically. In more realistic cosmological models the anisotropy decays much more rapidly owing to quantum processes which occur at early stages of the evolution.

Another homogeneous cosmological model which has been studied in detail is the Bianchi IX space (Misner, 1969a, b; Belinski *et al.*, 1970; Ryan, 1971, 1972; Doroshkevich *et al.*,

1973; Bogoyavlensky and Novikov, 1973). For simplicity, let us only consider the case where the matrix β is diagonal. The metric then takes the form

$$ds^2 = c^2 dt^2 - e^{-2\Omega}[e^{2(\beta_+ + \sqrt{3}\beta_-)}(\omega^1)^2 + e^{2(\beta_+ - \sqrt{3}\beta_-)}(\omega^2)^2 + e^{-4\beta_+}(\omega^3)^2], \quad (10.215)$$

where

$$\omega_1 = \sin\psi\, d\theta - \cos\psi\, \sin\theta\, d\varphi, \quad \omega^2 = \cos\psi\, d\theta + \sin\psi\, \sin\theta\, d\varphi \quad \text{and}$$

$$\omega^3 = -(d\psi + \cos\theta\, d\varphi)$$

are differential forms on the 3-sphere parametrized by Euler angles $0 \leqslant \psi \leqslant 2\pi$, $0 \leqslant \theta \leqslant \pi$, $0 \leqslant \varphi \leqslant 2\pi$. As in the previous case, we want to determine the dependence of β_+ and β_- on Ω. We are interested in the behaviour of the model near the singularity, for $\Omega \to \infty$.

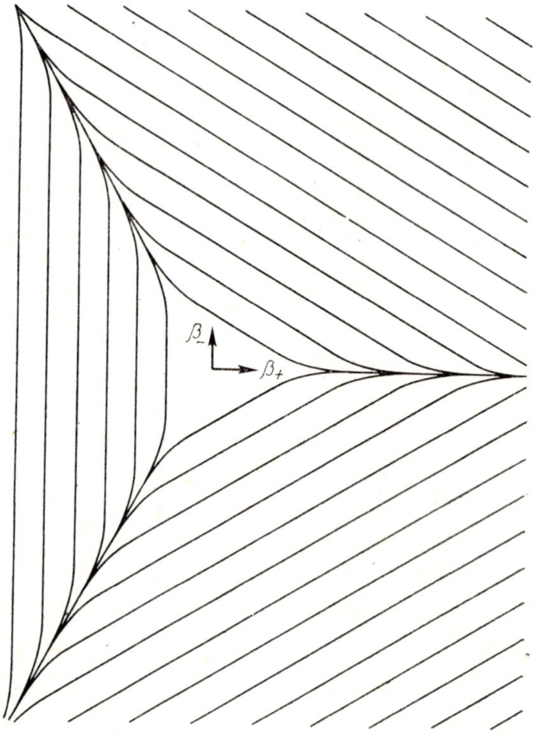

FIG. 10.16. Effective potential in the coordinates β_+, β_- (adapted from Misner, 1969 a).

The influence of radiation and matter will be neglected. The Hamiltonian

$$H^2 = p_+^2 + p_-^2 - e^{-4\Omega}(V(\beta_+, \beta_-) - 1) \quad (10.216)$$

depends on β_+ and β_- in a rather complicated way through the potential V (Fig. 10.16) given by

$$V(\beta_+, \beta_-) = 1 + \tfrac{2}{3}e^{4\beta_+}(\cosh 4\sqrt{3}\beta_- - 1) + \tfrac{1}{3}e^{-8\beta_+} - \tfrac{4}{3}e^{-2\beta_+}\cosh 2\sqrt{3}\beta_- . \quad (10.217)$$

To obtain a full analytical solution of this model seems an impossible task, but some interesting qualitative information about its behaviour can be derived quite easily. To this end, assume that the walls of the potential well $V(\beta_+, \beta_-)$ are perfectly elastic. In the Hamiltonian the potential is multiplied by $e^{-4\Omega}$ what implies that the walls of the potential barrier expand. The point describing the history of the universe in the plane (β_+, β_-) moves freely inside the potential well, and as long as it remains far from the walls, its velocity is unity. The walls of the well expand isotropically at the rate $\frac{d\beta_w}{d\Omega} \approx \frac{1}{2}$. The point representing the universe catches up with the walls and rebounds. The situation resembles the motion of a small ball enclosed in a triangular box. What law of reflection governs the bounces? The walls of the potential well expand and therefore the incidence and the reflection angles are not equal; it can be shown that they are related by the formula

$$\sin\theta_{out} = \frac{3\sin\theta_{in}}{5-4\sin\theta_{in}}. \tag{10.218}$$

Between collisions the Hamiltonian H is constant, but in each collision it jumps according to the rule

$$H_{in}\sin\theta_{in} = H_{out}\sin\theta_{out}, \tag{10.219}$$

where H_{in} and H_{out} are the values of H before and after the collision, respectively.

A different situation arises when after a few rebounds from the walls the point begins to move towards one of the corners of the well. The cross-section of the potential well is an equilateral triangle, and therefore the expansion rate of the corners is double that of the walls and is approximately unity. The point representing the universe moves at about the same velocity and needs a long time to catch up with the escaping edge. During this long period of free motion the potential can be neglected and the solution is given by relations similar to those in model I. If we represent the velocity components parametrically as

$$\begin{aligned} \frac{d\beta_+}{d\Omega} &= \tfrac{1}{2}(3p_3-1) = \frac{u^2+u-\frac{1}{2}}{u^2+u+1}, \\ \frac{d\beta_-}{d\Omega} &= \frac{\sqrt{3}}{2}(p_2-p_1) = \sqrt{3}\,\frac{u+\frac{1}{2}}{u^2+u+1}, \end{aligned} \tag{10.220}$$

where p_1, p_2 and p_3 are the Kasner parameters, then the directions of the edges are given by $u = -1, 0, \infty$. It is readily seen that for each of these values two of the Kasner parameters are zero and the third equals 1. The metric becomes

$$ds^2 = c^2dt^2 - \left(\frac{t}{\tau}\right)^2 dx^2 - dy^2 - dz^2, \tag{10.221}$$

and if we take into account that the universe is now closed and x changes only from 0 to 4π, we can see at once that after a sufficiently long period of time a light ray running along the x-axis will have made its way round the whole universe.

Using this fact, Misner (1969a) has advanced the hypothesis of what he has termed "great mixing". Misner assumes that the initial distribution of matter was highly anisotropic and inhomogeneous. But, from the early times, light could travel round the universe in all directions, transmitting information and energy over arbitrary regions and gradually smoothing out any inhomogeneities. It is this mixing, suggests Misner, that has led to the observed large scale isotropy and homogeneity of the present universe.

This so-called *Mixmaster model*, attractive as it is, has proved very difficult to defend. A very detailed analysis of light propagation in Bianchi IX spaces carried out by Grishchuk *et al.* (1971) indicates that in practice light will not have enough time to travel round the whole universe in any direction even once. And thus the mystery of the observed isotropy of the large-scale distribution of matter remains unexplained.

Literature

ALLER, L. H. (1958) *The Abundances of the Elements in the Sun and Stars*, in *Encyclopedia of Physics*, ed: Flügge, S., vol. L1, Astrophysics II: Stellar Structure, Springer-Verlag, Berlin.

ALLER, L. H. (1961) *Abundance of the Elements*, Interscience Publishers, New York.

ALPHER, R. A. and HERMAN, R. C. (1949) *Phys. Rev.*, **75**, 1089.

BARDEEN, J. M. (1980) *Phys. Rev.* **D22**, 1882.

BARROW, J. D. (1976) *Mon. Not. Roy. Astr. Soc.* **175**, 359.

BARROW, J. D. (1977) *Nature* **267**, 117.

BARROW, J. D. and MATZNER, R. A. (1977) *Mon. Not. Roy. Astr. Soc.* **181**, 719.

BELINSKII, V. A., KHALATNIKOV, I. M. and LIFSHITZ, E. M. (1970) *Adv. Phys.* **19**, 525.

VAN DEN BERGH, S. (1975) *The Extragalactic Distance Scale*, in *Galaxies and the Universe*, eds.: Sandage, A., Sandage, M. and Kristian, J., Vol. IX of Stars and Stellar Systems, gen. ed.: Kuiper, G. P., The University of Chicago Press, Chicago.

VAN DEN BERGH, S. (1978) *Galaxy Haloes and the Missing Mass Problem*, in *Structure and Properties of Nearby Galaxies*, eds.: Berkhuijsen, E. M. and Wielebinski, R.; D. Reidel Publ. Comp., Dordrecht.

BERKHUIJSEN, E. M. and WIELEBINSKI, R. (eds.) (1978) *Structure and Properties of Nearby Galaxies*, D. Reidel Publ. Comp., Dordrecht.

BIANCHI, L. (1897) *Mem. Soc. Ital. delle Sci.* **11**, 267.

BINNEY, J. (1974) *Mon. Not. Roy. Astr. Soc.* **168**, 73.

BLAAUW, A. and SCHMIDT, M. (eds.) (1965) *Galactic Structure*, Vol. V of Stars and Stellar Systems, gen. ed: Kniper, G. P., The University of Chicago Press, Chicago.

BOGOYAVLENSKY, O. I. and NOVIKOV, I. D. (1973) *Sov. Phys. J.E.T.P.* **37**, 747.

BONDI, H. (1961) *Cosmology*, Cambridge University Press, Cambridge.

BONNOR, W. B. (1957) *Mon. Not. Roy. Astr. Soc.* **117**, 104.

BOYNTON, P. E. and PATRIDGE, R. B. (1973) *Astrophys. J.* **181**, 243.

CA LITZ, R., FRAUTSHI, S. and NAHM, W. (1973) *Astron. Astrophys.* **26**, 171.

DANESE, L. and DE ZOTTI, G. (1977) *Rivista del Nuovo Cimento* **7**, 277.

DANESE, L. and DE ZOTTI, G. (1978) *Astron. Astrophys.* **68**, 157.

DANZIGER, I. J. (1970) *Ann. Rev. Astron. Astrophys.* **8**, 161.

DICKE, R. H., PEEBLES, P. J. E., ROLL, P. G. and WILKINSON, D. T. (1965) *Astrophys. J.* **142**, 414.

DOLGOV, A. D. and ZELDOVICH, YA. B. (1981) *Rev. Mod. Phys.* **53**, 1.

DOROSHKEVICH, A. G. (1972) *Sov. Astron.* **16**, 986.

DOROSHKEVICH, A. G., LUKASH, V. N. and NOVIKOV, I. D. (1971) *Sov. Phys. J.E.T.P.* **33**, 649.

DOROSHKEVICH, A. G., LUKASH, V. N. and NOVIKOV, I. D. (1973) *Sov. Phys. J.E.T.P.*, **37**, 739.

DOROSHKEVICH, A. G., SUNYAEV, R. A. and ZELDOVICH, YA. B. (1974) *Galaxy Formation in the Fried-*

man Models of the Universe, in *Confrontation of Cosmological Theories with Observational Data*, ed.: Longair, M., D. Reidel Publ. Comp., Dordrecht.

DYSON, F. J. (1979) *Rev. Mod. Phys.* **51**, 447.

ELLIS, J., GAILLARD M. K., and NANOPOULOS, D. V. (1980) *Grand Unification and Cosmology* in *Unification of th Fundamental Particle Interactions*, eds. Ferrara, S., Ellis, J., Van Nieuwenhuizen, P., Plenum Publishing Corporation, New York.

FALL, M. S. (1979a) *Rev. Mod. Phys.*, **51**, 21.

FALL, S. M. (1979b) *Nature* **281**, 200.

FIELD, G. B. (1969) *Rivista del Nuov Cimento* **1**, 87.

FIELD, G. B. (1975) *The Formation and Early Dynamical History of Galaxies*, in *Galaxies and The Universe*, eds.: Sandage, A., Sandage, M. and Kristian, J., Vol. IX of Stars and Stellar Systèm, gen, ed.: Kuiper, G. P., The University of Chicago Press, Chicago.

FIELD, G. B. and SHEPLEY, L. C. (1968) *Astrophys. Space Sci.* **1**, 309.

FRI DMAN, A., *Z. Physik* **10**, 377 (1922); **21**, 326 (1924).

GAMOW, G. (1948) *Nature*, **162**, 680.

GOTT, J. R. (1977) *Ann. Rev. Astron., Astrophys.* **15**, 235.

GOTT, J. R., GUNN, J., SCHRAMM, D. N. and TINSLEY, B. (1974) *Astrophys. J.* **194**, 543.

GREENBERG, O. W. (1978) *Ann. Rev. Nucl. Part. Sci.*, **28**, 327.

GRISHCHUK, L. P., DOROSHKEVICH, A. G. and LUKASH, V. N. (1971) *Sov. Phys. JETP* **34**, 1.

GUNN, J. E. (1977) *Observational Tests in Cosmology*, in *Décalages vers le Rouge et Expansion de l'Universe – l'Evolution des Galaxies et ses Implications Cosmologiques*, eds:. Balkowski, G. and Westerlund, B. E., CNRS Paris

GUNN, J. E. (1978) *The Friedmann models and optical observations in cosmology*, in *Observational Cosmology*, eds:. Maeder, A., Martinet, L. and Tammann, G., Geneva Observatory, Sauverny.

GUNN, J. E. and TINSLEY, B. M. (1975) *Nature* **257**, 454.

HAGEDORN, R. (1965) *Nuovo Cimento Suppl.* **3**, 147.

HARRISON, E. R. (1967) *Rev. Mod. Phys.* **39**, 862.

HARRISON, E. R. (1973) *Ann. Rev. Astron. Astrophys.* **11**, 155.

HAWKING, S. W. (1966) *Astrophys. J.* **145**, 544.

HAYASHI, C. (1950), *Progr. Theor. Phys.* **5**, 224.

HODGE, P. W. (1966) *Galaxies and Cosmology*, McGraw-Hill, New York.

HOHL, F. and HOCKNEY, R. W. (1969) *J. Comp. Phys.* **4**, 367.

HU, B. L. (19 0) *Elementary Particle Physics and Cosmology*, in Proceedings of the First Theoretical Particle Physics Conference, Academy of Science Press, Peking.

JAUNCEY, D. L. (ed.) (1977) *Radio Astronomy and Cosmology*, D. Reidel Publ. Comp., Dordrecht.

JEANS, J. (1902) *Phil. Trans. Roy. Soc.* **199A**, 49.

JONES, B. J. T. (1976) *Rev. Mod. Phys.* **48**, 107.

KASNER, E. (1921) *Am. J. Math.* **43**, 217.

KERR, F. J. and WESTERHOUT, G. (1965) *Distribution of Interstellar Hydrogen in Galactic Structure*, eds.: Blaauw, A., Schmidt, M., Vol. V of *Stars and Stellar System*, gen. ed.: Kuiper, G. P., The University of Chicago Press, Chicago.

KOLB, E. W. and WOLFRAM, S. (1980) *Nucl. Phys.* **B172**, 224.

KOMPANEETS, D. A., LUKASH, V. N. and NOVIKOV, I. D. (1982) *Sov. Astron.* **26**, 259.

KUNDT, W. (1968) *Recent Progress in Cosmology* Springer Tracts in Modern Physics 47, 111.

KUNDT, W. (1971) *Survey of Cosmology. Is Our World Implied by Thermal Equilibrium in the Hadron Era*? Springer Tracts in Modern Physics, Vol. 58.

LIFSHITZ, E. M. (1946) *J. Phys. USSR*, **10**, 116.

LIFSHITZ, E. M. and KHALATNIKOV, I. M. (1963) *Adv. Phys.* **12**, 185.

LIN, C. C. (1967) *Ann. Rev. Astron. Astrophys.* **5**, 453.

LIN, C. C. (1975) *Theory of Spiral Structure*, in *Structure and Evolution of Galaxies*, ed.: Setti, G., D. Reidel Publ. Comp., Dordrecht.

LIN, C. C. nd SHU, F. H. (1968) *Density Wave Theory of Spiral Structure*, in *Astrophysics and Genaral Relativity*, Vol. II; eds.: Chreten, M., Deser, S. and Goldstein, J., Gordon and Breach New York.

LIN, C. C. and SHU, F. H. (1972) *Theory of Galactic Spirals*, MIT Press, Cambridge, Mass.

LINDE, A. P. (1979) *Rep. Progr. Phys.* **42**, 389.

LONGAIR, M. S. (1971) *Rep. Progr. Phys.* **34**, 1125.

LONGAIR, M. S. (ed.) (1974) *Confrontation of Cosmological Theories with Observational Data*, D. Reidel Publ. Comp., Dordrecht.

LONGAIR, M. S. (1978) *Radio Astronomy and Cosmology*, in *Observational Cosmology*, eds.: Maeder, A., Martinet, L. and Tammann, G., Geneva Observatory.

LUKASH, V. N. (1981) *Sov. Phys. J.E.T.P* **52**, 807.

LUKASH, V. N., NOVIKOV, I. D., STAROBINSKY, A. A. and ZELDOVICH, YA. B. (1976) *Nuovo Cimento* **35B**, 293.

MACCALLUM, M. A. H. (1971) *Cosmological Models from a Geometrical Point of View*, in *General Relativity and Cosmology*, ed.: Sachs, R., Academic Press Inc., New York.

MACCALLUM, M. A. H. (1979a) *The Mathematics of Anisotropic Spatially-Homogeneous Cosmologies*, in *Physics of the Expanding Universe*, ed.: Demiański, M., Springer-Verlag, Berlin.

MACCALLUM, M. A. H. (1979b) *Anisotropic and Inhomogeneous Relativistics Cosmologies*, in *General Relativity: An Einstein Centenary Survey*, eds.: Hawking, S. W. and Israel, W., Cambridge University Press, Cambridge.

MAROCHNIK, L. S. and SUCHKOV, A. A. (1974) *Sov. Phys. Uspekhi* **17**, 85.

MC KELLAR, A. (1941) *Publ. Dom. Astrophys. Obs. Victoria B. C.* **7**, 251.

MILNE, E. A. (193) *Quart. J. Math. Oxford* **5**, 64.

MILNE, E. A. and MC CREA, W. H. (1934) *Quart. J. Math. Oxford* **5**, 73.

MISNER, C. W. (1967a) *Nature* **214**, 40.

MISNER, C. W. (1967b) *Astrophys. J.* **151**, 431.

MISNER, C. W. (1969a) *Phys. Rev. Lett.* **22**, 1071.

MISNER, C. W. (1969b) *Phys. Rev.* **186**, 1319.

NOVIKOV, I. D. and ZELDOVICH, YA. B. (1967) *Ann. Rev. Astron. Astrophys.* **5**, 627.

NOVIKOV, I. D. and ZELDOVICH, YA. B. (1973) *Ann. Rev. Astron. Astrophys.* **11**, 387.

OORT, J. H., KERR, F. J. and WESTERHOUT, G. (1958) *Mon. Not. Roy. Astr. Soc.*, **118**, 379.

OZERNOI, Z. M. (1978) *The Whirl Theory of the Origin of Structure in the Universe*, in *The Large Scale Structure of the Universe*, eds.: Longair, M. S. and Einasto, J., D. Reidel Publ. Comp., Dordrecht.

PARISKI, YU. N. (1973) *Sov. Astron.* **17**, 291.

PARISKI, YU. N. (1978) *Search for Primordial Perturbations of the Universe, Observations with Rotan-600 Radio Telescope*, in *The Large Scale Structure of the Universe*, eds.: Longair, M. S. and Einasto, J., D. Reidel Publ. Comp., Dordrecht.

PARKER, L. (1976) *The Production of Elementary Particles by Strong Gravitational Fields*, in Proceedings of the Symposium on Asymptotic Properties of Space-Time, ed.: Witten, L., Plenum Publ., Corp., New York.

PARTRIDGE, R. B. (1979a) *Cosmological Anisotropies in the Microwave Background*, in *Physics of the Expanding Universe*, ed.: Demiański, M., Sprin er-Verlag, Berlin.

PARTRIDGE, R. B. (1979b) *Observational Cosmology*, Marcel Grossmann Meeting on General Relativity, ed.: Ruffini, R., North-Holland Publ. Comp., Amsterdam.

PEEBLES, P. J. E. (1965) *Astrophys. J.* **142**, 1317.

PEEBLES, P. J. E. (1966) *Astrophys. J.* **146**, 542.

PEEBLES, P. J. E. (1968) *Astrophys. J.* **153**, 1.

PEEBLES, P. J. E. (1971) *Physical Cosmology*, Princeton University Press, Princeton.

PEEBLES, P. J. E. (1980) *The Large Scale Structure of the Universe*, Princeton University Press, Princeton.

PEEBLES, P. J. E. and YU, J. T. (1970) *Astrophys. J.* **162**, 815.

PEIMBERT, M. (1975) *Ann. Rev. Astron. Astrophys.* **13**, 113.

PENZIAS, A. A., (1978) *American Scientist*, **66**, 291.

PENZIAS, A. A. and WILSON, R. W. (1965) *Astrophys. J.* **142**, 419.

POOLEY, G. G. and RYLE, M. (1968) *Mon. Not. Roy. Astr. Soc.* **139**, 515.

PRESS, W. H. (1980) *Physica Scripta* **21**, 702.

Press, W. H. and Vishniac, E. T. (1980) *Astrophys. J.*, **236**, 323; **239**, 1.

Rees, M. (1971) *Some current ideas on galaxy formation*, in *General Relativity and Cosmology*, ed.: Sachs, R., Academic Press, New York.

Rees, M. (1975) *Formation of Galaxies, Radio Sources and Related Problems*, in *Structure and Evolution of Galaxies*, ed.: Setti, G., D. Reidel Publ. Comp., Dordrecht.

Rees, M. (1978) *Nature* **275**, 35.

Reeves, H., Audouze, J., Fowler, W. A. and Schramm, D. N. (1973) *Astrophys. J.* **179**, 909.

Rindler, W. (1956) *Mon. Not. Roy. Astr. Soc.* **116**, 662.

Rohlfs K. (1977) *Lectures on density wave theory*, Springer-Verlag, Berlin.

Rowan-Robinson, M., Negroponte, J. and Silk, J. (1979) *Nature* **281**, 635.

Ryan, M. P. (1971) *Ann. Phys.* **65**, 506 (1971a); *Ann. Phys.* **68**, 541 (1971b).

Ryan, M. P. (1972) *Ann. Phys.* **70**, 301.

Ryan, M. P. and Shepley, L. (1975) *Homogeneous Relativistic Cosmologies*, Princeton University Press, Princeton.

Ryle, M. (1968) *Ann. Rev. Astron. Astrophys.* **6**, 249.

Sandage, A. R. (1961) *Astrophys. J.* **133**, 355.

Sandage, A. R. (1970) *Physics Today*, February, p. 34.

Sandage, A. R. (1975) *The Redshift*, in *Galaxies and the Universe*, eds.: Sandage, A., Sandage, M. and Kristian, J., Vol. IX of Stars and Stellar Systems, gen. ed.: Kuiper, G. P., The University of Chicago Press, Chicago.

Sandage, A., Sandage, M. and Kristian, J. (eds.) (1975) *Galaxies and the Universe*, Vol. IX of Stars and Stellar Systems, gen. ed.: Kuiper, G. P., The University of Chicago Press, Chicago.

Schramm, D. N. and Wagoner, R. V. (1974) *Physics Today*, December, p. 41.

Schramm, D. N. and Wagoner, R. V. (1977) *Ann. Rev. Nucl. Sci.* **27**, 37.

Sciama, D. (1971) *Modern Cosmology*, Cambridge University Press. Cambridge.

Searle, L. and Sargent, W. L. W. (1972a) *Comments Astrophys. Space Phys.* **4**, 59.

Searle, L. and Sargent, W. L. W. (1972b) *Astrophys. J.*, **173**, 25.

Setti, G. (ed.) (1975) *Structure and Evolution of Galaxies*, D. Reidel Publ. Company, Dordrecht.

Shu, F. H. (1970) *Astrophys. J.* **160**, 89.

Silk, J. (1968) *Astrophys. J.* **151**, 459.

Silk, J. (1974) *Spectrum of Density Perturbations in Expanding Universe*, in *Confrontation of Cosmological Theories with Observational Data*, ed.: Longair, M., D. Reidel Publ. Comp., Dordrecht.

Smoot, G. F., Gorenstein, M. Y. and Muller, R. A. (1977) *Phys. Rev. Lett.* **39**, 898.

Steigman, G. (1979) *Ann. Rev. Nucl. Part. Sci.* **29**, 313.

Sunyaev, R. A. and Zeldovich, Ya. B. (1970) *Astrophys. Space Sci.* **7**, 3.

Sunyaev, R. A. and Zeldovich, Ya. B. (1980) *Ann. Rev. Astron. Astrophys.* **18**, 537.

Taub, A. H. (1951) *Ann. Math.* **53**, 472.

Tayler, R. J. (1978) *Galaxies: Structure and Evolution*, Wykeham Publications Ltd., London.

Thaddeus, P. (1972) *Ann. Rev. Astron. Astrophys.* **10**, 305.

Tinsley, B. M. (1977) *Physics Today*, June, p. 32.

Tinsley, B. M. (1979) *Cosmology and Galactic Evolution*, in *Physical Cosmology*, eds.: Balian, R., Audouze, J. and Schramm, D., North-Holland Publ. Comp., Amsterdam.

Tinsley, B. M. and Larson, R. B. (eds.) (1977) *The Evolution of Galaxies and Stellar Population*, Yale University Observatory, New Haven.

Toomre, A. (1969) *Astrophys. J.* **158**, 899.

Trimble, V. (1975) *Rev. Mod. Phys.*, **46**, 755.

Turner, M. S. and Schramm, D. N. (1979a) *Nature*, **279**, 303.

Turner, M. S. and Schramm, D. N. (1979b) *Physics Today*, September, p. 42.

Valentijn, E. A. and van der Laan, H. (1978) *Extended Radio Emission in Clusters of Galaxies: Recent Westerbork Observations*, in *The Large Scale Structure of the Universe*, eds.: Longair, M. S., Einasto, J., D. Reidel Publ. Comp., Dordrecht.

de Vaucouleurs, G. (1959) in *Handbuch der Physik*, ed.: Flügge, S., Vol. 53, Springer-Verlag, Berlin.

DE VAUCOULEURS, G., (1979a) *The Scale of the Universe and the Hubble Constant*, lectures at the Symposium "Problems of the Cosmos", Rome, 2 th Sept.
DE VAUCOULEURS, G. and BOLLINGER, G. (1979b) *Astrophys. J.* **233**, 433.
WAGONER, R. V. (1973) *Astrophys. J.* **179**, 343.
WAGONER, R. V. (1979) *The Early Universe*, in *Physical Cosmology*, eds.: Balian, R., Audouze, J., Schramm, D., North-Holland Publ. Comp., Amsterdam.
WAGONER, R. V., FOWLER, W. and HOYLE, F. (1967) *Astrophys. J.* **148**, 3.
WEBSTER, A. S. (1977) *The Statistical Analysis of Anisotropies*, in *Radio Astronomy and Cosmology*, ed.: Jauncey, D. L., D. Reidel Publ. Comp., Dordrecht.
WEINBERG, S. (1971) *Astrophys. J.* **168**, 175.
WEINBERG, S. (1972) *Gravitation and Cosmology*, J. Wiley, N. Y.
WEINBERG, S. (1979) *Phys. Rev. Lett.*, **42**, 850.
WHITE, S. D. M. and REES, M. J. (1978) *Mon. Not. Roy. Astr. Soc.* **183**, 341.
WILSON, R. W. (1979) *Science* **205**, 866.
WILSON, M. L. and SILK, J. (1981) *Astrophys. J.* **243**, 14.
WOODY, D. P. and RICHARDS, P. L. (1979) *Phys. Rev. Lett.* **42**, 925.
ZELDOVICH, YA. B. (1967) *Sov. Phys. Uspekhi* **9**, 602.
ZELDOVICH, YA. B. (1979a) *Cosmology and the Early Universe*, in *General Relativity, An Einstein Centenary Survey*, eds.: Hawking, S. W., Israel, W., Cambridge University Press, Cambridge.
ZELDOVICH, YA. B. (1979b) *Creation of Particles by Gravitational Field*, in *Physics of the Expanding Universe*, ed.: Demiański, M., Springer-Verlag, Berlin.
ZELDOVICH, YA. B. (1979c) *Cosmological Microwave Background Radiation and Formation of Galaxies*, in *Physics of the Expanding Universe*, ed.: Demiański, M., Springer-Verlag, Berlin.
ZELDOVICH, YA. B., ILLARIONOV, A. F. and SUNYAEV, R. A. (1972) *Sov. Phys. JETP* **35**, 643.
ZELDOVICH, YA. B. and NOVIKOV, I. D. (1983) *Relativistic Astrophysics*, Vol. 2: *Structure and Evolution of the Universe*, The Chicago University Press, Chicago.
ZELDOVICH, YA. B. and SUNYAEV, R. A. (1969) *Astrophys. Space Sci.* **4**, 285.
DE ZOTTI, G. (1979) *Constrains on the Possible Distortions of the Cosmic Background Radiation Spectrum*, in *Physics of the Expanding Universe*, ed.: Demiański, M., Springer-Verlag, Berlin.

APPENDIX

SOME CONSTANTS USED IN ASTROPHYSICS

Speed of light c	$2.997924 \cdot 10^{10}$ cm s^{-1}
Planck constant $h=2\pi\hbar$	$6.626176 \cdot 10^{-27}$ erg s
Boltzmann constant k	$1.380622 \cdot 10^{-16}$ erg K^{-1}
Gravitational constant G	$6.6720 \cdot 10^{-8}$ cm^3 g^{-1} s^{-2}
Electron charge e	$4.803248 \cdot 10^{-10}$ esu
Electron mass m_e	$9.109534 \cdot 10^{-28}$ g
Proton mass m_p	$1.672648 \cdot 10^{-24}$ g
Atomic mass unit	$1.660565 \cdot 10^{-24}$ g
Stefan–Boltzmann constant σ	$5.67032 \cdot 10^{-5}$ erg cm^{-2} s^{-1} K^{-4}
Thomson cross-section σ_T	$6.652453 \cdot 10^{-25}$ cm^2
Astronomical unit	$1.4960 \cdot 10^{13}$ cm
Parsec pc	$3.0856 \cdot 10^{18}$ cm
Light year	$9.4605 \cdot 10^{17}$ cm
Solar mass $M_\odot$	$1.989 \cdot 10^{33}$ g
Solar radius $R_\odot$	$6.9598 \cdot 10^{10}$ cm
Solar luminosity	$3.90 \cdot 10^{33}$ erg s^{-1}

SUBJECT INDEX

OTHER TITLES IN THE SERIES IN NATURAL PHILOSOPHY